Rural Non-Agricultural Employment in India

An Analysis of its Determinants and Impact on Poverty and Inequality

About the Author

R.R. Biradar is a faculty in Economics at Karnatak University, Dharwad since 1999. Previously, he worked at the Centre for Multi-disciplinary Development Research, Dharwad and the Institute for Social and Economic Change, Bangalore. He has published several research articles in reputed journals, edited books and reports. He has been the recipient of "Sanjay Thakur Young Labour Economist Award" conferred by the Indian Society of Labour Economics for the year 2005 for his research contribution on : Growth of Employment in ICT Sector in India: Emerging Issues and Challenges. In 2006, he received the "Sir Ratan Tata Visiting Fellowship" by the Institute for Social and Economic Change, Bangalore in collaboration with the Ratan Tata Trust. The Fellowship was awarded to carry out a project on "Incidence of Poverty among Social Groups in Rural India: Why Do SCs/STs Experience Greater Incidence of Poverty?"

In 2007, he visited China under India-China Cultural Exchange programme jointly sponsored by ICSSR, New Delhi and CASS, China for a study on "Human Development, Poverty and Inequality in India and China; Emerging Evidence and Issues."

He is directing a UGC Major Research Project on 'Poverty Alleviation through Micro-Credit Programme of SHGs: A Study of Four NGOs in South India' during 2007-09.

Rural Non-Agricultural Employment in India

An Analysis of its Determinants and Impact on Poverty and Inequality

R.R. Biradar

CONCEPT PUBLISHING COMPANY, NEW DELHI-110059

ISBN-13: 978-81-8069-625-1 **ISBN-10: 81-8069-625-1**

First Published 2009

Published and Printed by

Concept
Since 1974

Concept Publishing Company
Regd. Office:
A/15-16, Commercial Block, Mohan Garden
New Delhi-110059 (India)
Phones : 25351460, 25351794, *Fax* : 091-11-25357109
Email : publishing@conceptpub.com,
website: www.conceptpub.com

Editorial Office:
H-13, Bali Nagar, New Delhi-110 015, India.

Cataloging in Publication Data--*Courtesy:* D.K. Agencies (P) Ltd. <docinfo@dkagencies.com>

Biradar, R. R., 1968-
Rural non-agricultural employment in India : an analysis of its determinants and impact on poverty and inequality / R.R. Biradar.
p. cm.
Originally presented as the author's thesis (Ph. D.)--Karnatak University, 2005.
Includes bibliograph.cal references (p.).
Includes index.
ISBN 13: 9788180696251 ISBN 10: 8180696251

1. Rural poor--Employment--India. 2. Labor supply--India. 3. Manpower policy, Rural--India. 4. Poverty--India. 5. Income distribution--India I. Title.

DDC 331.110954 22

Foreword

Dr. Biradar's analytical work on rural non-agricultural employment (RNAE) provides useful insights into the role historically played by the rural non-agricultural sector in the process of growth. He has marshalled empirical evidence from different groups of countries to highlight this role. "The mass of satistical data and the evidence put forth in this study," Dr. Biradar hopes, "helps policy makers and development practitioners in deciding about the type of policy intervention needed to tackle simultaneously the triadic problems of educated youth unemployment, poverty and inequality in developing economies like India." In this sense, Dr. Biradar's study is an important contribution to the corpus of empirical literature on the subject.

The mass of empirical evidence presented in the book, no doubt, yields valuable insights into the role of the rural non-agricultural sector in development but it is doubtful whether we can derive any specific policy-inputs. What is important is that this theme had acquired a contemporary relevance in the current discourses on development which emphasise the concept of "Inclusive Growth". The title of the Draft Approach Paper to the Eleventh Five-Year Plan — "Towards Faster and More Inclusive Growth" reflects this re-discovered wisdom. Inclusive growth is perhaps a new label to the rather old-fashioned concept of "broad-based and decentralised growth" which some of us have been advocating for many years.

Inclusive growth needs to be contextualised. Even if the 8 or 9 percent GDP growth targeted in the Eleventh Plan is sustained, and export growth is maintained at, say, 15 or 20 percent, the organised sectors like large and medium industries, the services sector, etc., cannot do much to tackle effectively the massive problems of unemployment and poverty. In other words, employment generated autonomously by the normal process of growth will fall far short of requirments. The policy thrust will have to be, perforce, on broad-based and decentralised growth, if the macroeconomic objectives of reducing unemployment and poverty are to be achieved. Viewed from this perspective, inclusive growth has become a growth and development imperative: growth, because a high GDP growth of say 8 or 9 per cent can be sustained only if other sectors or segments of the economy, which have been sluggish because of a number of reasons including policy neglect, can be activated; development, because this is perhaps the best route by which the bulk of the poor can be provided with livelihood and food security. I have spelt out this theme in my recent book "*Inclusive Growth: Development Perspectives in Indian Economy*" (Academic Foundation, New Delhi, 2007).

In this context the contemporary experience of China is illuminating. China, whose macroeconomic parameters were almost comparable to those of India, provides the most fascinating example of such inclusive growth. Today, China's manufacturing power is reshaping the world economy. The World capital labour ratio has nearly halved because of the incorporation of its labour force in the global economy. Analytically, what is important to highlight, is that this impressive growth in manufacturing would not have been able, by itself, to make a significant dent on the levels of unemployment and/or poverty. It was the vibrant rural non-farm sector which played a critical role in the success story of drastically reducing poverty and employment. China's Township and Village Enterprises (TVEs) provided increasing job opporunities outside agriculture, thereby augmenting sources of rural household income. The phenomenal growth of this sector is demonstrated by the fact that its contribution to GDP rose to nearly one-third of GDP in 2004, from practically zero in 1952.

Facilitating inclusive growth in India by making the rural non-agricultural sector vibrant is a far more complicated task, involving micro planning, evolving area specific solutions in the light of the resource endowments of the concerned region or area, and promoting participation of a number of actors in the task—panchayati raj institutions, Central and State Governments and NGOs, SHGs etc. Moreover, all those involved in the devleopment of the Indian economy — banks, cooperatives, development finance institutions, NGOs and administrators will have to rethink their own strategies in response to the challenge which the expansion of the rural non-agricultural sector poses. I do hope Dr. Biradar will carry forward his study to tackle these issues so that he can provide concrete policy inputs.

Mumbai **N.A. Mujumdar**
13 November, 2007

Preface

Development economists like Lewis or Fei and Ranis have argued that the solution for rural development and disguised unemployment lies in shifting surplus labour from subsistence sector — agriculture to the urban capitalist sector — manufacture. As the ability of urban capitalist sector to absorb a reasonable degree of labour is limited, the growth of rural non-agriculture sector has become an important aspect of the development paradigm to combat the problems of unemployment, poverty and inequality in most developing countries like India.

A considerable decline in the share of agriculture in the national income and marginal decline in the share of agricultural workers resulted in a sharp decline in the productivity per worker in agriculture and a substantial increase in the non-agricultural sector in rural areas. This created an abundant scope for non-agricultural investment, production and employment for educated youth. Consequently, the structure of the labour market and employment in rural areas began to undergo several changes. Among others, the withdrawal of labour force (in the age groups of 5-9, 10-14, 15-19 and 20-24 years) in favour of schools/colleges to acquire higher educational qualifications, skills and knowledge and the shift of workers in favour of the non-agricultural sector were found to be the important ones. An analysis exploring the factors contributing to a sharp decline in WPRs in the 1990s and the growth process of RNAE and its impact on poverty and income inequality forms the core of this book.

The decline in WPRs mainly came from: (i) an increase in unemployment rates among illiterates and educated persons up to primary level and among the youth and old-aged people; (ii) withdrawal of workers from the labour market in favour of schools/ colleges and corresponding rise in SPRs; and (iii) the increase in the ratio of population neither in the labour market, nor in schools/

colleges. The withdrawal of young labour force from the labour market is expected to have a beneficial impact on the economy in terms of human capital accumulation and thereby improving the efficiency in resource allocation as well as utilisation, resulting in a faster and more inclusive growth in the long-run.

Occupational diversification in favour of the rural non-agricultural sector has always been considered as an important strategy to tackle the problems of unemployment, poverty and inequality in most developing countries like India. Since the 1970s, especially when the sustained decline in poverty began to occur, the growth of RNAE has been on the gradual rise in varying degrees by gender, activities and areas. It was found to be much faster in respect of males as compared to females. The growth of male RNAE was mainly determined by growth related factors, whereas in respect of female RNAE, it was mainly poverty related factors. The residual sector hypothesis, as advanced by Vaidyanathan, seems to be untenable in respect of male RNAE, while it remains tenable in respect of female RNAE. The growth of male RNAE seemed to have greater poverty reduction effects than that of female RNAE. It is due to the fact that activities resulting largely from growth induced factors appear to be more productive, remunerative and highly paid as compared to activities resulting largely from distress induced/poverty-compelled.

The extent of poverty and income inequality reduction tended to depend *inter alia* on the type of activity status in the rural non-agricultural sector in which the poor were involved. The distribution of income from cultivation, followed by rural non-agricultural regular employment, was found to be most inegalitarian. On the contrary, the distribution of incomes from rural non-agricultural wage employment, rural non-agricultural self-employment and agricultural wage employment was more egalitarian.

It has been found that cultivation, followed by rural non-agricultural regular employment and non-agricultural self-employment contributed the most to total income inequality. The income from wage employment in the agricultural and non-agricultural sectors contributed negatively to overall income inequality; and the contribution of the former was higher than that of the latter.

In view of a gradual decline in the share of public expenditure for poverty alleviation and social security programmes, the promotion of RNAE can be considered as a strategic approach for combating the problems of unemployment, poverty and inequality in rural areas. The strategic approach for promotion of RNAE forms the core of this book.

The book is organised in seven chapters. Chapter 1 presents the research design and approach covering the relevance of RNAE, research issues and database and limitations. Chapter 2 provides an issue-wise thorough review of literature on the growth process of RNAE at different levels (state/regional, district/taluk and village/household) and its impact on poverty and inequality. Several research issues are also identified for further research. The growth of employment and unemployment by age groups and level of education during the pre-reform and reform periods in India is analysed in Chapter 3. This chapter basically attempts to explore the factors contributing to a sharp decline in WPRs in the 1990s. Chapter 4 presents the trends and patterns of male and female RNAE by age groups, level of education and size of landholdings during the pre-reform and reform periods in India. The determinants of male and female RNAE are analysed in Chapter 5. The impact of RNAE on poverty and income inequality is examined in Chapter 6. In the last Chapter a summary of the findings and important policy prescriptions are presented.

This book will be useful to researchers, policy makers and dvelopment practitioners in deciding the type of policy intervention needed to tackle the problems of unemployment, poverty and income inequality in rural areas of developing countries like India.

R.R. Biradar

Dharwad
December 2007

Email: rrbiradar2008@gmail.com

Acknowledgements

This book is based on my Ph.D thesis, "Rural Non-Agricultural Employment in India: An Analysis of its Determinants and Impact on Poverty and Inequality" submitted to Karnatak University, Dharwad in 2005. Pursuance of research on issues relating the dynamics of labour market and employment structure and its impact on distributional aspects rather calls for versatile research inputs being obtained from many academicians and also unstinted support from many well-wishers. During the course of my research work, I have received continuous advice, guidance and assistance from my research guide, academicians, colleagues, friends and officials of many institutions, for which it is my great privilege to acknowledge their contributions.

I express my sincere and profound gratitude to my research guide Dr. S.T. Bagalkoti, Prof., Department of Economics, Karnataka University Dharwad for his unstinted, constant encouragement, intellectual research inputs and valuable suggestions during the entire course of my research work. I am most grateful to all my teachers and colleagues in the Department of Economics for their support and encouragement for my research work.

I place on record the blessings and unstinted support extended by my *Guruji* Dr. M. S. Chandakavate, Principal (Retd.), G. P. Porwal College, Sindagi, Bijapur District in shaping my research and teaching career. I express my sincere gratitude to Dr. D. Rajasekhar, Professor at the Institute for Social and Economic Change (ISEC), Bangalore who has been the instrumental in shaping my research career, without his scholarly research guidance, the present work would have not seen the light of the day.

I am thankful to Prof. N.A. Mujumdar for his kindness to go through the manuscript and write a foreword, discussing several issues dealt in this book.

I am greatly indebted to Prof. P. R. Panchamukhi, Prof. G.K. Chadha, Prof. R. S. Deshpande, Prof. Alak Sharma, Dr. R. P. Mamgain, Prof. Errol D'Souza, Prof. R.M. Girji and Dr. P.M. Honaker; Dr. S.K. Melakar and Dr. Ramu Sajjan for their critical comments and suggestions on earlier version of my work presented in various conferences and seminars, which have immensely helped me in finalising the present work. I am thankful to Prof. G.B. Sajjan, Principal (Retd.), Bijapur and Prof. Parthasarthy, Bangalore for their valuable suggestions on an earlier version of the book.

In pursuance of my research work for Ph.D degree, I have been greatly benefited from several libraries at Karnataka University, Dharwad; ISEC, Bangalore; JNU, New Delhi; CMDR, Dharwad; Central Statistical Organisation, New Delhi; Planning Commission, New Delhi; Ministry of Agriculture and Co-operation, New Delhi; NIRD, Hyderabad and NSSO, West Bengal. Therefore, I am thankful to the officials of all these libraries, who extended their kind co-operation in collection of data and literature. I am deeply indebted to my wife Shaila and son Rakshit for their co-operation in completion of the work. It is needless to mention that the author alone should be held responsible for any flaws or shortcomings.

I dedicate this work to my parents and teachers with all due reverence.

R.R. BIRADAR

CONTENTS

Abbreviations

BC	:	Backward Caste
CDS	:	Current Daily Status
CPIAL	:	Consumer Price Index for Agricultural Labourers
CV	:	Coefficient of Variation
CWE	:	Casual Wage Employment
CWS	:	Current Weekly Status
GAI	:	Gross Area under Irrigation
GCA	:	Gross Cropped Area
GDP	:	Gross Domestic Product
HYVs	:	High Yielding Variety of Seeds
ILO	:	International Labour Organisation
ITI	:	Industrial Training Institute
NCAER	:	National Council for Applied Economic Research
NSA	:	Net Sown Area
NSSO	:	National Sample Survey Organisation
PCIA	:	Per Capita Income from Agriculture
PHIA	:	Per Hectare Income from Agriculture
RFL	:	Rate of Female Literacy
RML	:	Rate of Male Literacy
RNAE	:	Rural Non-Agricultural Employment
RSE	:	Regular Salaried/Wage Employment
SC	:	Scheduled Caste
SE	:	Self-employment
SNA	:	System of National Accounting
SPR	:	Student Population Ratio
ST	:	Scheduled Tribe
UN	:	United Nations
UPS	:	Usual Principal Status

UPSS	:	Usual Principal and Subsidiary Status
WB	:	World Bank
WPR	:	Work Participation Rate
WTO	:	World Trade Organisation

1 Research Issues and Approach

The Background

Diversification of occupational structure in favour of the non-agricultural sector has received a renewed importance in the development paradigm of many developing countries like India, as it has poverty reduction and income equalising effects. Occupational diversification may be due to distress in agriculture (distress diversification) and/or due to the vigorous growth in agricultural and non-agricultural sectors (inter-sectoral growth-led diversification). In fact, both types of processes may be at work simultaneously (Basant 1998, 1994; Biradar 2003; Biradar and Bagalkoti 2004; Biradar 2006a). Although distress and growth-led forces are operating, the dominant one is required to be identified, as they have differential levels of impact on poverty and income inequality reduction (Biradar and Bagalkoti 2004; Biradar and Rajasekhar 2007).

In this regard, the growth process of industrialised countries illustrates a certain pattern of diversification in production as well as occupational structure over time. Several studies (Fisher 1939; Clark 1940; Kuznets 1966; Chenery and Syrquin 1975) on the growth process of the present industrialised economies have revealed that as economic growth takes place, the primary sector largely consisting of agriculture tends to decline in its relative contribution to the national output even though its absolute contribution increases. The growth of the primary sector contributes to the expansion of the secondary or industrial sector through its multiple linkages. Once the economy becomes industrialised, that is, when the stage at which the labour force, released from the primary sector, particularly from agriculture, would be productively utilised is reached, the share of the secondary sector tends to decline in its relative importance. At still higher levels of economic growth, the share of the tertiary sector consisting largely of services of all kinds begins to rise. There will also be corresponding changes in the occupational distribution of the workforce in each sector.

This pattern of economic growth observed in the so-called industrialised economies of the world, however, has not been apparently occurring in India. There has been a rapid transition from agriculture to the service sector, bypassing manufacturing sector. Inter-sectoral growth movements in India during the last five decades reveal that while the share of agriculture in national income has been steadily declining, the share of the service sector has witnessed a rapid increase and that of the secondary sector has increased at a much slower pace than what was expected during the same period (Nagaraj 1990; Bhalla 1997; Madheswaran and Dharmadhikary 2000). Before the country reached a stage at which it can be called an "Industrialised Economy", the share of the service sector in national income had begun to rise by a greater extent[1]. This is contrary to the experience of the present industrialised economies such as the USA and UK.

As regards employment structure, it is observed that change in the occupational structure of the workers has not matched a shift in the production structure (Nagaraj 1990; Bhalla 1997; Parthasarathy, Shameem and Reddy 1998; Bhatt 2003). While there has been a continuous decline in the share of agriculture in national output during the last five decades, there was no decline in the share of employment until 1970. The beginning of the 1970s, however, started showing a sustained decline in the share of workers in agriculture and a corresponding increase in the rural non-agricultural sector. At the all India level, for instance, while the share of workers in agriculture declined from 74 percent in 1972-73 to 60 percent in 1999-2000 (NSSO 1981, 2001), the share of income from agriculture consistently declined from about two-thirds to a little over one-fourth of the national income during the same period (GoI 2003a). The pace of diversification in national output was found to be much faster than that of diversification in the occupational structure, especially in rural India. However, the pattern of occupational shift in favour of the rural non-agricultural sector was more clearly evident in respect of male workers than female workers. Within the rural non-agricultural sector, the increase in the share of employment in the service sector was much higher than that in the manufacturing sector; the bulk of the increase was explained by an increase in the proportion of workers in casual wage employment during the same period.

Since the 1970s, especially when the sustained decline in poverty has been accompanied by a shift of workers away from agriculture in favour of the rural non-agricultural sector, the growth of rural non-agricultural employment (RNAE)[2] has occupied an important place in the development paradigm. A considerable decline in the share of agriculture in the national income and marginal decline in the share of agricultural workers have resulted in a sharp decline in productivity per worker in agriculture and a substantial increase in non-agricultural sector in rural areas (Nagaraj 1990; Unni and Rani 2000; Madheswaran and Dharmadhikary 2000). This created abundant scope for investment and thereby rapid increase in production and employment opportunities in the rural non-agricultural sector in India. Consequently, the growth of RNAE has increased in varying degrees across gender, activities and states over time.

In view of the commitment by development planning of generating 'ten million employment opportunities per year over the 10th Plan period' (GoI 2002c) and 'halving poverty by 2015'— one of the Millennium Development Goals as set at the UN General Assembly in 2000 (WB 2001)—the need for a "sustained developmental strategy" which has greater unemployment and poverty reduction-effects, is felt increasingly important in the recent past. As agriculture and urban organised manufacturing sectors have virtually ceased to employ more labourers, the growth of the rural manufacturing sector has acquired a greater importance in the knowledge-based economy of the New Millennium. The important question, however, is whether the rural non-agricultural sector can bear the burden of providing sustained employment to the growing labour force, especially for educated youth.

Relevance of RNAE in India : An Overview

The relevance of RNAE can be understood from the development models formulated by Lewis (1954) and Fei and Ranis (1961) and many other development economists like Chenery and Syrquin (1975), Sanghvi (1969), Rosenzweig (1988), Lanjouw and Lanjouw (1995), and Leones and Feldman (1998). They point out that the solution for rural development and reduction of disguised unemployment lies in shifting surplus labour from the subsistence-agriculture-sector to the urban capitalist–manufacturing-sector. The ability of the urban organised capitalist sector to absorb a reasonable degree of labour force being limited, the growth of RNAE has become quite relevant for its multiple benefits. In the wake of several changes in the macroeconomic policies across the globe; the promotion of RNAE has become an important strategy for combating unemployment, reduction of poverty and inequality in most of the developing countries.

A cursory glance at the growth of RNAE and its impact on poverty and income inequality in most of the developing countries of Latin America and the Caribbean, Asia and Africa brings forth the relevance of RNAE in India. A few studies, regardless of the period and the level of sectoral diversification, have revealed that the proportion of workers in rural non-farm activities was relatively higher in Latin America and the Caribbean than in Asia and Africa (Hazell and Haggblade 1993; Lanjouw and Lanjouw 1995; Reardon, Berdegue and Escobar 2001). The growth of non-agriculture was found to be second in importance to agriculture in terms of their contribution to national income in several countries of Asia, Africa and Latin America. Evidence from Asia and African countries clearly shows that an inverse relationship existed between the size of landholdings and the share of non-agricultural income to total household income, which implies that for the smallest landholding categories in each country, non-agricultural income accounted for a major share of household income (Samuel 1979; Liedholm and Kilby 1989). Studies in Asian and African countries suggest that the Gini coefficient of income tended to decline when income from the rural non-agricultural sector was included in total household income (Samuel 1979; Leidholm and Kilby 1989). It follows that the growth of RNAE has income equalising effect. The growth of employment in agricultural and non-agricultural sectors tended to move in opposite directions and income earned from non-agricultural sources had a complementary effect on net agricultural income (Leidholm and Kilby 1989; Anderson and Leiserson 1980). The expansion of RNAE tended to smoothen and stabilise household income over different seasons of agriculture in a year[3]. The growth of RNAE also reduced rural-to-urban migration, minimising the adverse effects of urbanisation (Samuel 1979).

It has also been found that, in Latin America and the Caribbean, the poor households derived their major share of income from RNAE, which contributed to a decline in the incidence of poverty (Ferreira and Lanjouw 2001; Ruben and Van den Berg 2001; Reardon, Berdegue and Escobar 2001; Lanjouw 2001; Corral and Reardon 2001) and also reduction of inequality in the distribution of income (Janvry and Sadoulet 2001). A micro-level study by Adams (1994) in Pakistan and Leones and Feldman (1998) in northern Leyte, Philippines have also pointed out that income from non-agricultural sources had an income-equalising effect. Another category of studies, on the contrary, has found that the expansion of RNAE has raised income inequality among rural households. Reardon, Delgado and Matlon (1992) in Burkina Faso; Collier, Radwan and Wangwe (1986) in Tanzania; and Matlon (1979) in Nigeria, for instance, have found that the non-farm income raised the inequality in income distribution. Elbers and Lanjouw (2001) in Ecuador have observed that highly productive RNAE sharply raised the average income and reduced poverty but widened income inequality as compared to low productive ones. Escobal (2001) in rural Peru has also found that non-agricultural income, particularly from self-employment, followed by wage employment, contributed significantly to an increase in income inequality. Similarly, in rural China, a household level study by Kung and Lee (2001) has also observed that an increase in the share of non-agricultural income aggravated the degree of income inequality. Most of the studies in Asia, Africa and Latin America and the Caribbean have clearly shown that the growth of RNAE contributed to a decline in the incidence of poverty but the outcome of its impact on income inequality was found to be divergent.

In India, in view of growing population pressure on agriculture, declining land frontiers, deteriorating employment elasticities in agriculture and organised urban manufacturing sector[4] and excess supply of labour force, on the one hand; and increasing investment opportunities due to liberalisation, privatisation and globalisation, on the other; the importance for sustained growth of RNAE has been well recognised, especially since the 1990s, at a time when "educated youth unemployment" started swelling in its worst forms. The relevance of sustained growth of RNAE can also be observed in the context of several changes witnessed in agriculture. It may be noted that while the possibilities of raising net sown area (NSA) are gradually declining; the efforts to raise productivity through yield-enhancing technologies and double cropping have been constrained due to inadequate public investment in irrigation, flood control, drainage and dryland technology. In the recent past, even private investment in agriculture has gradually declined due to a sharp decline in public investment, particularly in irrigation, water management practices and dryland technology. The decline in capital formation in labour-intensive agricultural activities, and a consequent rise in mechanisation of agriculture, has shrivelled up the potentiality of employment growth in agriculture. The increasing number of small and marginal farmers and declining average size of operational landholdings have further tightened the investment opportunities and thereby livelihood avenues in agriculture. Consequently, a large number of small and marginal farmers and agricultural labourers are finding it increasingly difficult to get sustained and adequate employment from agriculture. In the absence of adequate employment growth, the excess concentration of population in agriculture has also led to a reduction or stagnation in the growth rate of daily real wage earnings in agriculture.

A large number of studies in India (see, Chapter 2), employing secondary data at the state/regional, district and village/household levels, have clearly illustrated that there had been an occupational diversification in favour of rural non-agricultural sector in varying degrees across gender, activities and areas. Further, the growth of RNAE was attributed to several factors originating within and/or outside agriculture in rural as well as urban areas. These factors were broadly categorised under growth related factors (inter-sectoral linkage hypothesis), distress-induced factors (residual sector hypothesis), factors external to agriculture but within the rural economy and factors originating from the urban economy.

The studies under review (see, Chapter 2) have shown that growth related factors such as the value of agricultural income per hectare/agricultural worker (cultivators and agricultural labourers) as an index of agricultural growth, degree of commercialisation, cropping intensity, mechanisation of agriculture, proportion of area under irrigation, consumption of chemical fertilizers and irrigation pumpsets energized, contributed to the growth of RNAE. On the other hand, the distress-induced/poverty-compelled factors such as person-day unemployment rates, proportion of people living below the poverty line, proportion and average size of small and marginal landholdings, share of landless labour households to total number of households, share of rural agricultural labourers to total number of labourers, land-man ratio and daily real wage rates in agricultural and non-agricultural occupations were also found to be important factors contributing to the growth of RNAE. Factors external to agriculture but within the rural economy such as rate of literacy (human resource development), infrastructure, per capita public expenditure for rural development, etc., were also determinants of RNAE. While the factors within agriculture and rural economy largely contributed to the growth of RNAE; factors external to the rural economy, i.e., originating from the urban sector such as degree of urbanisation and urban agglomeration, defined in terms of proportion of workers in urban manufacturing sector, as shown by several studies at the regional/district/taluk levels also emerged as important factors contributing to the growth of RNAE, especially for males.

As regards the impact of RNAE on the incidence of poverty, a few studies, particularly at the village/household level, have revealed that the poorer households, namely, small and marginal farmers, landless agricultural labourers and rural artisans derived their major sustenance from the rural non-agricultural sector, thereby contributing to a decline in poverty (Unni 1997a; Rajasekhar and Biradar 1998; Lanjouw and Shariff 2004). Studies examining the impact of RNAE on income inequality have found divergent results (Unni 1997a; Birthal 1997). An increase in non-agricultural income raised the degree of inequality in the distribution of rural income and ownership of land (Unni 1997a). It has been argued that if the macroeconomic policies push the workers into low productive non-farm jobs, the level of inequality in rural income distribution is more likely to increase. However, another household level study by Birthal (1997) in three villages of Moradabad district in Uttar Pradesh has observed a different pattern. The distribution of income from non-agricultural wage employment and all other non-agricultural jobs was more unequally distributed as compared to livestock and agriculture. An additional increase in agricultural income, followed by livestock and salaried jobs, contributed to a further increase in inequality in the distribution of income; whereas an increase in incomes from business, art crafts, non-agricultural and agricultural wage employment (mainly from non-agricultural sources) tended to reduce it. Thus, the impact of rural non-agricultural activities on income inequality was found to be divergent within the country.

Research Issues

As evident from the review of literature (see, Chapter 2), little attention is paid to explore a satisfactory explanation to the sharp decline in work participation rates (WPRs) in the 1990s. The extent of male and female participation in RNAE by different age groups, level of educational attainments and size of landholdings during the pre-reform and reform periods have scarcely been studied. Examining the participation of male and female workers in different activity statuses of rural non-agricultural sector (self-employment, regular and casual) is also another area of research. Most of the studies have attempted to identify the factors contributing to the growth of only male RNAE as a whole. As the factors that determine the growth of male RNAE are different from those of female RNAE, exploring the factors determining the extent of female participation in RNAE is also quite essential for policy intervention. Moreover, there are few studies attempting to explore the factors contributing to the growth of sub-activities in the rural non-agricultural sector for males and females separately; estimating the elasticities of important determinants of male and female RNAE as a whole and sub-activities; and, identifying what are the growth-led and distress-induced RNAE and also sub-activities for males and females. Because, the activities that are largely growth induced appear to be more productive, remunerative and highly paid and hence have had a greater beneficial impact on poverty reduction, as compared to the activities that are largely distress-induced/poverty-compelled[5]. The dominant forces behind the occupational diversification in favour of the rural non-agricultural sector are required to be identified.

The extent of poverty and income inequality reduction, *inter alia*, depends on the type of activity status in the rural non-agricultural sector in which the poor are involved. Issues relating to the nexus between RNAE and rural poverty, the extent of impact of RNAE on poverty reduction, and the poverty alleviation approach at the macro level have received little scholarly attention. More importantly, at the macro level, an analysis examining the extent of impact of RNAE as a whole and sub-activities on income inequality requires to be attempted. The need of the hour, however, is not only to explore the cause and nature of RNAE, but also its impact on poverty, income inequality and unemployment. The analysis of these issues would certainly help policy makers and development practitioners in deciding the type of policy intervention needed to tackle the long-standing problems of unemployment, poverty and inequality in India, especially in rural areas.

Against this background, the author has attempted to review the studies relating to different aspects of RNAE in India at different levels; analyse the dynamics of employment-unemployment by gender, age groups, level of education and activity status in rural and urban areas; examine the quality of rural employment by gender, age groups, level of education in rural and urban areas; explore the factors contributing to a sharp decline in WPRs in the 1990s both in rural and urban areas; examine the trends and patterns of RNAE by gender and activities across major states and by age groups, level of education and size of landholdings at the all India-rural level; explore the factors contributing to the growth of RNAE; measure the impact of RNAE on poverty and income inequality in rural areas; and offer policy prescriptions for sustainable growth of RNAE so as to address the problems of educated unemployment, poverty and income inequality in a holistic approach.

Database and Limitations

Nature and coverage of data

The study is based on secondary data collected from the various quinquennial surveys on employment and unemployment by the National Sample Survey Organisation (NSSO). Till date, NSSO has conducted seven quinquennial surveys[6] [the first 27th round survey (1972-73) followed by the 32nd round (1977-78), 38th round (1983), 43rd round (1987-88), 50th round (1993-94), 55th round (1999-2000) and the 61st round (2004-2005) surveys]. These surveys are the only source for comprehensive, detailed and also more comparable data on nation-wide and state-wise employment-unemployment figures. In all the quinquennial surveys, three distinct reference periods, namely, a long period, usually the last one year, one week and one day were employed and known, respectively, as usual principal status (UPS)[7], current daily status (CDS)[8] and current weekly status (CWS)[9]. In order to capture the different dimensions of employment-unemployment, these three (UPS, CDS and CWS) different activity statuses of an individual person have been employed since the first quinquennial survey in 1972-73. While the first activity status is expected to capture the employment-unemployment situation of a permanent nature, the last two are expected to capture the seasonality in employment-unemployment situation.

The data on workforce participation by gender, activity and unemployment rates for sixteen major states, namely, Andhra Pradesh, Assam, Bihar, Gujarat, Haryana, Himachal Pradesh, Karnataka, Kerala, Madhya Pradesh, Maharashtra, Orissa, Punjab, Rajasthan, Tamil Nadu, Uttar Pradesh and West Bengal were collected for the period 1972-73 to 1999-2000. In view of limited access to the data for 2004-05, the 61st round of NSSO, the present study has been limited to the period between 1972-73 and 1999-2000.

The data on workforce participation in agricultural and non-agricultural sectors by age groups, level of education and size of landholdings have also been collected at the all India level due to non-availability of the data at the state level. Since the time series data are not available, the period from 1983 to 1987-88 is regarded as the pre-reform period and 1993-94 to 1999-2000 as the reform period for examining the dynamics of employment and unemployment across major states and at the all India level. In some cases where data either for 1983 or 1987-88 are not made available, the period from 1983 or 1987-88 to 1993-94 is also considered as the pre-reform period. The sixteen major states are chosen as they have, by and large, similar socio-economic and physiographic features.

To explore the important determinants of RNAE, other data on per hectare/capita net income from agriculture, rural per capita non-food consumption expenditure, daily real wage rates of male and female agricultural labourers in agricultural and rural non-agricultural occupations, rate of literacy, proportion of area under non-foodgrains, area under irrigation, per hectare fertiliser consumption, degree of mechanisation of agriculture, proportion of people living below the poverty line, proportion and average size of small and marginal holdings, proportion of landless households, Gini ratios of operational landholdings, degree of urbanisation, urban agglomeration, rural infrastructure and the like have also been collected from various government documents/reports for the selected states for 1999-2000. Appropriate adjustments have also been made in some of the data for those states which have been affected due to the formation of new states, namely, Uttaranchal, Jharakhand and Chhattisgarh. The data on income shares from different sources of agriculture and non-agriculture in rural areas for the selected states for the year 1993-94 have been collected from NCAER (Lanjouw and Shariff 2004).

Tools of data analysis

In the present study, to analyse the trends and patterns of RNAE across gender activities in major states, simple averages, percentages, ratios, compound annual growth rates, and to measure the extent of variability in the proportion of male and female workers in RNAE across major states, the coefficients of variation have been used. To identify the important determinants of RNAE for males and females and estimate the elasticities of them, log-linear regression model has been formulated. The composite indices for rural infrastructure and mechanisation of agriculture have been constructed using the principal component analysis. The Gini coefficient is used to estimate the extent of inequality in the distribution of income from different sources including RNAE. For decomposing the income inequality by income sources, the decomposition technique by coefficient of variation has been employed.

Why is the NSS source of data preferred?

The NSS data are found to be more consistent than those of the Census. It has been found that in the 1951 Census of India, 'income' was used as the criterion to ascertain the activity status of a person. Accordingly, the population was divided into three categories, namely, 'self-supporting', 'earning dependent' and 'non-earning dependent'. The activity status of a person depended on his/her income earning status. Those who earned enough to support themselves were categorised as 'self-supporting', those who earned an income but not enough to support themselves entirely were considered as 'earning dependents', while those who earned nothing and were completely dependent on others were labelled 'non-earning dependents'. The census of 1961 saw a departure from the 1951 census. In the place of income, the concept of work was employed for the first time to ascertain the activity status of a person. The population was thus divided into two categories under 'workers' and 'non-workers'. The 1961 census, therefore, merely divided the entire population into these two categories of 'workers' and 'non-workers', the unemployed and those out of the labour force (NSS categories) being clubbed together into the 'non-worker' category. Therefore, the comparison of the data on employment-unemployment situation between the 1951 and 1961 censuses becomes erroneous.

The census of 1971 was also modelled on the lines of the 1961 census. In the 1971 census, those identified as workers were termed as 'primary workers', whereas those found to be primarily non-workers but who worked marginally, their secondary work was termed and recorded as 'secondary workers'. The 1971 census adopted the concept of main activity (i.e., activity not necessarily work) in which the person engaged himself/herself most of the time. This was a big shift from the 1961 census where a person was deemed to be a worker if he/she had worked for the major part of the working season or for the last fortnight preceding the date of enumeration. The effect of the 1971 classification was that many people who did work of an intermittent nature were likely to have reported under non-work as their main activity; one would assume that if the 1961 definition were to be applied most of these people would have been categorised as workers.

The 1981 census saw the streamlining and standardisation of not only the concept used to ascertain an individual's activity status but also the reference periods. The population was now divided into three categories under 'main workers', 'marginal workers' and 'non-workers'. Those who pursued 'economic activities' of one nature or another, for more than 183 days in a year, were identified as 'main workers'. Those who worked for less than 183 days were designated as being 'marginal worker' and those who did not engage in any form of activities were considered as being 'non-workers'. The concept of main workers introduced in the 1981 census, and used till date, is similar to the concept of primary workers used in the 1971 census and that of the secondary workers to the marginal workers.

The concept of economic activity used in the 1981 and 1991 censuses was found to be more restrictive than in the 2001 census. The economic activities included in the 1981 and 1991 censuses were all market and non-market activities, excepting for: (1) growing of crops, vegetables, flowers and other crops if done exclusively for home consumption; and (2) own account production of fixed assets. In the 2001 census, however, the concept of work was expanded to include the production of milk for domestic consumption, while the processing of primary commodities for own consumption and own account production of fixed assets continued to lie outside the scope of the definition adopted to identify workers. The term 'economic activity' as used in the NSSO surveys, includes all market activities. Of the non-market activities, the two included activities are: (1) all activities relating to the agricultural sector that result in agricultural production for own consumption; and (2) construction of own houses, roads, wells, etc., production of tools and machinery for household enterprise and construction of private or community facility free of charge. The concept of work as used by the Census till 1991 was more restrictive than that of the NSSO, but it was less restrictive in the 2001 census as it included an activity in which many rural women were more likely to be engaged. The scope of the concept of economic activity in the NSSO surveys continues to be wider than that of the Census. Moreover, unlike the Census, the NSSO measures different dimensions of employment structure by using three reference periods, namely, one year for UPS, one day for CDS and one week for CWS.

Further, the NSSO conducts its surveys systematically through well-trained enumerators who are permanent employees of the Organisation. Data collection in the field is supervised meticulously at every stage. Information collected through household schedules is, therefore, more complete and hence credible. Whereas in the case of the Census, information has been collected through the appointment of honorary enumerators, who are employees of other departments of the government, not of the Census organisation. They have neither training nor skills, and more importantly, they lack inclination to carry out the census work, as this work is in addition to their departmental work. Lack of involvement of enumerators in the census work is thus supposed to affect the quality of the census data (Visaria 1996; Deshpande 2003). The Census data since 1981 till date are found to be more comparable in terms of concept, reference period and coverage. In order to delineate the dynamics of structural changes in rural employment, in general, and RNAE, in particular, over a period of time, therefore, the data published by NSSO covering three decades were employed.

Comparability problems across the quinquennial surveys

The data published by the NSSO seem to be inter-temporally more consistent than those of the Census. Despite overall consistency across the quinquennial rounds, some minor inconsistencies do emerge. In the 27th round and the rounds prior to that (7th to the 10th), the 'usual status' concept was interpreted in a manner by which a person who remained unemployed for a long period in the past but became employed during the latter part of the period and if his/her status of being employed was likely to continue in the future, was identified as being 'employed' even though the person had remained unemployed during the major part of the long period (usually a year). This was so because the 'priority rule' was applied even to determine the activity status of an individual, whereby the working status received priority over the unemployment status. Opposed to this, a person in the same situation was categorised as being 'unemployed' in subsequent rounds beginning from the 32nd till the 55th. As such, the estimate of the employed would have been lower and that of the unemployed higher in the 27th round, had the revised definition, used in the latter rounds, been applied to ascertain employment levels during the 27th round.

The employment estimate of the 27th round, thus, would not be comparable with those of all the subsequent rounds *in toto*. Because, this again did not impose any minimum period of work to qualify as a worker; it merely attempted to record the enduring aspect of a person's activity pattern. The 32nd round also saw the concept of 'usually employed' being introduced. A person categorised as a 'worker' on the basis of his/her principal status was referred to as 'principal status worker' or 'main worker'. A person categorised as a 'non-worker' (according to the principal/usual activity status) who pursued some gainful activity in a subsidiary capacity was considered to be usually working in a subsidiary capacity and was referred to as 'subsidiary status worker' or 'marginal worker'. 'Principal status worker' and 'subsidiary status worker' taken together constituted 'all workers' or 'usual principal + subsidiary status' according to the usual status classification. Thus, the 'usually employed (US)' could be disaggregated into those employed in a 'principal' category/status (UP) and those employed in a 'subsidiary' category/status (SS).

As mentioned above in the 27th round, a person was considered as working if he/she pursued some gainful activity (not for a major part) during the latter part of a long period of time in the past and was likely to continue in the future as a 'worker'. The estimates of usual status workers based on the 27th round data included both categories of 'workers', those pursuing gainful activity in 'principal' capacity and also those pursuing gainful activity in 'subsidiary' capacity (those who got employed later). In other words, both 'principal status' and 'subsidiary status' workers together constituted the 'usual status' workers in the 27th round. Therefore, the estimates of 'workers' from the 27th round onwards would then be comparable to the principal (main) and subsidiary (marginal) status workers (main plus marginal) taken together in the subsequent rounds. The concepts, definitions and reference periods used in all the subsequent quinquennial rounds starting from the 32nd to the 50th are more or less identical enough to make comparison of broad employment estimates over time. Some changes have occurred over the years in the survey reports with respect to the disaggregated presentation of data by age composition, level of education and social background of the population surveyed. Hence, inter-temporally comparability of the data on RNAE is restricted.

Organisation of the Book

The book is organised into 7 chapters. Chapter 1 presents the introduction and research methodology covering relevance of RNAE, research issues—objectives, database and limitations. Chapter 2 sets forth a comprehensive issue-based review of literature on RNAE and identifies a wide range of issues for research. An analysis relating to the dynamics of employment and unemployment in India, focusing on the growth of employment, changes in WPRs by gender, age groups and level of education, the quality of employment and unemployment situation, is provided in Chapter 3. The data on trends and patterns of RNAE by gender, activities, age groups, level of educational attainments and size of landholdings during the pre-reform and reform periods are analysed in Chapter 4. In Chapter 5, the determinants of RNAE and elasticity of the important determinants are analysed. The impact of RNAE on poverty and income inequality has been examined in Chapter 6. The last Chapter contains a summary of the findings and important policy prescriptions.

NOTES

1. A phenomenal increase in the share of the service sector in the national income is regarded as "excess growth" by Bhattacharya and Mitra (1990). They point out that income from the service sector has been growing much in excess of the demand generated for the services by the commodity sector.
2. It includes all the non-agricultural activities such as mining and quarrying; manufacturing activities of cottage and small-scale rural industries; public utilities; construction; trade and commerce; hotel, restaurant business; transport, storage and communication; personal and community services; and the seasonal and contractual non-agricultural activities available within villages or in near by towns. These activities are also termed as rural non-farm activities (RNFAs) (Biradar and Rajasekhar 1998; Biradar 2006a).
3. An analysis of the monthly income fluctuations of 424 households in Thailand revealed that the variability of total household income was substantially less than the variability of net farm income over the year (Leidholm and Kilby 1989; Anderson and Leiserson 1980).
4. Employment elasticity with respect to agricultural output, based on CDS basis, has been on the decline. Although the employment elasticity declined in the earlier period, the decline was especially sharp in the 1990s. Employment elasticity in agriculture steeply declined from 0.87 to 0.01 during the period between 1983 to 1987-88 and 1993-94 to 1999-2000 (GoI 2002b).
5. It is due to the fact that daily real income or wage earnings are more likely to be higher in the growth-induced rural non-agricultural activities than in the distress-induced as it depresses the real income or wage rates due to distress (Sharma 2001; Unni 1997b).
6. As per the recommendation of an Expert Committee on Unemployment Estimates headed by Prof M.L. Dantawala set up by the Planning Commission, New Delhi, in 1968, the NSSO methodology underwent drastic changes in the manner in which employment surveys were being conducted till around early 1970s. NSSO, thereafter, embarked on five-yearly employment surveys based on a large-scale sample at both states and central levels as against the earlier thin sample surveys.
7. Principal status activity of a person is defined as the activity status on which a person spends relatively longer time (major time criterion) during the 365 days preceding the date of survey. A person who pursues some economic activity in a subsidiary capacity is considered as subsidiary status worker. These two groups, viz., principal status and subsidiary status

workers together constitute the usual status workers (US = UP + SS) (NSSO 2001).

8. Current weekly status of a person is the activity status obtaining for a person during a reference period of 7 days preceding the date of survey. According to this status, a person is considered working or employed if the person was engaged for at least one hour on any one day of the previous week in any economic activity (employed). On the other hand, a person who had not worked for even one hour on any single day of the week, but had been seeking or had been available for work at any time for at least one hour during the week is considered 'seeking/available for work' (unemployed) (NSSO 2001).
9. Current daily activity status of a person is determined on the basis of his/her activity status on each day of the reference week. According to this, a person is considered working or employed for the entire day if he/she had worked for 4 hours or more during the day. If a person had not been engaged in any work even for one hour, but had been seeking or available for work for 4 hours or more he/she is considered unemployed for the entire day (NSSO 2001).

2 Rural Non-Agricultural Employment in India

A Review of Evidences and Issues for Research

Introduction

An analysis of the employment situation in India, *prima facie,* reveals that there has been a decline in employment growth rate, quality of employment, WPRs and, more importantly, shift of workers away from agricultural to non-agricultural sector, especially in rural areas. Very few studies (Chadha 2001a; Biradar and Bagalkoti 2001; Chadha and Sahu 2002; Hirway 2002) have attempted to explore the possible reasons for the observed decline in WPRs, particularly in the 1990s. They show that the decline in WPRs was partly due to an increase in the student-population ratio (SPR). Nevertheless, satisfactory explanations need to be explored. Occupational diversification of workers in favour of the rural non-agricultural sector has assumed considerable importance in the development paradigm, as it has "welfare implications" in terms of poverty alleviation and income-equalising effects.

In this chapter, an attempt has been made to review the important studies on RNAE focusing on the trends and patterns, determinants and its impact on poverty and income inequality and identify research gaps for the present study. A large number of studies have appeared on various aspects of RNAE at the all-India, state, region/district and taluk levels drawing upon the data from the NSSO and the Census. The basic objectives of most of the studies based on secondary data relate to the analysis of trends and patterns and factors contributing to the growth of RNAE. A number of primary studies have also been carried out to analyse the economics of RNAE, to identify the socio-economic background under which the workers take up jobs in the rural non-agricultural sector and to examine the impact of RNAE on poverty and income inequality.

Diversification of Rural Employment

The economic growth process of the present industrialised countries illustrates a certain pattern of sectorial diversification in production as well as occupational structures over time. Several studies (Fisher 1939; Clark 1940; Kuznets 1966; Chenery and Syrquin 1975) attempting to explore the growth process of the present industrialised economies reveal that as economic growth takes place, the primary sector, largely consisting of agriculture, tends to decline in its relative contribution to the national output even though its absolute contribution increases. The growth of the primary sector leads to an increase in the relative share of secondary or industrial sector to the national output through its multiple linkages. Once the economy becomes industrialised, the share of the secondary sector starts declining in its relative importance, and that of the tertiary sector consisting largely of services of all kinds begins to rise. There would also be corresponding changes in the occupational distribution of work-force in each sector.

This pattern of economic growth observed in the industrialised economies of the world has not been apparently occurring in most developing countries like India. A cursory glance at the rural non-farm activities in the developing countries of Latin America, Asia and Africa would be of importance before looking into the Indian context. In many developing countries of Latin America, Asia and Africa, a large number of rural households, in general, and the poor, in particular, depend on RNAE as a major source of employment and income. Anderson and Leiserson (1980) have assessed the extent and importance of non-agricultural activities in rural areas and towns in terms of their contribution to the output, employment and income earnings of rural households[1]. The National Census data across 15 developing countries have revealed that the rural non-farm activities were the primary source of employment and income earnings for approximately one-fourth of the rural labour force (one-third including labour force in rural towns) and an important source of secondary earnings during the slack season of agriculture for the small and landless farmers in most of the Latin American, Asian and African countries (Anderson and Leiserson 1980; Leidholm and Kilby 1989) The proportion of workers in rural non-farm activities was found to be relatively higher in Latin America than in Asia and Africa (Hazell and Haggblade 1993; Lanjouw and Lanjouw 1995; Reardon, Berdegue and Escobar 2001). A quick glance at the share of workers in rural non-farm activities within these regions would be quite interesting.

In Latin America, the proportion of workers in non-agricultural activities in rural settlements has estimated at 26 to 28 percent during 1970-1990 (Hazell and Haggblade 1993). A recent review study by Reardon, Berdegue and Escobar (2001) in Latin America and the Caribbean has estimated that the rural non-agricultural sector provides a little over one-third of employment in rural areas[2]. A considerable regional variation also existed in the proportion of workers in RNAE. For instance, while the proportion of male workers varied from 18 percent in Bolivia to 57 percent in Costa Rica; in respect of females, it varied from 16 percent in Bolivia to 93 percent in Panama in 1997. In most of the countries of Latin America and the Caribbean, the participation of female workers in RNAE was found to be more than that of their male counterparts. The proportion of male workers in RNAE was estimated at 32 to 47 percent, whereas in respect of females, it was estimated at 60 to 80 percent in 1997 (Reardon, Berdegue and Escobar 2001). The simple average data for 12 countries of Latin America and the Caribbean have revealed that the share of income from RNAE was estimated at 46 percent. The share of income from RNAE, however, varied from 22 percent in Honduras in 1997 to 68 percent in Haiti in 1996 (Reardon, Berdegue and Escobar 2001).

In Asia, where diversification appeared to be slower than in Latin America and the Caribbean and faster than in Africa, the proportion of workers in RNAE varied considerably across space and gender. Hazell and Haggblade (1993) and Lanjouw and Lanjouw (1995) have estimated that in most of the 11 countries of Asia, the RNAE accounted for 30 to 50 percent of rural employment during

1980-1990. The proportion of workers in RNAE, however, varied from as low as 11 percent in 1980 and 20 percent in 1986 in China to 47 percent in 1966 and 67 percent in 1980 in Taiwan. Unlike in the case of Latin America and the Caribbean, the proportion of male workers was found to be higher in RNAE than that of female workers in Asia.

In Africa, the proportion of workers in RNAE was estimated at 14 percent (Hazell and Haggblade 1993). A review study by Reardon, Berdegue and Escobar (2001) estimated that the share of workers in RNAE was around 18 percent with considerable regional variation. In Africa, Western Nigeria accounted for the largest share of workers, estimated at 60 percent, whereas Rwanda accounted for the lowest at 5 percent in 1978. Although the male workers seemed to have accounted for a larger share in RNAE, a significant gender gap was quite discernible in most of the African countries[3]. The simple average data on income earnings from RNAE in Africa was approximately estimated at 45 percent of the rural income (Reardon, Stamoulis, Cruze, Balisacan, Berdegue and Banks 1998). In most of the countries of Latin America and the Caribbean, Asia and Africa, the diversification of production structure seemed to be much faster than that of occupational structure.

Sectoral patterns of RNAE

Although there had been a consistent decline in the contribution of income from agriculture to the national income, there was hardly any decline in the share of labour force involved in agriculture until 1970. Surprisingly, even after two decades of development experience, the share of labour force in agriculture continued to be around 70 percent. Since the 1970s, however, the rural employment structure has started diversifying in favour of the non-agricultural sector with varying degrees across gender, areas and sub-sectors. Since then, a large number of studies have been carried out to analyse the different aspects of RNAE using the NSS and Census data published at different points of time.

To begin with, an attempt was made by Krishnamurty (1984) to analyse the changes in the proportion of workforce involved in agricultural and non-agricultural occupations, both in rural and urban areas, drawing upon the Census data for 1971 to 1981 and the NSS data for 1972-73 to 1977-78. The results of the Census data indicated a greater structural shift away from agriculture towards manufacturing than those of the NSS. But the NSS data indicated a structural shift away from agriculture to the service sector (Krishnamurty 1984:2128).

A study by Vaidyanathan (1986) has attempted to examine the pattern of labour use in agricultural and non-agricultural occupations in rural areas by using the NSS data for 1972-73 to 1983. The proportion of rural workers (US), both male and female, in RNAE had increased; the increase in respect of males was, however, higher than that of females. Going by different activities in the rural non-agricultural sector, the increase of workers in construction, transport, trade and services seemed to be higher than that of the manufacturing sector (Vaidyanathan 1986: A-131). Basant and Kumar (1989) have analysed the various aspects of RNAE at the macro level, drawing upon the NSS data for 1972-73 to 1983 and the Census data for 1961 to 1981 and have also compared the results of NSS data with those of the Census.

The study has found that, though the NSS and Census data showed a constant increase in the proportion of male workers in RNAE, the former showed a relatively greater increase of male workers than the latter[4]. Both sets of data showed an increase in the share of workers in manufacturing and tertiary sectors; however, a greater increase took place in the tertiary sector as compared with the secondary sector, particularly in respect of male workers. Within the secondary sector, it was observed that the share of non-household manufacturing had increased at the expense of household manufacturing for both male and female workers.

The rise in the share of male workers in the manufacturing sector was greater as per the NSS data, compared to the Census. In relative terms, the Census data showed a greater shift of workers in favour of the tertiary sector than the NSS data; but both suggested a faster growth of employment in the service sector than in the manufacturing one in rural areas. Contrary to the observation made by Krishnamurty (1984) using the Census data, Basant and Kumar using the NSS data have arrived at the conclusion that the share of tertiary sector increased at a faster rate than that of the secondary sector[5].

Dev (1990) has analysed the changes in distribution of workers involved in sub-activities of agricultural and non-agricultural sectors in rural areas, using the NSS data for 1972-73 to 1983 at the all India level. The study has found that an increase in the share of tertiary sector was higher than that of secondary sector (Dev 1990). Unlike as in the case of male workers, an increase in the share of female workers in the secondary sector was higher than that of the tertiary sector. Another study by Unni (1991) has observed that the proportion of workers in RNAE increased from 14.3 percent in 1961 to 18.6 percent in 1983; the increase was greater in respect of males as compared to females. Bhalla (1993), using the Census data from 1961 to 1991, has found that, by and large, there had been a gradual increase in the share of workers in RNAE[6]; the increase was greater among the rural male *vis-à-vis* female workers, particularly during 1981 to 1991.

Studies by Vaidyanathan (1994), Basant and Kumar (1994) and Visaria (1995), drawing upon the NSS data covering relatively a longer period, that is, from 1972-73 to 1987-1988, have observed the following three broad trends in RNAE. Firstly, the proportion of workers in RNAE had increased: the trend was more clearly evident among male workers than among female workers. Secondly, within the rural non-agricultural sector, the increase in the share of the tertiary sector was higher than that of the secondary sector. Finally, the bulk of the increase in RNAE was explained by an increase in the proportion of casual workers.

In another study, Bhalla (1997), using the Census data has computed a diversification index[7] at the state and district levels for 1961, 1971 and 1981 and found that rural India experienced, over the two decades up to 1981, a deepening of workforce diversification—an increasing complexity and balance within the non-cultivating segment—rather than a significant widening of rural workforce diversification. Of the three broad sectors of rural economy, the tertiary sector diversified the fastest, the secondary the second fastest, while the primary sector as a whole scarcely diversified at all (Bhalla 1997). Drawing upon the NSS data from 1972-73 to 1993-94, Chadha

(1997) and Parthasarathy, Shameem and Reddy (1998) observed two types of phases in RNAE expansion in the early 1990s. The expansion of RNAE was observed for both male and female workers during 1977-78 to 1987-88, but it decelerated during 1987-88 to 1993-94. The deceleration was far more serious in respect of female workers. The early decade of 1990s appeared to have broken the usual pace of occupational diversification towards the rural non-agricultural sector. The female workers seemed to be going back to agriculture and letting their male counterparts to take up jobs in the rural non-agricultural sector. It was also noted that the proportion of male and female workers in manufacturing (US) declined sharply and in the tertiary sector continued to grow in the 1990s (Chadha 1997).

Studies by Chadha (2001a, 2002), and Chadha and Sahu (2002), drawing upon the NSS data (US) covering the period from 1972-73 to 1999-2000, have found that, in rural areas, the proportion of male workers in RNAE increased, but not at the same pace as it did in the pre-1987-88 years. The inter-sectoral shifts were relatively more in the 1970s and 1980s than in the 1990s. The study has pointed out one clear trend of the early 1990s, i.e., a substantial "slow down" of the process of weaning away of rural male workers from agriculture and its complete halt, if not reversal, in the case of rural females[8]. The service sector grew relatively much faster than the manufacturing sector for rural male workers, whereas the manufacturing sector grew much faster than the service sector for rural female workers in the 1990s (Chadha 2001a, 2002; Chadha and Sahu 2002).

Spatial patterns of changes in RNAE

The proportion of workers in RNAE is not uniform across the states and over time. A considerable variation has existed in the magnitude and rate of increase in RNAE. The spatial pattern of diversification is, therefore, designed to review the important studies identifying the states that experienced a rapid diversification of employment in favour of the rural non-agricultural sector. Vaidyanathan (1986) has found considerable variations in the proportion of workers in RNAE across sixteen major states and year-to-year fluctuations. Another study

(Dev 1990), using the NSS data for fifteen major states in 1977-78, has also found considerable regional disparity in the share of workers in RNAE. Similarly, Unni (1991) has found that the regional variation across the NSS regions was significant and the coefficient of variation was quite higher for females (84.6 percent) than that for males in the rural non-agricultural occupations (37.4 percent).

A couple of studies by Chadha (1997) and Parthasarathy, Shameem and Reddy (1998) have also pointed out that the proportion of male workers in RNAE showed a continuously increasing trend in a majority of the states, while a declining trend of growth was noticed in a few states during 1987-88 to 1993-94. Most of the states, however, showed a significant increase in the proportion of male workers in RNAE during 1983 to 1987-88. A rapid increase in the share of male workers in 1987-88 over 1983 was largely attributed to the drought of 1987-88. Evidence for drought-driven RNAE was also explored by Parthasarathy, Shameem and Reddy (1998).

Chadha (2001a, 2002) has computed the compound annual growth rates of workforce in agricultural and non-agricultural activities. The employment growth rates in agricultural and non-agricultural sectors at the all India level and in most of the states appeared to have improved during the reform period (1993-94 to 1999-2000) as compared to the pre-reform period (1983 to1993-94) (Chadha 2001a: 510). The compound annual growth rates estimated by the same author at the all India level using the Economic Census data, however, did not support the trends observed from the NSS UP+SS data (Chadha 2001a: 512).

Seasonal patterns of changes in RNAE

The seasonal pattern of RNAE is designed to explore whether employment in agricultural and non-agricultural sectors in rural areas tends to vary systematically over different sub-rounds within the NSSO survey year. In view of the declining average size of landholdings and deteriorating employment elasticity in agriculture, on the one hand, and the increasing number of small and marginal farmers and agricultural labourers, on the other, a larger proportion of rural workers are unable to find adequate and sustained employment in agriculture all through the year. A large chunk of rural workers, therefore, started to seek supplementary jobs in the rural non-agricultural sector during the slack season of agriculture. Consequently, the proportion of workers in RNAE started fluctuating significantly season-to-season within a year.

An attempt, therefore, was made by Basant and Kumar (1989) to measure the extent of seasonality in agricultural and non-agricultural employment. The study has found that, at the aggregate level, employment in the rural non-agricultural sector seemed to have moved in the direction opposite to employment in agriculture during different sub-rounds of the NSSO survey year. Some proportion of rural workers appeared to have found work in the rural non-agricultural sector when adequate employment was not available in agriculture. This does not, however, mean that all the non-agricultural employment in rural areas was necessarily residual in nature because some of the workers were engaged in casual wage as well as in self-employment of the rural non-agricultural sector throughout the year. Some proportion of casual workers seemed to have shifted back and forth between agricultural and non-agricultural pursuits to stabilise their annual household income in rural areas. But, the fluctuations over sub-rounds were not too large; change from one to another exceeded not more than 5 percentage points. The growth of employment in agricultural and non-agricultural sectors tended to move very systematically over sub-rounds in a year and it rather varied counter-cyclically. The seasonal fluctuations in RNAE remained significant even in irrigated areas (WB 1978).

Marginalisation or subsidiarisation of RNAE

The term 'marginalisation' is used to mean an increasing proportion of workers in subsidiary occupations, resulting from lack of employment in the main occupation throughout the year and/or low productive nature of the main occupation, be it in agriculture or non-agriculture. Marginalisation of RNAE occurs only when the workers engaged in agriculture or rural non-agriculture as their main occupation are unable to find adequate work so as to engage themselves in their main occupation for the full length of the year and increasingly rely on subsidiary occupations in the rural non-agricultural sector. A large proportion of the workforce do not have steady activity either because of the increasing seasonality of agricultural employment and/or because of the increasing unorganised and informal nature of their work pattern, which invariably requires them to undertake more than one activity at different times or seasons within the span of a year.

With a gradual decline in the quality of employment in rural areas (defined in terms of increasing proportion of workers in casual wage employment either in agriculture or non-agriculture), the process of subsidiarisation, particularly in respect of females, has been accelerated since the 1990s. In this regard, to begin with, a study by Basant and Kumar (1989) found that as compared to the 1961 census, the 1971 census reported a smaller proportion of workers engaged in secondary activity[9]. In both the census estimates, and in most of the activities categorised as main, it was the male workers who reported a relatively higher incidence of secondary occupation as compared to their female counterparts. A majority of them were involved in non-household manufacturing activities, trade and services, as casual labourers (Basant and Kumar 1989:37).

According to the NSS data, the process of subsidiarisation of rural employment in the non-agricultural sector increased significantly, particularly among casual agricultural workers in 1983 over 1977-78. Among agricultural workers, the proportion of casual workers reporting RNAE as their subsidiary occupation was not only higher than that of the self-employed and regularly employed workers but also increased from 7 percent to 17 percent for males and from 4.1 percent to 10.5 percent for females during 1977-78 to 1983. In the case of rural non-agricultural workers, the proportion of casual workers reporting agriculture as their subsidiary occupation was not only higher than that of the self-employed and regularly employed workers, but it also increased from 9.2 percent to 12.6 percent in respect of males and from 7.4 percent to 23.7 percent in respect of females during the same period. Although the extent of workers reporting agriculture as their subsidiary occupation among the causal non-agricultural workers in rural areas was quite large, the pace of increase was higher among the male casual agricultural workers (seeking subsidiary works in the non-agricultural sector) and also among the female casual non-agricultural workers (seeking subsidiary occupation in agriculture) during the same period (Basant and Kumar 1989:39).

Quality of rural non-agricultural employment

The quality of employment in the rural non-agricultural sector can be measured in terms of casualisation of non-agricultural employment, the incidence of poverty among non-agricultural workers and the incidence of unemployment among non-agricultural households as compared to agricultural households. If the proportion of such incidences is either higher or increasing over time in non-agriculture *vis-à-vis* agriculture, it means that the quality of RNAE is either low or deteriorating. Further, the educational background of workers in the rural non-agricultural sector can also be taken to measure the extent of quality of employment. An attempt has been made to review the studies covering the above aspects of rural non-agricultural sector in India.

Casualisation of employment

It is evident from the studies reviewed in the earlier section that there had been a gradual shift of rural workers away from agricultural to non-agricultural sector; the process of which seemed to have slowed down, on the one hand, and the pace of casualisation of RNAE seemed to have accelerated, on the other, in the 1990s as compared to the 1980s. In this regard, a couple of studies by Basant and Kumar (1989, 1998) have attempted to analyse the NSS data on workforce distribution by activity status and employment structure for 1972-73 to 1987-88 at the all-India level. The studies have concluded that there had been a gradual increase in the proportion of casual workers employed both in agricultural and non-agricultural sectors during 1972-73 to 1987-88 (Basant and Kumar 1998:83).

In the case of agriculture, while a gradual decline occurred in the share of self- and regular employment for males followed by females, a marginal increase took place in the share of casual employment, particularly in respect of males, during 1972-73 to 1987. In respect of rural non-agricultural sector, on the other hand, a corresponding, if not one-to-one, increase took place in the share of workers in self- and regular employment for both male and female workers. The share of male and female workers in casual RNAE increased much faster than that of self- and regular RNAE and even causal employment in agriculture during the same period. Although the degree of casualisation was smaller in the rural non-agricultural sector, it increased rapidly since 1983 and more so in 1987-88.[10]

Incidence of unemployment

Dev (1990) has estimated the incidence of unemployment[11] in principal occupation both in agricultural and non-agricultural sectors by using the NSS data on person-days unemployment (CDS) in 1977-78 at the all India level. The study has pointed out that the incidence of person-days unemployment was higher in non-agriculture as compared to agriculture and severe among females as compared to their male counterparts at the all India level in 1977-78 (Dev 1990:1531). The incidence of person-days unemployment for all persons in almost all the sub-sectors of non-agriculture was higher than that of agriculture, excepting for trade and services. Further, the incidence of person-days unemployment for females was higher in transport, followed by public utilities (electricity, gas and water), manufacturing, services and trade as compared to their male counterparts during the same period. There was a concentration of person-days unemployment for all persons in construction, transport, mining and quarrying and other areas.

Incidence of poverty

Dev (1990), using the data on incidence of poverty in agricultural and non-agricultural sectors for 1977-78, has observed that in contrast to the incidence of unemployment, the incidence of poverty among households in the rural non-agricultural sector was lower than in agriculture. Going by employment status, although the incidence of poverty was highest for casual labour in both agricultural and non-agricultural sector, the latter had a relatively lower incidence of poverty. The regular wage employment in agriculture and self-employment in the non-agricultural sector were found to be second important activities which had a greater incidence of poverty. Agricultural self-employment and non-agricultural regular employment had a much lower incidence of poverty. Going by gender, like in the case of incidence of unemployment, the female workers experienced a greater share of poverty incidence as compared to their male counterparts during the same period (Dev 1990).

The study has also attempted to explore the degree of variation in poverty incidence in agricultural and non-agricultural occupations across states in rural areas. In the case of the non-agricultural occupations, it varied from 12 percent in Punjab and Rajasthan to 51 percent in West Bengal. The incidence of poverty in the rural non-agricultural occupations was higher than in agriculture in Punjab and Haryana, whereas it was quite higher in agriculture as compared to what it was in the rural non-agricultural sector in the rest of the states under study. Further, the incidence of poverty was highest in the manufacturing sector in Bihar, lowest in Punjab; for construction, it was highest in Orissa, lowest in Punjab; and for trade and transport, it was highest in West Bengal, lowest in Punjab. Among all the states, Punjab witnessed a lower incidence of poverty in almost all the non-agricultural activities (Dev 1990:1533).

Factors Contributing to the Growth of RNAE

Growth of employment in the rural non-agricultural sector results from a variety of factors emanating from within as well as outside agriculture and external to the rural economy. Above all, the growth and distress related factors were found to be dominant forces in the promotion of RNAE. Serious attention has to be accorded if the growth of RNAE is the result of sustained squeezing of labour from agriculture due to lack of employment growth rather than as a response to a vigorous and dynamic growth of the rural non-agricultural sector. In the process of creating employment in the rural non-agricultural sector both types of forces (growth and distress) may be operating simultaneously. For promoting productive and sustained growth of RNAE and hence a rapid decline in poverty and inequality, the dominating forces need to be explored in the areas with different levels of development. An attempt has been, therefore, made in this section to review a series of studies at different levels, identifying the factors which contributed to the growth of RNAE.

A variety of factors have been identified in the literature to explain the variations in the proportion of workers in RNAE at different levels. For analytical purpose, these factors have been grouped into the following broad categories : agricultural growth-led diversification (inter-sectoral linkage hypothesis); distress diversification (residual sector hypothesis); factors external to agriculture such as rural infrastructure, human capital and per capita public expenditure for rural development; and factors external to the rural economy such as degree of urbanization and urban agglomeration. The studies attempting to identify the factors which determine the extent of variation in the proportion of workers in RNAE have been categorised into two main groups, based on the nature and coverage of data: studies based on secondary data at the state and district/taluk levels; and studies based on primary data covering a large and thin sample size at the household level.

Studies Based on Secondary Data

Agricultural growth-led diversification (Inter-sectoral linkage hypothesis)

Growth of RNAE mainly depends on agricultural prosperity through its multiple inter-sectoral linkages. In the early 1970s, Mellor and Lele (1973) and Mellor (1976) postulated that due to the adoption of green revolution technologies in agriculture and consequent increase in agricultural productivity and income levels of the farmers as well as agricultural labourers, a "virtuous cycle" could emerge between agricultural intensification and non-agricultural activity through "inter-sectoral linkages" including consumption and production linkages. The consumption linkages operate, based on Engel's law. That is, with increasing levels of income of the farmers and agricultural labourers, the relative share of consumption expenditure on non-food items tends to increase and that of food items declines, thereby stimulating the growth of RNAE through consumption-linkages.

The growth of RNAE would also take place through backward and forward production linkages, emanating from agricultural prosperity. The backward production linkages from agriculture to non-agriculture in terms of the demand of agriculturists for inputs such as ploughs, tools, fertilizers, electricity, machinery equipments, distribution and repair services of agricultural implements and the like stimulate the growth of RNAE. Similarly, the forward production linkages through the processing of agricultural commodities such as paddy, tobacco, oil seeds, sugarcane, etc., also stimulate the growth of RNAE. Further, the growth of the rural non-agricultural sector would, in turn, raise agricultural productivity through investment of surplus back into agriculture, and consequent growth of rural non-agricultural sector. A virtuous spiral of inter-sectoral linkage-led growth, thus, would be vigorously operating both in agricultural and non-agricultural sectors in rural areas.

Several empirical studies have documented the power of farm and non-farm linkages in Asia, Africa and Latin American countries. Johnston and Kilby (1975), based on evidence from Asia, have argued that a development strategy focused on small farms would generate rapid, equitable and geographically dispersed growth because of labour-intensive linkages with the rural non-farm economy. Based on evidences from India, Pakistan and Taiwan, they have highlighted the potential importance of production linkages through the demand from small farmers for fertilizers, equipments, repair services provided by rural blacksmiths and construction inputs. Hirschman's (1958) early indictment of agriculture as a low-linkage, underpowered engine has erred because it ignored these important agricultural growth-led linkages.

Haggblade, Hazell and Brown (1989), based on the evidence from rural sub-Saharan Africa and Asia, have argued that the production and consumption linkages resulting from agricultural growth stimulated demand for rurally supplied non-farm goods and services[12]. Farm-non-farm linkages were found to be larger in Asia as compared to Africa. Consumption linkages were much stronger than production linkages in most of the Asian and African countries. Consumption expenditure by peasant families provided an important stimulus to the growth of RNAE. Raising farm labour productivity, therefore, was important not only because it permitted the release of labour from agriculture to non-agricultural pursuits but also because it boosted up per capita income to the levels that enabled consumption diversification in favour of non-food items (Haggblade, Hazell and Brown 1989:1191). While no directly comparable studies were available for Asia, studies relating to India (Rangarajan 1982) and South Korea (Adelman 1984) showed a stronger linkage effect from agricultural growth in Asia than in Sub-Saharan Africa[13]. However, in relative terms, consumption linkages were much stronger than those of production and even stronger than those estimated in Asia.

A similar observation was made by Janvry and Sadoulet (1993), which said that consumption linkages exerted the strongest influence in Asian countries than that in Latin American countries. They further argued that the difference was due to more unequal distribution of land and income in Latin America. This implied that a few landowners benefited from the bulk of income effects of agricultural growth. These large landowners were often absentees, so that they did not demand locally produced goods, and the level of their income was such that they tended to prefer luxurious items met by urban industries or imports.

Although production and consumption linkages attracted most of the initial interest in exploring the "inter-sectoral growth linkages" (Mellor and Lele, 1973; Johnston and Kilby 1975; Mellor 1976), more recent investigations have identified a third important linkage, known as "labour market interactions". This refers to the shifting of workers from low skilled, low productive, dead-end, lowly paid works to high-skilled, high investment, more productive, better paid works within the rural non-agriculturai occupations. This happens especially when the daily real wage rates tend to increase rapidly in agriculture, as a result of which the opportunity cost of the workers involved in the rural non-agricultural occupations tends to rise. Consequently, the workers are more likely to change their occupation from lowly paid jobs to high paid jobs in the rural non-agricultural sector. Though the evidence on this line of development is scanty, a couple of studies (Hossain 1988; Ahmed and Hossain 1990) carried out in Bangladesh reveal that increasing agriculture wages raised the opportunity cost of labourers in non-agricultural occupations, due to which the workers started changing their activity in favour of relatively better paid non-agricultural jobs in rural areas.

Consumption Linkages

A. Studies at the State/Regional level

Consumption and production linkage effects, along with others, on the growth of RNAE have been explored by many studies using the data at the state level published by the NSSO and the Census. Not all studies have found a strong association between consumption linkage and the growth of RNAE. For instance, based on the Census data Vyas and Mathai (1978) have argued that agricultural growth could not stimulate the growth of rural non-farm economy. According to them, skewed income gains in agriculture limited consumption linkages, and inadequate rural infrastructure restricted the ability of rural firms to supply the required input and consumer goods (Vyas and Mathai 1978). Similarly, Sankaranarayan (1980) has attempted to test the inter-sectoral linkage hypothesis by considering rural per capita consumption expenditure and the value of crop output per agricultural worker as indirect and direct linkages of agricultural prosperity, respectively. However, no strong correlation between these linkages and the proportion of workers in RNAE was obtained. He has concluded, therefore, that the differences in agricultural prosperity, by and large, did not explain the variations in the share of workers in RNAE.

Vaidyanathan (1986) has considered the level of employment in rural non-agricultural activity to be a function of (a) the level of rural demand for non-agricultural goods and services produced locally; (b) the level of extra demand for rural products (services); and (c) location, share and technology of activities catering to these demands. The first one depended on agricultural prosperity (defined as per capita income of the agricultural classes and the degree of inequality in its distribution). A strong positive association was observed between the proportion of workers in RNAE and crop output per head of agricultural population, and a strong negative association between the share of all and male workers in RNAE and the Gini ratio for the concentration of operational landholdings. This implies that increased income per capita resulting from high crop output and egalitarian distribution of landholdings stimulated the growth of RNAE through consumption linkages. On the contrary, Dev (1990) at the state level and Unni (1991) at the NSSO regional level, have found that there was an insignificant positive impact of crop output per agricultural worker on the growth of workers in RNAE, particularly for females. But, they have found a significant impact of crop output per hectare (land productivity) on the share of workers in RNAE, particularly for males. Agricultural surplus acted as a source of capital for investment in rural non-agricultural pursuits (production linkage) rather than the source of expenditure on non-food items (consumption linkages).

In respect of females, Unni (1991) has found that with increasing levels of income and consumption expenditure, the participation of females in RNAE tended to decline and with contraction of income earning potentiality, the participation tended to rise. The females diversified their occupation in favour of RNAE only when there was distress in agriculture, reflected in terms of lower real wage earnings or low consumption potentiality (Unni 1991). Parthasarathy, Shameem and Reddy (1998) have found that in times of severe distress in agriculture (low foodgrains production) due to drought, both males and females, and more so in respect of females, were more likely to diversify their activities in favour of rural non-agricultural sector[14].

Although the gross value of agriculture per hectare played a relatively greater role in determining the level of RNAE, there had been a time lag to convert the growth impulses generated in agriculture into the rural non-agricultural employment opportunities (Parthasarathy, Shameem and Reddy 1998). Another study by Gill and Sawhney (2003) has also found that there was a strong positive association between per capita net state domestic product (NSDP) and male RNAE, and a weak one with respect to total RNAE. The male per capita gross state domestic product (GSDP) originating from agriculture had a strong positive impact on all and male RNAE. Thus, from the above analysis it appears that it was the productivity of agriculture per hectare and not per worker that contributed a lot to the growth of RNAE, in general, and male RNAE, in particular, mostly through production linkages. Secondly, in times of agricultural distress females were found to diversify their activities in favour of rural non-agricultural sector. This means that higher levels of income or consumption expenditure seemed to inhibit the participation of female workforce in RNAE.

B. Studies at the district and taluk levels

The review results obtained at the district and taluk levels present divergent growth processes of RNAE. One category of studies (Shukla 1991,1992; Hazell and Haggblade 1991; Srivatsava and Dubey 2002) has argued that the growth of agriculture through consumption linkages had a strong positive impact on the growth of RNAE. Another category of studies (Jayaraj 1989; Singh 1994; Narayanamoorthy, Rodrigues and Phadnis 2002) has argued that agricultural growth, through increased demand for non-farm goods and services, had still a positive association with the level of RNAE, but it was not statistically significant. Some other studies (Harriss 1987; Chandrasekhar 1993; Eapen 1994, 1995; Verma and Verma 1995; Chand 1996) have argued that the growth of agriculture has not resulted in the growth of RNAE, refuting Mellor's inter-sectoral linkage hypothesis. This category of studies intends to discover the factors external to agriculture and rural areas that contribute to the growth of RNAE.

A couple of studies by Shukla (1991,1992) in Maharashtra have found that a strong positive association existed between NSA per agricultural worker (as a proxy for average rural farm income and consumption) and the proportion of workers in RNAE. An U-shaped relationship was in evidence with regard to the distribution of landholdings. The greater the share of the smallest and largest size of landholdings in total number and area of operational landholdings, the higher was the proportion of workers in RNAE. The proportion of operational landholdings and area in lower (5 to 20 hectares) and higher size of landholding categories (20+ hectares) were considered as proxies for consumption linkages and these were found to have a strong negative effect with the former and a strong positive effect with the latter on the growth of RNAE. Hazell and Haggblade (1991), covering 85 districts spread over 12 states in India, have argued that the growth of RNAE was primarily the result of agricultural growth by raising the income levels of the farmers and daily wage earnings of agricultural labourers, resulting in an increased demand for non-farm goods and services.

At the NSS region level, Srivatsava and Dubey (2002), considering the rate of poverty reduction and average agricultural wage rate as an index of demand for non-farm goods and services, have argued that reduction in poverty incidence and an increase in the average wage rate in agriculture stimulated the growth of RNAE through an increased demand for non-farm goods and services. They have concluded that in better-off regions, the growth of RNAE was demand-driven, and in poorer regions it was the result of distress diversification.

Another category of studies argues that the growth of agriculture positively contributed to the growth of RNAE through its consumption linkages, but statistically not significant, implying that there were other factors, which appeared to be more important than the factors within agriculture in the promotion of RNAE. For instance, a study by Jayaraj (1989) in Tamil Nadu has considered rural per capita gross cropped area (GCA) and proportion of irrigated area to total cropped area as proxies for rural per capita agricultural income and has found an insignificant positive association between the proportion of male workers in RNAE and rural per capita agricultural income, implying weak consumption as also production linkages from agriculture to RNAE in Tamil Nadu.

In Uttar Pradesh, a study by Singh (1994) in order to explore the factors contributing to the growth of RNAE within and outside agriculture, selected the following variables: (1) level of agricultural development and rural prosperity (value of agricultural output per capita, value of agricultural output per hectare and per capita rural consumption expenditure); (2) degree of inequality in income/asset ownership (Gini coefficient of consumption expenditure, the proportion of people below poverty line and Gini coefficient of landholdings); (3) degree of commercialisation; (4) degree of mechanisation of agriculture; (5) demographic pressure and extent of landlessness; (6) average size of the village; and (7) degree of urbanisation. The study has found that out of the indicators measuring agricultural development and rural prosperity, the value of agricultural output per capita was positively, but insignificantly, associated with the proportion of workers in RNAE, whereas the value of agricultural output per hectare and per capita rural consumption expenditure were positively and significantly associated.

In respect of the second set of indicators, which were partial reflections of consumption linkages, captured through the Gini coefficient ratio for consumer expenditure and landholdings, had a strong and weak positive impact on the growth of RNAE, respectively. The proportion of people living below the poverty line had a strong negative impact on the proportion of workers in RNAE, implying that in poor districts lack of demand inhibited the growth of RNAE. The study has concluded that higher rural incomes, coupled with greater inequalities, raised the level of demand for non-agricultural products, resulting in the growth of RNAE (Singh 1994:206). As the value of agricultural output per capita had a weak effect on the proportion of RNAE, it appeared that although the growth of RNAE was in part demand-driven, there was little evidence to believe that the entire demand resulted only from agricultural growth (as rural per capita consumption expenditure is a function of incomes from agricultural as well as non-agricultural occupations) had a strong positive impact on the growth of RNAE. The growth of RNAE was, thus, partially explained by the consumption linkages originated from agricultural growth, indicating the importance of other factors outside agriculture. Narayanamoorthy, Rodrigues and Phadnis (2002), using the census data for 256 districts, spreading over 13 states, have found a positive but weak association between the value of agricultural output per hectare and the proportion of workers in RNAE.

A study by Harriss (1987) in North Arcot district of Tamil Nadu in India[15], however, has found that the local economy was not growing on the lines of Mellor's inter-sectoral linkage hypothesis. The increasing integration, penetration and concentration of national markets and allocation of public sector resources were the important agents of growth of RNAE (Harriss 1987:40). A study by Chandrasekhar (1993) in West Bengal has found that the compound growth rate of agricultural output had an insignificant positive impact, while the compound growth rate of agricultural output per capita had a weak negative impact on the proportion of male workers in RNAE. The cropping intensity, density of population and WPRs also played no significant role in determining the level of male RNAE. The study, however, has claimed that the persistent rise in the proportion of workers in RNAE in total rural employment was not the "fall out" of the dynamism but rather of inadequate diffusion of the dynamism of the green revolution. The relationship between the growth of agriculture and RNAE appeared to be non-linear (Chandrasekhar 1993).

Eapen (1994) in Kerala has found that a rapid increase in the proportion of workers in RNAE was the result of a sustained increase in the demand for goods and services produced in the non-farm sector, particularly for non-farm services and construction inputs. This growth in demand for such non-farm goods and services was attributed to a rapid increase in the Gulf remittances and daily wage rates in the rural non-agricultural, particularly construction, and agricultural sectors. It was rather the consumption linkages that had a profound impact on the growth of workers in RNAE; however, it was not led by the sustained and vigorous growth of agriculture but by a rapid inflow of remittances and increase in the daily wage rates in non-agricultural and agricultural occupations. This was also reflected in terms of an inverse relationship between the farm size and the proportion of workers in RNAE, implying that with increasing levels of income from agriculture, the proportion of workers in RNAE tended to decline.[16]

In another study, Eapen (1995) has clearly found no significant and specific association (negative in 1981 and positive in 1991) between the proportion of male workers in RNAE and the value of per capita agricultural output. He has concluded that the growth of male workers in RNAE was led not by agricultural growth but by out-migration and remittances from Gulf countries (Eapen 1994, 1995). A study, carried out by Verma and Verma (1995) in the Eastern region of the country, has found that there had been a negative but insignificant association between the proportion of male RNAE and the average value added in agriculture per hectare; a positive but weak association in respect of average value added in agriculture and allied activities per agricultural worker. Rural income per capita had a negative but insignificant impact on the proportion of male workers in RNAE. The share of non-food items in total rural consumption was negatively and significantly associated with the proportion of male workers in RNAE. The study has claimed that the inter-sectoral linkage hypothesis was untenable and the residual sector hypothesis was more applicable. That is, the growth of agriculture played hardly any role in determining the level of RNAE through its increased demand for non-food items in the eastern part of the country (weak consumption linkages) (Verma and Verma 1995).

Chand (1996) in Himachal Pradesh has found that per capita output had a strong negative impact on the growth of RNAE. The study has concluded that empirical evidence based on district level data provided support to the inverse relationship between agricultural growth and the percentage of workers in RNAE. The primary data collected from 225 households spread over three panchayats, representing different levels of development in Solan district, however, lent credence to the inter-sectoral linkage hypothesis which suggested a strong positive association between agricultural progress and RNAE (Chand 1996: 850). Synthesizing the divergent results obtained at the macro and micro levels in Himachal Pradesh, the study has concluded that agricultural growth initially might reduce the proportion of workers in RNAE, but as agricultural growth got perpetuated and broad based, it resulted in a faster growth of RNAE as compared to total rural employment. This observation based on primary data is similar to the observation made by Chandrasekhar (1993) in West Bengal and Chadha (1986, 1994) in Punjab and Haryana. The regional/district level studies clearly show that the direction of impact of agriculture on the growth of RNAE depended upon, *inter alia*, the stage of agricultural prosperity of the region/ district. There would not be a monotonic relationship between the growth of agriculture and non-agriculture over time.

Production linkages (Forward and Backward)

A. Studies at the State/Regional level

Sustained growth of RNAE depends on, *inter alia,* the magnitude of backward and forward production linkages from agriculture to non-agriculture. Sankaranarayan (1980) has formulated the alternative hypothesis that while agricultural growth (value of crop output per worker) could not adequately explain variations in the proportion of workers in RNAE, the degree of commercialisation of agriculture explained variations in the proportion of workers in RNAE. The degree of commercialisation was captured by two indicators, namely, exports from agriculture to non-agriculture (forward linkages) and percentage of cash purchase to total consumer expenditure (both food and non-food items) in rural areas (backward linkages). The proportion of the area under non-food crops to total cropped area was used to measure the extent of exports from agriculture. But, this indicator would not capture the share of food crops that were marketed. The percentage of cash purchase to total consumer expenditure in rural areas was used to measure the imports into agriculture. Both these indicators were found to have a strong positive impact on the proportion of workers in RNAE, implying that agricultural growth through its backward and forward production linkages strongly stimulated the growth of RNAE.

A study by Vaidyanathan (1986) has also hypothesised that the degree of commercialisation was indirectly sought to be captured by an increasing function of the level of crop output per head of agricultural population, inequality in the distribution of operational landholdings and the proportion of area under non-foodgrains. While the crop output per capita had a significant positive, and inequality had a strong negative impact, on the proportion of workers in RNAE, the proportion of area under non-foodgrains gave mixed results. On the basis of NSS data (1977-78), the proportion of area under non-foodgrains was not significantly associated with RNAE. According to the Census data (1971), the two variables showed a strong positive relationship.

Vaidyanathan himself has admitted that we do not have a direct measurement of commercialisation of agriculture and that the gross crop output was an incomplete and crude measurement of net income of the agricultural classes. Before arriving at any conclusion about the effect of agricultural prosperity and the degree of commercialisation on RNAE, he has advocated further refinement in the variables capturing commercialisation of agriculture (Vaidyanathan 1986). Papola (1987) has hypothesised that agricultural growth through its direct and indirect effects contributed to the growth of employment in the industrial sector in rural areas.

The direct impact of agricultural growth on rural industrial activity was by way of supplying more raw materials (forward linkages) and creating more demand for agricultural inputs and services (backward linkages). The indirect impact of agricultural growth was observed through an increase in the consumption demand for non-farm goods and generating surplus for investment. He has, however, observed that the direct impact in terms of input-supplying and output-using linkages was limited. More evidence on the indirect impact of agricultural growth on the growth of manufacturing activities was observed. He has concluded that the effects of consumption linkages from agricultural growth on rural industries were relatively better than those of production linkages (Papola 1987).

Similarly, a study by Unni (1991) has hypothesised that predominance of non-food crops in the cropping pattern of a region would have direct and indirect positive effects on RNAE. The direct impact on non-agricultural activities was by supplying raw materials for processing and other industrial activities and the indirect impact was through inter-linkages between output, credit and labour markets. The impact of a non-food predominant cropping pattern on the growth of rural non-agricultural activities was sought by capturing the two indicators, namely, the percentage of non-food crops to total cropped area and the ratio of non-food crops to total quantum of cereals.

Unni (1991), using the NSS regional level data, has found an insignificant association between the share of non-food crops to total cropped area (as a proxy for commercialisation) and the share of workers in RNAE. The ratio of non-food crops to total cereals— the other indicator to measure commercialisation of agriculture— was found to have no direct relationship with RNAE. But, it was negatively and significantly associated with the proportion of female workers in RNAE. The growth of agriculture with non-foodgrains dominating cropping pattern in a region appeared to inhibit rather than facilitate the growth of female RNAE. In this case, forward production linkages from agriculture to non-agriculture seemed to be insignificant. In respect of backward production linkages, the variables such as the use of oil engine, electric pump sets and tractors per hectare and fertilizer consumption per hectare did not show any significant relationship with the proportion of workers in RNAE. She has found that, in relative terms, forward production linkages were more effective than backward production linkages in the promotion of RNAE.

Parthasarathy, Shameem and Reddy (1998), using the proportion of area under non-foodgrains as a proxy for commercialisation of agriculture, have found a strong positive impact on all, but insignificant on male, RNAE. This impact appeared to have sharply reduced in the case of all RNAE and increased in the case of male RNAE in 1993-94 over 1987-88, but both of them were not statistically significant. This implies that the importance of commercialisation in the promotion of RNAE declined in respect of female workers and increased in respect of male workers over time. The study has concluded that the impact of commercialisation on male RNAE was statistically not significant both in 1987-88 and 1993-94 (weak forward production linkages) (Parthasarathy, Shameem and Reddy 1998:150-1). A study by Kaur (2002) has also found a positive association between the percentage of area under non-food crops and the percentage of workers in RNAE; but, again, statistically not significant (Kaur 2002: 974). The agricultural growth-led RNAE through production linkages was, however, partially substantiated at the state level. Thus, studies at the state level indicate the existence of weak production linkages from agriculture to non-agriculture. But, forward production linkages were more operative than backward production linkages.

B. Studies at the district and taluk levels

Several studies have also been carried out to examine the impact of backward and forward production linkages on the growth of RNAE for both males and females, using the district and taluk level data for a state. From the review it appears that there are two important categories of studies that have attempted to test the production linkage hypothesis, i.e., the growth of agriculture stimulates the growth of non-agricultural employment through its backward and forward production linkages. One category of studies at district and taluk levels (Harriss 1987; Jayaraj 1989; Shukla 1991, 1992; Singh 1994; Eapen 1995; Rajasekhar 1995; Rajasekhar and Biradar 1998; Sharma, Kumar and Sharma 1999; Eapen 2001; Narayanamoorthy, Rodrigues and Phadnis 2002; Biradar 2007) argues that the commercialisation of agriculture continued to have a positive impact on the growth of RNAE, but still statistically insignificant. Another category of studies at the district level (Chandrasekhar 1993; Mecharla 2002; Basant and Parthasarathy 1998; Singh 1993), on the other hand, has discovered that there was an inverse relationship between the degree of commercialisation and the proportion of workers in RNAE, but statistically not significant.

A study by Harriss (1987) in North Arcot district of Tamil Nadu has found that production activities were more important in terms of number of units, capital invested and employment created than consumption activities in the promotion of RNAE. The study has also found that forward production linkages were more important than backward production linkages (Harriss 1987:44). The increased production of paddy and groundnuts in North Arcot district was accompanied by a growth in the activities which had production linkages; both backward (inputs) and forward (processing) production activities seemed to have important non-local linkages.

Jayaraj (1989) in his study in Tamil Nadu has considered the proportion of non-foodgrain crops output to total crop output and the concentration ratio of landholdings as partial measurements of commercialisation. He has hypothesised that the concentration of land with large farmers led to an increase in market dependence as: (1) the proportion of crops marketed increased either due to an increase in the proportion of area devoted to commercial crops or due to a decline in the proportion of crops used for own consumption; and (2) the use of purchased inputs increased, leading to an increase in the demand for non-farm services. But the relationship between RNAE and concentration ratio of landholdings was found to be non-linear. The demand for consumer goods of rural origin, which is a function of income generated in the agricultural sector and its distribution, might tend to decline with increasing inequality in the distribution of income. The results obtained at the district level, however, provided no conclusive evidence on the hypothesised relationship, perhaps due to the smallness of the sample size.

At the taluk level estimates, the proportion of area irrigated by wells to total irrigated area was used in place of the percentage of non-foodgrains crop output to total crop output as it did not yield significant results at the district level for capturing commercialisation of agriculture. The concentration ratio of landholdings and the proportion of area irrigated by wells to total irrigated area were found to have a positive impact on RNAE. The degree of commercialisation, defined in terms of proportion of well irrigation, stimulated the growth of RNAE at the taluk level (Jayaraj 1989).

Shukla (1992), in her study in Maharashtra, chose the following indicators to explore production linkages (backward and forward) from agricultural to non-agricultural sector : the value of output of major crops per hectare of NSA, ratio of area under total food crops and sugar crops to total NSA, ratio of total area under oil seeds and non-food crops to total area under crops, average yield of all cereals, foodgrains (cereals and pulses) and cotton which would have a direct bearing on forward production linkages. The proportion of gross irrigated area (GIA) to GCA, number of irrigation pump sets and tube-wells energized, number of tractors and fertilizer consumption were considered as important indicators for backward production linkages from agricultural to non-agricultural sector. While the ratio of area under non-food crops and fibre crops to total area under all crops and average yield of sugarcane were negatively related to the proportion of workers in RNAE, the ratio of area under food crops, oilseeds and the average yield of all cereals and pulses had a strong positive impact on the growth of RNAE. The average yield of cotton also contributed to it.

With regard to input using linkages, utilization of fertilizers and agricultural credit in general contributed to the growth of RNAE, while the use of tractors seemed to inhibit it. While there had been a strong positive association between the share of workers in RNAE and the GIA as a percentage of cropped area, the intensity of pumpsets and tubewells was negatively related to it (Shukla 1992:1479). The backward or input linkages of agriculture were weak, whereas the forward or output linkages, particularly crops like cereals, pulses and oilseeds dominating, were found to be more strong in the growth of RNAE. The low impact of commercial crops on the growth of RNAE was due in part to the poor agro-processing linkages. Shukla (1992:1486) has concluded that production linkages were weaker than consumption linkages in the growth of RNAE in Maharashtra.

A study by Singh (1994) in Uttar Pradesh has illustrated similar linkage effects on the growth of RNAE. The backward production linkages from agricultural to non-agricultural sector were captured through the use of tractors (per 1,000 hectares of NSA), pumpsets (per 1,000 hectares of NSA) and electricity consumption (per hectare of NSA) and the forward production linkages were captured through the proportion of area under non-food crops. The indicators used to measure the backward production linkages of agricultural growth had a strong positive correlation with RNAE, particularly in developed (western region) regions, as compared to the relatively backward (central and eastern) regions. Among these, consumption of electricity exercised a greater positive influence on the growth of RNAE, even in backward regions (Singh 1994:203). In the case of forward production linkages, the proportion of area under non-food crops was positively associated with the proportion of workers in RNAE, but statistically it was not significant in the state as a whole. Going by the level of development in the state, however, a strong positive impact was discernible in developed (western) regions, whereas in the case of relatively backward regions, a weak negative (central) and positive (eastern) impact was observed on the proportion of workers in RNAE. In only developed regions, thus, backward production linkages, as against forward production, from agriculture turned out to be more effective in the growth of RNAE. While there were several factors outside agriculture, the consumption linkages appeared to have exercised a strong positive influence on the growth of RNAE as compared to production linkages in Uttar Pradesh.

A few studies (Rajasekhar 1995; Rajasekhar and Biradar 1998; Biradar 2007) in Karnataka have found that the commercialisation of agriculture, defined as the proportion of area under non-foodgrains to GCA, had a positive impact on the growth of RNAE. The extent of positive impact, however, declined sharply in 1991 over 1981, implying that the magnitude of RNAE resulting from agricultural growth, through its production linkages, tended to deteriorate and enhance the importance of factors outside agriculture (Rajasekhar 1995; Rajasekhar and Biradar 1998). Similarly, a study by Sharma, Kumar and Sharma (1999) in Himachal Pradesh has shown that a strong positive association existed between the proportion of area under non-foodgrains and the level of RNAE. The impact of commercialisation of agriculture on the growth of RNAE in rural Himachal Pradesh seemed to have come down in 1991 over 1981. The backward production linkages were weak due to industrial backwardness in most of the cities, which were in fact trading centres and administrative units in nature (Sharma, Kumar and Sharma 1999: 261).

A study by Eapen (2001) in Kerala, using the 1971 Census data at the village level has found that the proportion of area under irrigation to total cultivated area was positively associated with the proportion of workers in RNAE (Eapen 2001:81). Higher proportion of area under irrigation, including the use of pumpsets, new variety of seeds, mechanical threshers, chemical fertilizers, repair services and so on, seemed to have stimulated the growth of RNAE mostly through backward production linkages[17]. Narayanamoorthy, Rodrigues and Phadnis (2002) have found a strong positive association between the proportion of irrigated area to GCA (proxied for degree of commercialisation) and the proportion of workers in RNAE. The extent of association between these two variables increased in 1991 over 1971.

Contrary to the above empirical investigations, another category of studies (Singh 1993; Eapen 1995; Basant and Parthasarathy 1998; Mecharla 2002) have argued that the degree of commercialisation does not invariably contribute to the growth of RNAE. For instance, in Haryana, a study (Singh 1993), considering the proportion of area under non-foodgrains as a proxy for commercialisation of agriculture, has found that there was a negative impact on the growth of RNAE in rural Haryana. There were many factors originating outside agriculture such as degree of urbanisation, population density, density of road length, rate of literacy and number of manufacturing units located in urban centres (urban agglomerations), which exerted significant influence on the diversification of rural economy in favour of RNAE. As most of the area was irrigated in Haryana, paddy and wheat were largely cultivated and marketed for commercial purpose, but they were not generally included in the list of non-foodgrains. Consequently, the proportion of area under non-foodgrains was negatively associated with RNAE in Haryana.

Similarly, a study by Eapen (1995) in Kerala, viewing the proportion of non-food crops to GCA, as a proxy for commercialisation of agriculture, has indicated that the degree of commercialisation did not positively contribute to the growth of RNAE. The extent of negative contribution, however, sharply declined in 1991 over 1981, implying that the diversified consumption pattern in favour of superior and processed agro-commodities started to influence RNAE through commercialisation of agriculture. The study has found that consumption linkages were more operative than production linkages in rural Kerala (Eapen 1995).

In rural Gujarat, to capture the commercialisation of agriculture and its impact on the growth of male RNAE, Basant and Parthasarathy (1998) selected the following important indicators, namely, (1) percentage of area under irrigation, (2) percentage of area under non-food crops, (3) horsepower of mechanized equipments (tractors, electric motors and oil engines) per unit of GCA, and (4) number of tractors per unit of GCA. They have found that most of the above explanatory variables, excepting for percentage area under non-food crops, had a positive impact on the level of male RNAE both at district and taluk levels with varying levels of significance, and most of them were significant only at the taluk level. The proportion of area under non-food crops contributed negatively to the growth of RNAE at the district as well as the taluk level. The number of tractors per unit of GCA was the only factor, among others mentioned above, which had a strong positive influence on the growth of male RNAE at district followed by taluk level. A study by Mecharla (2002) in Andhra Pradesh, considering the percentage of non-food cropped area to GCA has also shown an insignificant negative impact on the growth of RNAE, more so in the case of female workers. This was, however, not reflected in terms of close positive association between the proportion of irrigated area and the proportion of workers in RNAE. He has pointed out that there was a conceptual problem with regard to commercialisation, i.e., a large segment of food crops, which are more likely to be processed and marketed, was not included in the list of non-food crops (Mecharla 2002: 813).

Distress Diversification (Residual Sector Hypothesis)

A. Studies at the State/Regional level

The validation of residual sector hypothesis has been hotly debated in the literature. In the absence or lack of agricultural growth-led consumption and/or production (forward and backward) linkage effects on RNAE or non-existence of a mechanism for conversion of agricultural growth impulses into non-agricultural employment opportunities or inability to absorb conventionally reasonable size of labour force in agriculture, the growth of RNAE can be termed as "distress diversification" into unproductive, lowly paid, low status, dead-end non-agricultural jobs. When labourers are not absorbed fully or under-employed in agriculture, the rural non-agricultural sector acts as a "sponge" or "sink" for employment. Such a spill-off of excess labour force from agriculture to the rural non-agricultural sector has been put forward as the "residual sector hypothesis" by Vaidyanathan (1986). This implies that in the absence of improvement in agricultural productivity or prosperity of agriculture, labourers are pushed into unproductive, lowly paid, low status, dead-end non-agricultural jobs.

In order to test the hypothesis, he used NSS person-day (CDS) and US unemployment rates to capture the excess labour problem in agriculture. A strong positive association was observed between the person-day unemployment rates and the proportion of workers in RNAE. Many scholars, however, questioned the validity of taking NSS person-day unemployment rate as a proxy for excess labour supply in agriculture and thereby measuring the distress situation in agriculture and thereby testing the residual sector hypothesis. Does the positive association between the proportion of workers in RNAE and person-day unemployment rate really indicate a distress situation in agriculture? Is the growth of RNAE due to distress in agriculture necessarily in a residual sector?

The residual sector hypothesis put forward by Vaidyanathan (1986) received considerable attention from many scholars (Bhalla 1997; Dev 1990; Unni 1991). Bhalla (1997) has identified two kinds of distress diversification in which rural non-agricultural activities were found to be residual labour force absorbers. The first was the case of supplementary workers who had no main occupation, but were engaged in subsidiary occupation to supplement household income. Such workers were included in subsidiary status of occupation according to NSS estimates. The second was the case of persons with a main occupation who also engaged in a secondary occupation, but were not included in UPS. The workforce engaged in these categories of employment status seemed to be due to distress diversification. In both these cases of distress diversification, the wage rates in the rural non-agricultural occupations (or returns to family labour) were likely to be lower than the prevailing wage rates or even below subsistence. For the 56 NSS regions, Dev (1990) has attempted to test the residual sector hypothesis using the person-day unemployment rate. A strong positive association existed between the person-day unemployment rate and the proportion of workers in RNAE, lending support to the residual sector hypothesis. However, the validity of the residual sector hypothesis has been questioned by Dev (1990) in the context of positive association between agricultural productivity and unemployment rate[18]. The evidence on residual sector hypothesis, however, was not clear with the person-day unemployment rate.

Similarly, Unni (1991) has also questioned the validity of the residual sector hypothesis. At the 56 NSS regional levels, she has found a strong positive association between the person-day unemployment rate and the proportion of workers in RNAE. Whether such growth of RNAE could be regarded as a residual sector has been a matter of intense debate at different levels of academic discussions. She has argued that the NSS unemployment rates captured only open and visible unemployment, which was likely to be higher in agriculturally developed regions. In such areas, there would be in-migration expecting more and better employment opportunities than in agriculturally backward regions. Consequently, the proportion of wage dependent households and casual labourers was likely to be higher in agriculturally developed regions, further raising the reported unemployment rates.

The percentage of casual workers was high in agriculturally developed regions due to low land-man ratio and inequality in land distribution. As a result, the proportion of people reporting unemployment would also be higher. If this were to be the case, to the extent that the proportion of non-agricultural workers was higher in agriculturally developed regions, the unemployment rate would be positively associated with the proportion of workers in RNAE. The unemployment rate was significantly and positively associated with the index of agricultural development and agricultural productivity per hectare, which were also strongly and positively associated with the proportion of workers in RNAE. A strong positive association between the unemployment rate and the proportion of workers in

RNAE in agriculturally developed regions, therefore, would not be surprising, but expected. Unni (1991) further argued that the higher unemployment rate should have reduced the non-agricultural wage rate relative to the agricultural wage rate. The results obtained for the NSS 56 regions also showed that there was no correlation between the proportion of workers in RNAE and the ratio of non-agricultural wage to agricultural wage. In fact, there had been a strong positive correlation between the male non-agricultural wage rate and the proportion of agricultural workers, implying that there was no distress in agriculture (Unni 1991:117).

Unni has argued that the NSS unemployment rate would not be a good proxy to test the residual sector hypothesis for cross-section data. Hence, she considered two variables, namely, the proportion of landless rural labour households and the proportion of people living below the poverty line to capture the distress situation in agriculture and a lowly paid non-agricultural jobs and thereby examining the validity of the residual sector hypothesis.

In regions with a higher proportion of landless labour households and a higher incidence of poverty, the growth of RNAE would depend upon whether there was a demand for non-agricultural goods and services. If these regions were also highly impoverished, then lack of demand might inhibit, rather than facilitate, the growth of RNAE. She has found a strong negative association between the proportion of male and total workers in RNAE and the proportion of people in poverty in developed regions followed by backward regions. This implies that in regions with a higher incidence of poverty, the proportion of male workers in RNAE was lower. This could be due to lack of demand for non-agricultural goods and services in agriculturally backward regions.

In agriculturally developed regions, where unemployment rates were higher, the proportion of male workers in RNAE was also higher, and the proportion of the poor was also lower. In agriculturally developed regions, since the proportion of workers reporting as unemployed was higher, their entry into RNAE was also not necessarily a distress phenomenon. The positive association between RNAE and unemployment rate cannot be, however, treated as a confirmation of the residual sector hypothesis.

She has also found a strong negative association between the proportion of male workers in RNAE and the proportion of landless rural labour households. When the proportion of landless rural labour households was squared, it pointed to a strong positive impact on male RNAE. The relationship between the proportion of male workers in RNAE and the proportion of squared landless rural labour households constituted the "U-shaped curve". This implies that beyond a certain level, regions with higher proportion of landless rural labour households also have a higher proportion of workers in RNAE. A rapid increase in the proportion of landless rural labour households seemed to have created an excess labour supply situation, leading to distress diversification into RNAE. The activities in which they were involved, however, need not be residual in nature. Some of them might be engaged in productive, relatively remunerative non-agricultural jobs, as the incidence of poverty has a strong negative association with RNAE. The growth of RNAE in agriculturally developed regions was, therefore, not inhibited by lack of demand as in the case of regions with a higher incidence of poverty (as there was a strong negative association between the index of agricultural development and the incidence of poverty).

Further, the real wage rates prevailing in agricultural and non-agricultural occupations do not lend support to the residual sector hypothesis. The residual sector hypothesis implies that the high unemployment rates should reduce the non-agricultural wage rates, relative to the wage rates prevailing in agriculture. She has found no correlation between the proportion of workers in RNAE and the wage rates in non-agricultural occupations relative to those in the agricultural sector in 56 NSS regions. An increase in the real wage rates in agriculture shows an excess demand for labour force over supply of labour in this sector. This means that there is no pushing out or compulsion for agricultural labourers to spill over to RNAE, creating a distress induced employment situation. If the growth of RNAE were to be distress diversification : firstly, the real wage rates in agriculture should have declined indicating less demand for labour over supply; secondly, excess labour inflow into the rural non-agricultural sector should have also reduced the real wage earnings absolutely; and finally, the wage rates in the rural non-agricultural occupations should have been lower than those in agricultural occupations (Unni 1997b; Biradar and Bagalkoti 2001). In another study, Vaidyanathan (1994), therefore, changed his argument, maintaining that distress diversification hypothesis was no longer tenable in the context of rising real wage rates. After examining the data, he found that the residual sector hypothesis tended to become valid only when there was a stagnation or decline in the real wage rates in agricultural occupations. It should, however, be noted that distress movement of labour force into the rural non-agricultural sector due to lack of employment elsewhere cannot completely be ruled out.

Parthasarathy, Shameem and Reddy (1998), using the CDS unemployment rate for persons, have attempted to test the residual sector hypothesis for the data collected for 15 major states. They have found that there was a positive association between the proportion of workers, particularly for males, in RNAE and the CDS unemployment rate and arrived at the conclusion that the growth of RNAE was in the residual sector (Parthasarathy, Shameem and Reddy 1998:151). A strong negative association existed between the changes in the share of male workers in RNAE and the changes in the incidence of poverty between two periods (1972-73 to 1977-78 and 1983 to 1993-94). The decline in the incidence of poverty was attributed not to the growth of RNAE, but to a rise in public expenditure (Parthasarathy, Shameem and Reddy 1998:148). A study by Kaur (2002) considered the following variables to capture the distress situation in agriculture : (1) unemployment rate according to CDS; (2) density of population; (3) land-man ratio; and (4) number of marginal landholdings. The unemployment rate, density of population and marginal landholdings had a positive, and the land-man ratio a negative, impact on the proportion of workers in RNAE. He concluded that the growth of RNAE was distress-induced (Kaur 2002:974). A study by Srivatsava and Dubey (2002) has attempted to analyse the distress diversification phenomenon in 71 comparable NSS regions using the incidence of poverty as a proxy for distress situation. They found a negative correlation between the incidence of poverty and the proportion of workers in RNAE. The negative correlation further improved in 1999-2000 over 1983, implying that the higher incidence of poverty discouraged the growth of RNAE due to lack of demand for non-farm goods and services. A weak positive association, however, existed between the change in the incidence of poverty and the change in the proportion of workers in RNAE. The possible explanation is that even in regions where poverty reduction was insignificant, the share of RNAE increased, lending support to Vaidyanathan's residual sector hypothesis. They have concluded that the growth of RNAE in the better-off regions was demand-induced; in the poorer regions, it was an outcome of distress diversification (Srivatsava and Dubey 2002:7501).

B. Studies at the district and taluk levels

Since the data on unemployment rates are not available below the state level, the district/taluk level studies have attempted to test the residual sector hypothesis by considering distress related factors such as lower land-man ratio; higher proportion of landless rural labour households, agricultural labourers, small and marginal landholdings; lower size of landholdings; out-migration of the workforce; higher incidence of poverty; and lower daily real wage rates in agricultural and non-agricultural occupations.

A study by Verma and Verma (1995) has attempted to test the residual sector hypothesis *vis-à-vis* the inter-sectoral linkage hypothesis in the eastern part of the country. The high unemployment rate for males (CDS) and the incidence of poverty were taken to represent the distress phenomenon. They have observed that while the inter-sectoral linkage effect was weak, the residual sector hypothesis was more applicable in the eastern part of the country. While none of the proxies used for inter-sectoral linkage hypothesis, and the factors outside agriculture, yielded significant and expected results, the unemployment rate (CDS) had a significant positive impact on the proportion of male workers in RNAE. The incidence of poverty was negatively associated with male RNAE, but statistically not significant. The negative association might be due to lack of demand and absence of infrastructure facilities in the eastern region with a higher incidence of poverty (Verma and Verma 1995:428). If non-agricultural activities were further visualized in terms of traditional and non-traditional sub-sectors, the validation of residual sector hypothesis was more meaningful in backward areas with a preponderance of traditional non-agricultural sector in which hired labour was hardly used and rather carried out as a means of family subsistence.

Eapen (1995) in Kerala has found that while the proportion of marginal landholdings was significantly and positively associated with RNAE, particularly for females, the proportion of agricultural labourers had an insignificant negative impact on RNAE, particularly for males. A strong negative correlation was found between the proportion of male workers in RNAE and land-man ratio, implying that regions with lower land-man ratio had a higher proportion of male workers in RNAE. The impact of surplus labour in agriculture on the growth of RNAE was captured by another regional level study by Singh (1994) in Uttar Pradesh. The study has attempted to seek validation for the residual sector hypothesis at the regional level by considering the following indicators capturing mainly the population pressure on land: (1) agricultural worker per thousand hectare of net sown area; (2) proportion of agricultural labourers to total workforce; and (3) employment intensity of agricultural workers. These variables reflect the overall demand and supply situation of rural agricultural labourers.

It was hypothesised that the higher the value of these variables, the lower would be the labour absorptive capacity of agriculture and greater the need for seeking employment outside agriculture. While the proportion of agricultural labourers to total workforce and the employment intensity of agricultural workers were significantly and positively related to the proportion of workers in RNAE, the agricultural worker per thousand hectares of NSA was negatively and significantly related to it. This observation provided support to the distress diversification hypothesis that the workforce unable to get adequate work in agriculture would be forced to find jobs in the rural non-agricultural sector so that it acted — on the analogy of the "informal sector" in urban areas — as a "sponge" for the excess labour force in rural areas. It was, however, not necessarily a residual sector, as the incidence of poverty showed a strong and negative association with RNAE.

A few studies (Rajasekhar 1995; Rajasekhar and Biradar 1998; Biradar 2007) in Karnataka have also attempted to examine the distress diversification hypothesis. They have found that the proportion of marginal landholdings was positively, and land-man ratio and proportion of agricultural labourers were negatively, associated with the proportion of female workers, followed by all and male workers, in RNAE, but statistically not significant. They have, therefore, concluded that the growth of male RNAE was better explained by growth related factors, whereas the growth of female RNAE was better explained by distress induced factors. Basant and Parthasarathy (1998), in their study in rural Gujarat, regarded casualisation of labour in agriculture as a factor reflecting compulsion to seek employment in the rural non-agricultural sector. The share of agricultural labourers in the total workforce and the Gini index for land distribution were used to measure the extent of casualisation in agriculture. They found a positive but very insignificant association between the proportion of male workers in RNAE and the percentage of agricultural labourers and the Gini index for land distribution. This showed that although distress situation was in operation, it did not severely cause a distress diversification of workers into the rural non-agricultural sector (Basant and Parthasarathy 1998: 246).

A study by Sharma, Kumar and Sharma (1999) in Himachal Pradesh has found that the Gini ratio of operational landholdings was positively associated with RNAE, but statistically not significant. An important observation was that the incidence of poverty was positively and significantly associated with RNAE, which lent support to the residual sector hypothesis (Sharma, Kumar and Sharma 1999:261-2). In Andhra Pradesh, a study by Mecharla (2002) has also attempted to test the distress diversification hypothesis at the district level. The average size of landholdings and the proportion of people living below the poverty line were taken as representative of distress diversification. He found a negative but insignificant association between the proportion of workers in RNAE and the average size of landholdings and the incidence of poverty. Regions with a higher incidence of poverty had a smaller proportion of workers in RNAE due to lack of demand for non-agricultural goods and services. The growth of modern RNAE was explained by growth related factors, whereas the traditional RNAE was explained by distress related factors such as higher proportion of small and marginal farmers and landless labourers (Mecharla 2002: 817).

Contributing factors external to agriculture

The growth of RNAE was also determined by factors external to agriculture but within the rural economy. Rural infrastructure, particularly road length, telecommunications, banks, electricity consumption and human capital, government programmes, etc., were also important factors external to agriculture contributing to the growth of RNAE.

Rural Infrastructure

A. Studies at the State/Regional level

Availability of rural infrastructure such as road length, telecommunications, banks and electricity has also played a vital role in determining the magnitude of RNAE in a variety of ways. Expansion of roads, transportation and communication facilities certainly encourage specialisation and division of labour by rural households. The infrastructure facilitates exchange of goods and services between villages and small rural towns, and promotes trade, commerce and business in rural areas. Expansion of trade and commerce, marketing and distributional networks linking farm and non-farm sectors depend on the availability of transport and communication (Oshima 1994; Lane 1996; Islam 1997). The ease of transport and communications between rural and urban areas, however, confronts the rural enterprises with increased competition from goods produced in urban areas/

imports. This is especially true for rural small-scale and cottage industries that face competition from cheaper products produced by urban-based large-scale industries or imported from outside the country. Good infrastructure facilities also affect the taste and preferences of rural households and orient them towards urban products, especially in the case of rich households (Islam 1997). This may have an adverse impact on the growth of RNAE.

The extent of impact of production and consumption linkages, degree of urbanisation, urban agglomeration, rural-urban linkages and the like largely depends on the availability of infrastructure facilities both in terms of quantity and quality. It is, therefore, imperative to review the studies aimed at analysing the impact of rural infrastructure on RNAE at different levels. At the aggregate level, few studies have been carried out. A study by Kaur (2002) in 15 major states considering road length per square kilometre as an index of rural physical infrastructure found that there was a strong and positive association between the road length and the proportion of male workers in RNAE, implying that road networks contributed positively to the growth of male RNAE.

B. Studies at the district and taluk levels

In Maharashtra, Shukla (1991), using the proportion of villages with electricity as an index of infrastructure, found an insignificant negative association with the incidence of RNAE, but a strong positive association with the density of RNAE[19]. At the disaggregated level, specifying different proxies for different sub-sectors, she found a strong positive association between the proportion of workers in manufacturing, processing, servicing and repairing activities and the proportion of villages with a *pucca* road; between the proportion of workers in trade, hotels and financing-insurance-real estate-business services and the proportion of villages with infrastructure such as electricity, *pucca* road, post and telecommunication, education, medical and drinking water facilities. These infrastructure facilities, however, had a strong negative impact on the growth of employment in construction and transport-storage-communication (Shukla

1991:2590). A study by Singh (1994) in Uttar Pradesh, considering electricity consumption per thousand NSA as a proxy for rural infrastructure, found a positive impact on RNAE.

In hilly and backward areas, the impact of infrastructure might not be effective. Chand (1996) in Himachal Pradesh, for instance, found an insignificant negative correlation between the proportion of workers in RNAE and road density in 1991. However, when the tribal districts (Lahaul-Spiti and Kinnaur) were excluded, the relationship was positive and more significant (Chand 1996:843). These results, by and large, were consistent with the findings by Sharma, Kumar and Sharma (1999) in Himachal Pradesh during the same period. However, a study (Verma and Verma 1995) in the eastern part of the country, a region, more or less, with the same topographical features as Himachal Pradesh, found a positive association between the infrastructure index estimated by the Centre for Monitoring Indian Economy and the share of workers in RNAE in rural areas of eastern states, but statistically not significant. Mecharla (2002) has considered the number of banks per lakh population and total length of roads per hundred square kilometres of geographical area to assess the impact of infrastructure on the 1981-91 pooled data of RNAE in Andhra Pradesh. The number of banks per lakh population had an insignificant negative impact on RNAE. Going by gender, it had a strong negative impact on female RNAE and positive on male RNAE. A positive but insignificant association existed between the total road length per hundred square kilometres of geographical area and the proportion of workers in RNAE (Mecharla 2002:816). A study by Narayanamoorthy, Rodrigues and Phadnis (2002) has attempted to examine the impact of infrastructure, defined in terms of percentage of villages with a *pucca* road, on RNAE for 1971, 1981 and 1991. The results of multiple regression showed a strong positive association between the proportion of workers in RNAE and the percentage of villages with a *pucca* road; but the extent of impact appeared to have marginally declined in 1991 over 1971. They concluded that the growth of RNAE was an education-infrastructure-induced effect in most of the selected districts.

Human Capital

The rate of participation of workers in RNAE depends on, *inter alia*, the level and quality of their education, which can be considered as a surrogate to human capital (Biradar 2004). The earning levels of the workers also depend on the level and quality of education acquired by them. It has been argued that workers with higher educational attainments are more likely to secure better jobs in the rural non-agricultural sector. Chadha in his studies (1997, 2001a, 2004) has attempted to estimate the human capital base of workers involved in both agricultural and non-agricultural occupations. He has found that at the all India level, the proportion of illiterate workers, both males and females, was higher in agricultural than in non-agricultural occupations in rural areas. With increasing levels of educational attainment, the participation of workers tended to increase in the rural non-agricultural sector and decline in the agricultural sector. With increasing levels of educational attainment, the participation of workers in casual wage employment in the rural non-agricultural sector declined far more than in regular employment (Chadha 2001a, 2004). Similarly, a study by Kaur (2002) also found a strong positive association between the proportion of workers in RNAE and the rate of literacy.

B. Studies at the district and taluk levels

Jayaraj (1989), in his study in Tamil Nadu, has shown that the rate of literacy was significantly and positively associated with male RNAE both at district and taluk levels. Another study (Singh 1993) in rural Haryana has found that the rate of literacy played a strategic role in diversification of employment in favour of RNAE. A study in Kerala (Eapen 1994) has pointed out that a larger proportion of literate workers were involved in non-agricultural, as compared to agricultural, occupations. Further, the study showed that the proportion of workers with a better educational background increased rapidly in non-agricultural, as compared to agricultural, occupations in 1987-88 over 1977-78.

A couple of studies (Chand 1996; Sharma, Kumar and Sharma 1999) in Himachal Pradesh have also revealed that the rate of literacy had a strong positive impact on the growth of RNAE (particularly in non-tribal districts) (Chand 1996: 843). A study by Mecharla (2002) in Andhra Pradesh has found that the growth of literacy was positively associated with the growth of modern RNAE and negatively associated with the traditional RNAE. A couple of studies (Rajasekhar and Biradar 1998; Biradar 2007) in Karnataka have found a strong positive association between RNAE and the rate of literacy, especially in respect of male workers. A similar observation was also made by Narayanamoorthy, Rodrigues and Phadnis (2002).

Contributing factors external to rural economy (Non-local linkage hypothesis)

Although agricultural prosperity contributes to the growth of RNAE through internal factors like consumption and production linkages, factors external to the rural economy have come to occupy an important role in determining the magnitude of RNAE in the recent past (Jayaraj 1989; Unni 1991; Shukla 1991; Bhalla 1993). Factors such as the degree and growth of urbanisation, urban agglomeration, size of nearby villages/cities, public resource allocation for rural development and poverty alleviation programmes, macro level policy environment (economic reforms) and the like are also simultaneously influencing the growth of RNAE in varying degrees across gender and activities. This category of studies questions the strength and validity of inter-sectoral linkage hypothesis and examines the impact of non-local linkages on the growth of RNAE.

Degree of urbanisation

A. Studies at the State/Regional level

At the NSS regional level, Unni (1991) has attempted to estimate the impact of urbanisation, defined as the share of population in urban areas to total population, on the share of workers in RNAE. She found a strong positive impact of urbanisation on male RNAE and weak negative impact on female RNAE in 1977-78. Parthasarathy, Shameem and Reddy (1998) found a positive association between the degree of urbanisation and the proportion of workers in RNAE, but not statistically significant, suggesting that the degree of urbanisation had little influence on RNAE at the macro level (Parthasarathy, Shameem and Reddy 1998:149). The pace of urbanisation slowed down in the 1980s, particularly in 1987-88, the year of drought during which the degree of urbanisation did not produce any positive result due to little demand and production linkages with urban centres.

B. Studies at the district and taluk levels

At the state level, the data seemed to suggest that urbanisation had limited impact on the growth of RNAE. A large number of studies using the district/taluk level data, therefore, have attempted to examine the impact of growing urbanisation on the growth of RNAE. A study by Jayaraj (1989) has attempted to examine the impact of increasing urbanisation through "pull effect" and "push effect" on the magnitude of RNAE. "Pull effect" refers to the tendency among the rural workforce, in the initial stage of urban development, to migrate to the urban centres when better facilities and living conditions are available. This may lead to a decline in the proportion of workers in RNAE. On the other hand, "push effect" refers to the hesitancy of rural workers to stay in urban areas characterised by congestion, bad living conditions and expensive living standards, at a later stage of urban development. They would rather like to stay in rural areas and commute between rural and urban areas for work, as transport and communication facilities are relatively better now. Such process of changes resulting from the "supply-side effect" would lead to an increase in the proportion of workers in RNAE (Jayaraj 1989). On the other hand, the "demand-side effects" of urbanisation include expanding market support for goods and services, both in secondary and tertiary sectors, in the adjoining rural areas to satisfy non-local demands. This process may also lead to an increase in the proportion of workers in RNAE.

In Tamil Nadu, Jayaraj (1989) found a positive association between the degree of urbanisation and the proportion of male workers in RNAE at the district and taluk levels. Further, classifying the taluks on the basis of their degree of urbanisation, he has found that changes in the level of male RNAE were positively related to changes in the degree of urbanisation in taluks with a higher degree of urbanisation in the base year (push factor). The rural workers were hesitant to stay in more urbanised areas when better transport and communication facilities were available to commute between rural habitation and urban work place. A rapid growth in urbanisation, therefore, pushed the rural workers into the rural non-agricultural sector. The study has also found that the taluks or villages, which were located close to towns, had a larger proportion of workers in RNAE.

Similarly, Shukla (1991) in Maharashtra has assessed the impact of urbanisation on the magnitude of RNAE. The degree of urbanisation was significantly and positively related to the proportion of workers in RNAE. At the disaggregated level, urbanisation had a strong negative impact on the growth of employment in the manufacturing sector, particularly in the non-household segment (Shukla 1991). As there had been an increasing demand for manufacturing goods produced in urban centres resulting from changing consumption pattern in favour of urban brand, some rural labourers were pushed into the low-wage segment of rural non-agricultural economy, particularly the household sector. Most of the rural based manufacturing goods appeared to be substitutes for urban based manufacturing goods. A study by Singh (1993) in Haryana found a positive impact of urbanisation on the growth of RNAE. The impact of urbanisation on RNAE in areas with different levels of development indicated that in areas with agriculturally prosperous, well-developed network of transport, a high rate of literacy and presence of big industrial towns[20], the degree of urbanisation was positively associated with RNAE through consumption linkages. In areas which were relatively backward in terms of agriculture, social backwardness due to low rate of literacy and weak rural-urban linkages, the impact of urbanisation on RNAE was found to be very insignificant (Singh 1993:43).

Singh (1994) in Uttar Pradesh has arrived at a similar finding that the degree of urbanisation had a strong positive impact on the growth of RNAE. He also found different results in the areas classified with different levels of agricultural development (developed and backward). In agriculturally backward areas, where the rural demand for non-agricultural goods was low, the degree of urbanisation exercised a positive influence on RNAE. On the other hand, in agriculturally backward areas with a potentially dynamic rural economy, the urbanisation had a depressing effect on RNAE. This could be due to the fact that in the richer regions consumer tastes and preferences in favour of factory-produced goods seemed to be stronger (Singh 1994: 212). Similarly, a study by Eapen (1995) in rural Kerala also showed a strong positive correlation between the degree of urbanisation and male RNAE. Rajasekhar (1995) and Biradar (2007) in rural Karnataka, Verma and Verma (1995) in the eastern states and Basant and Parthasarathy (1998) in Gujarat have found that there was a weak positive association between the degree of urbanisation and the proportion of workers in RNAE, particularly in respect of males, implying that in districts with poor transport and communication facilities, the degree of urbanisation seemed to have an insignificant impact on the growth of male RNAE. The impact of urbanisation on RNAE was quite higher at the taluk level, as compared to the district level. For instance, a study in Gujarat (Basant and Parthasarathy 1998) found a significant positive impact of urbanisation on male RNAE at the taluk level, as compared to the district level. Further, the extent of impact varied with different levels of urbanisation across taluks in rural Gujarat. A strong positive influence of urbanisation on male RNAE was observed only in highly urbanised taluks (Basant and Parthasarathy 1998:251). A study by Mecharla (2002) in Andhra Pradesh has also attempted to assess the impact of urbanisation on male, female and all RNAE. The degree of urbanisation contributed positively to male RNAE, but it was statistically insignificant (Mecharla 2002:813).

At the NSS regional level, Srivastava and Dubey (2002:750-1) found a strong positive impact of urbanisation on the growth of RNAE, which increased in 1999-2000 over 1983, indicating an increasing role of urbanisation in determining the growth of RNAE. A strong positive association existed between changes in the level of urbanisation in 1999-2000 over 1983 and changes in the proportion of workers in RNAE during the same period. The push or spread effects or corridor effects of urbanisation seemed to have had a greater positive influence on the growth of RNAE.

A couple of studies (Chand 1996; Sharma, Kumar and Sharma 1999), however, found different results on the impact of urbanisation on RNAE. Chand (1996) in Himachal Pradesh found a strong negative correlation between the degree of urbanisation and the proportion of workers in RNAE in 1991. The pull effects appeared to have operated in those areas, which were known for tourism and industrial centres. Similarly, another study by Sharma, Kumar and Sharma (1999) has also found that the degree of urbanisation had a strong negative impact on RNAE in rural Himachal Pradesh; the extent of negative impact, however, declined in 1991 over 1981. In their view, such negative impact of urbanisation could be attributed to the mountainous character of the state, which made it impossible to commute from one's residence to the work place daily in rural areas. Thus, the extent and direction of urbanisation impact on RNAE tended to depend on the initial level of urbanisation, agricultural prosperity, topology and transport and communication networks.

Growth of urban agglomeration

Growth of urban agglomeration also influences the magnitude of RNAE in two ways. One is largely from the production side in manufacturing activities (backward linkages), which seeks to capture external scale economies. Regional industrialisation confers broad localisation benefits on similar activity in neighboring rural areas (Shukla 1991,1992). This happens as a greater ease of technology transfer, input availability and business sophistication make all employed resources more productive. This leads to the growth of ancillary industries in rural areas that provide support to the main industries located in urban areas and thereby creating a derived demand for labour in the rural non-agricultural sector. Another one is through what is described as "substitution effects". With the growth of the urban manufacturing sector, the influx of modern and superior goods produced in urban localities into rural areas certainly reduces the demand for locally produced conventional, as also to some extent non-conventional, goods (Haggblade, Hazell and Brown 1989). For instance, plastic items manufactured in urban localities tend to reduce the demand for locally produced pottery related items. As a result of such process, the direct demand for labourers in such non-agricultural activities may decline and to that extent there will be a decline in the growth of RNAE.

The urban agglomeration also contributes to the growth of RNAE through its backward production linkage effects. At the macro level, there is hardly any study capturing the impact of urban agglomeration on RNAE mostly due to data constraint. At the regional level, Shukla (1991,1992) in Maharashtra has attempted to assess the impact of urban agglomeration on RNAE. The average daily employment in factories per lakh district population was used to capture the impact of the manufacturing sector (proxied for urban agglomeration) on RNAE through its backward production linkages. The urban agglomeration had a strong positive impact on the growth of RNAE in rural Maharashtra. Eapen (1995) in Kerala also has attempted to examine the impact of rural-urban linkage characteristics on RNAE in 1981 and 1991 by considering the proportion of medium and small + medium towns to total number of towns and the proportion of population in large villages (above 5,000 persons) to total rural population. No significant association was found to exist between the characteristics of rural-urban linkages and the proportion of male workers in RNAE. The growth of such rural–urban linkages reduced the magnitude of male RNAE (Eapen 1995:637).

Studies Based on Primary Data

A large number of micro level studies based on village surveys at the household/individual level have mainly addressed the following important issues: What are the factors, whether growth or distress in agriculture, due to which the households diversified their occupational structure in favour of RNAE? What are the socio-economic characteristics of the households or workers against which they took up jobs in the rural non-agricultural sector? Which are the rural communities that benefited most from the expansion of RNAE? What are the sub-activities of the rural non-agricultural sector that contributed to a greater decline in poverty and inequality in the distribution of income or assets? What are the characteristics of growth and poverty-induced RNAE? What are the levels of investment, business transaction and margin associated with modern and traditional or growth and distress induced RNAE? The micro level studies which collect both quantitative and qualitative data at the levels of households and individuals can be classified into two important categories (Unni 1998). The first type of micro level studies are those covering a large number of villages and sample size, which do not necessarily focus on employment in the rural non-agricultural sector alone, but provide some analytical insights about the dynamics of different activities within the rural non-agricultural sector (Unni 1998). These studies are mainly descriptive and do not attempt any statistical analysis. Such studies are not many (Chadha 1994; Unni 1997a; Lanjouw and Shariff 2004). The second type of micro level studies are those covering a small number of villages and individual samples, which attempt to understand the rationale and processes of workforce participation in RNAE using statistical or econometric tools.

Based on the findings of detailed household surveys of 18 villages in three diverse development scenarios[21] Chadha (1984, 1994) argued that a rapidly growing and highly productive agriculture was capable of promoting a strong productive network of RNAE avenues within the village economy itself. The development experience of the Punjab state has provided empirical evidence to believe that the magnitude and nature of RNAE depends on, *inter alia,* the stage of (primitive, matured and post-matured) agricultural transformation of a region. He has also observed that the growth of agriculture alone does not automatically foster RNAE (Chadha 1984, 1994). Evidence from another study in six villages of West Bengal (Chandrasekhar 1993) has also illustrated that there was no monotonic association between the growth of agriculture and the proportion of workers in RNAE.

Analysing the relationship between the proportion of workers in RNAE and agricultural development (measured in terms of agricultural crop output per capita) in six villages with different levels of agricultural prosperity (Pre-Green Revolution, Green Revolution and Post-Green Revolution), Chandrasekhar (1993) has found that the average result of aggregation across state/district/region was a result of dominance of regions with a particular level of agricultural development over those that were less or more developed. He has argued that the persistent rise in the proportion of workers in RNAE was not the fall-out of dynamism but rather a combination of inadequate diffusion of that dynamism and the existence of mature Green Revolution areas and states which were losing out in terms of their degree of dynamism. A large segment of the area is yet to experience the Green Revolution phase for the better impact of broad-based agricultural development on RNAE (Chandrasekhar 1993:256). Describing the relationship between the growth of agriculture and RNAE, he has found that the share of workers in RNAE tends to decline (negative association) during the pre-Green Revolution phase and rise during the post-Green Revolution phase (positive association) as agricultural growth further sustains itself and picks up momentum.

A study by Chand (1996) in a cluster of villages in three panchayats representing different levels of agricultural development in Solan district of Himachal Pradesh has shown that backward panchayats had an insignificant proportion of workers in RNAE as their main occupation, whereas in panchayats with agricultural prosperity, a larger proportion of workers were found to be involved in service, trade and business related activities. Due to high adult literacy rate and attainment of a high level of agricultural development long ago in the developed areas, it became possible for a large proportion of younger labour force to undertake more attractive jobs outside agriculture. The differential impact of agricultural prosperity on RNAE was attributed to time lags to usher in growth dynamism in agriculture. The macro level evidence in Himachal Pradesh, however, suggests that there had been an inverse association between agricultural prosperity and RNAE.

It has been, therefore, argued that in the initial stages of agricultural growth, employment in agriculture grows faster than in non-agriculture and the pressure on the labour force to seek lowly paid non-agricultural jobs tends to reduce. Up to this stage, thus, an inverse association tended to exist between the proportion of workers in RNAE and the levels of agricultural development. As agricultural growth picks up and its base tends to expand further, and it ushers in prosperity through improved agricultural incomes, the linkage effect through increased demand for agricultural inputs and services and consumer goods tends to become strong enough to create employment in the rural non-agricultural sector more than in agriculture. Obviously, at this stage, it results in a positive association between agricultural development and RNAE (Chand 1996).

A household level study (Rajasekhar and Biradar1998) in four villages with different levels of agricultural performance in Tumkur and Shimoga districts of Karnataka, has indicated that while agriculturally prosperous villages had a larger proportion of workers in RNAE, particularly in trade, business and agro-processing activities (Kodur and Karehalli in Shimoga), agriculturally backward ones had a relatively smaller proportion of workers in RNAE. This employment may be characterised as distress type, lowly paid and seasonal (Polenahalli in Tumkur district). Further, the growth induced rural non-agricultural activities were characterised by a high level of initial investment, bank borrowings, working capital, employment generation and profit margin, as compared to the distress-induced RNAE (Rajasekhar and Biradar 1998).

Distress-driven movement of workers into RNAE cannot, however, be ruled out *in toto*. Several studies using primary data reveal that when agricultural growth is not able to provide employment throughout the year, a larger proportion of workers are being pushed into the rural non-agricultural sector for employment. A household study in Kot village of Kangra district of Himachal Pradesh (Saraswat, Singh and Daliya 1995) has found that increasing population pressure, marginalisation of landholdings (compulsive factors) compelled the workforce to seek jobs in the rural non-farm sector, which were less dignified and lowly paid (residual sector). A study by Shylendra and Thomas (1995) in the semi-arid region of Panchmahals district in Gujarat has revealed that there were two types of non-farm employment. One type of non-farm employment was artisan activities, services, trade and business, and white-collar jobs which are described as "non-farm proper" and the second type of activity was mainly seasonal and consisted of unskilled manual works known as "non-farm migratory".

The growth of "non-farm proper" activities was largely the result of pull factors, whereas the growth of "non-farm migratory" activities was the result of push factors. Growing population pressure and inability of rain-fed agriculture to absorb surplus labour force resulted in out-migration and forced the workers to seek jobs in less lucrative and undignified non-farm jobs. Therefore, initial unemployment and underemployment in agriculture pushed the workers into the lowly paid and seasonal non-farm jobs, known as "non-farm migratory". Three-fourths of the households in the villages surveyed resorted to seasonal migration, which was a widespread phenomenon in this tribal and agriculturally backward district of Gujarat.

Rajasekhar and Biradar's study (1998) in Karnataka has also found that in agriculturally backward areas as against agriculturally developed ones, the growth of RNAE was largely determined by distress-induced factors such as the proportion of small and marginal farmers and landless labourers, unemployment and underemployment, absence of technological breakthrough, little or no diversification of cropping pattern in favour of commercial crops, frequent droughts and the like. In such areas, the growth of RNAE was largely the result of distress related factors (distress diversification). A study by Nath (1996) in two villages of western Orissa has found that a limited diversification in the cropping pattern in favour of non-food crops resulted in a limited diversification of employment in favour of the rural non-agricultural sector. Basant (1998), surveying in rural Gujarat, has observed that there was a significant positive association between WPRs and the number of activities per worker and the proportion of workers without two-square meals (distress inducement) which was negatively associated with RNAE, particularly for males. A negative but insignificant association existed between male followed by female RNAE and the proportion of irrigated and cultivated area per agricultural worker. The proportion of landless households was significantly and positively associated with male RNAE, lending support to the distress diversification hypothesis (Basant 1998:156).

Sidhu and Toor (2002) have pointed out that there had been a negative but weak association between the proportion of workers in RNAE and labour and land productivity in agriculture, indicating that the areas with low earnings for labourers from agriculture and land productivity were more likely to have a higher share of workers in RNAE (distress diversification). Added to this, an inverse but insignificant association existed between the proportion of workers in RNAE and labour productivity in the rural non-agricultural sector, implying that in backward areas, activities in the rural non-agricultural sector were not lucrative but rather of a residual type (Sidhu and Toor 2002).

The foregoing analysis shows that the importance of consumption and production linkages resulting from agricultural prosperity has been well recognised by many scholars using both secondary and primary data as reviewed above. Another important linkage from agriculture to non-agriculture (i.e., if there is an increase in agricultural wage rate, the workers are more likely to shift their occupation from low productive to high productive RNAE), which is termed as the "labour market interactions" (as mentioned earlier in the review based on secondary data) has been advanced (Hazell and Haggblade 1991). A few primary studies (Chadha 1986; Bhalla, Sharma, Chadha and Kashyap 1989) have found that an increase in the demand for agricultural labour raised the wage rates in agriculture. Consequently, the opportunity cost of the labourers in the rural non-agricultural sector tended to increase. As a result, the workers started to shift from low productive to high productive occupations within the rural non-agricultural sector. The share of workers declined in very low-return household manufacturing and a parallel increase took place in higher-return modern small factories and services.

The growth of RNAE was also largely determined by the socio-economic characteristics of the rural workers. The proportion of workers in RNAE was influenced by the level and quality of education or number of schooling days, household productive assets, caste, age of the working members, gender and so on. Several studies (Shylendra and Thomas 1995; Unni 1996b, 1997a; Saleth 1996; Chand 1996; Basant 1998; Honakeri 2002; Lanjouw and Shariff 2004; Sidhu and Toor 2002) have well recognised the importance of educational status of the workers in determining the size and quality of RNAE. A study by Shylendra and Thomas (1995) in Mahudi village of Gujarat has revealed that persons with higher education had easy access to lucrative jobs in the rural non-agricultural sector. On the other hand, unskilled and illiterate workers seasonally migrated to nearby urban areas and engaged in unskilled and lowly paid non-agricultural jobs. A couple of studies by Unni (1996b, 1997a) in rural Gujarat have found that the participation of workers, particularly males, in better paid non-agricultural jobs increased with increasing levels of educational attainments. The earning levels of the workers in non-agricultural self-employed jobs were largely determined by the level of education of male workers (Unni 1997a). In other words, the workers with lower educational attainments were pushed into low-productive rural non-agricultural jobs.

A couple of studies by Saleth (1996) in Tamil Nadu and Chand (1996) in Himachal Pradesh have found that households with better education were more likely to participate in RNAE as compared to households with no or little education. A large-scale sample survey of 3,760 households spread over 30 villages in rural Gujarat (Basant 1998) has also identified the importance of education in promotion of RNAE. Another large-scale sample of household survey (Lanjouw and Shariff 2004), conducted across 16 major states in India, has also revealed that with increasing levels of education, the proportion of workers in casual RNAE declined and a corresponding increase took place in self- and regular/salaried RNAE. Low educational attainments coupled with no or little wealth and social status, restricted the access for the poor to the relatively more remunerative non-farm jobs, which had a significant poverty reduction effect (Lanjouw and Shariff 2004). A couple of household level studies in Tumkur and Shimoga

(Rajasekhar and Biradar 1998) and Gadag (Honakeri 2002) in Karnataka have shown that workers with more educational attainments gave up their family occupations and sought after modern growth-oriented and more competitive rural non-agricultural jobs. A primary survey of 340 households spread over 17 districts in Punjab by Sidhu and Toor (2002) has also revealed that the level of education had a positive impact on the growth of RNAE.

Further, like education, the asset endowment of a household, particularly land, cattle and wealth also tended to determine the level and nature of RNAE (Saraswat, Singh and Daliya 1995; Unni 1996b, 1997a; Saleth 1996; Nath 1996; Basant 1998; Lanjouw and Shariff 2004; Honakeri 2002; Sidhu and Toor 2002). A study by Saraswat, Singh and Daliya (1995) in Kot village of Kangra district in Himachal Pradesh has found that workers belonging to the medium and large-sized landholdings had a larger representation in relatively better non-agricultural jobs such as trade, business, doctor and laboratory. On the other hand, workers belonging to the small and marginal landholdings took up non-agricultural jobs, which were relatively less dignified, and lowly paid works, like non-agricultural wage employment, artisanship, band playing, mail and water carrier jobs. A detailed household sample survey conducted in rural Gujarat (Unni 1996b, 1997a) has pointed out that households with a very high value of land and other assets had more male representation and investment in high valued rural non-agricultural activities.

In respect of females, their participation in non-agricultural activities was encouraged not by the very high value of land but by the very high value of other assets. On the other hand, households without land and capital had a greater share of workers in low-productive, residual non-agricultural jobs. Similarly, a household survey in Western Orissa (Nath 1996) has also indicated that asset-holding households with easy access to land, non-land assets, capital and education combined their family activity with non-agricultural non-manual, permanent or semi-permanent jobs. The assetless households who lacked education, skill or training faced an entry barrier into more productive, highly paid non-agricultural jobs, but sold their labour in the agricultural, or lowly paid non-agricultural, employment market. A couple of studies in Karnataka (Rajasekhar and Biradar 1998; Honakeri 2002) have shown that households with more education, land and livestock assets were more likely to take up market-driven, high valued rural non-farm jobs in agriculturally prosperous villages, as compared to those in backward villages (Rajasekhar and Biradar 1998).

A study by Sidhu and Toor (2002) in Punjab, known as the "heartland" of India's green revolution, showed a strong negative association between RNAE and the average size of household, average household assets, cattle wealth per household and size of farms. Households which were resource-poor in terms of landed and livestock assets had a higher proportion of workers in RNAE as the labour absorption capacity of agriculture was limited; on the other hand, households which were resource rich had a lower proportion of workers in RNAE.

The magnitude and nature of RNAE was also largely determined by the caste of the household members (Chandrasekhar 1993; Kumar 1996; Rajasekhar and Biradar 1998; Unni 1997a; Lanjouw and Shariff 2004). A study in West Bengal (Chandrasekhar 1993) has pointed out that the workers belonging to upper castes accounted for a major share in RNAE, as compared to lower caste ones (SCs/STs) in the total workforce. A household survey (Kumar 1996) in Bhuvel in Kheda district of Gujarat has showed that a larger proportion of workers belonging to upper castes and rural artisan households took up non-agricultural jobs in business, trade and artisan-based activities, whereas workers belonging to SC/BC households sought after less dignified non-agricultural jobs like casual employment in construction, hotel and the like. A large-scale household survey by Unni (1997a) also yielded similar observation that the male workers with low caste background had an insignificant representation in dignified and high quality RNAE, as compared to upper caste household members. A similar observation has been reported by a study in Karnataka (Rajasekhar and Biradar 1998): participation of workers in RNAE declined with the declining social status of the workers. The lower caste of the workers acted as an entry barrier or constraint to take up certain non-agricultural jobs, and it was more so in backward villages (Tumkur district). A nation-wide large-scale sample survey by Lanjouw and Shariff (2004) has also pointed out that the workers belonging to SC/ST households, who possessed little human capital and assets, faced additional constraints to get jobs in the rural non-agricultural sector which had a greater poverty reduction effect. The age and gender of the workers were also other important factors, which determined the extent and nature of RNAE. The younger male workers were more likely to take up rural non-agricultural jobs, whereas the older male workers were more likely to engage in traditional agriculture (Unni 1996b; Saleth 1996). Even with the increasing age of the workers, the level of income from RNAE also increased (Lanjouw and Shariff 2004). It has been argued that the male workers were more likely to involve in RNAE and the female workers in agriculture and allied activities, mostly vacated by male folk (Chandrasekhar 1993; Unni 1996b; Rajasekhar and Biradar 1998; Lanjouw and Shariff 2004). The proportion of workers in RNAE tended to increase with a large number of household members and the increasing number of active female members (Saleth 1996).

The proportion of workers in RNAE was also determined by the availability of rural infrastructure like roads, telecommunication facilities, electricity, banking and the like (Rajasekhar and Biradar 1998; Honakeri 2002; Sidhu and Toor 2002). A high density of population, the size of the village and its closeness to nearby towns (Saraswat, Singh and Daliya 1995; Unni 1996b; Basant 1998; Rajasekhar and Biradar 1998; Honakeri 2002) and size of population in nearby towns (Basant 1998) also contributed positively to the growth of RNAE, particularly in respect of male workers. Government policies in terms of expenditure for rural development programmes appeared to have made an insignificant positive impact on the growth of employment in the rural non-agricultural sector (Chandrasekhar 1993; Nath 1996; Honakeri 2002). Though many of these programmes were meant for the benefit of the poor, adequate employment in the rural non-agricultural sector had not been generated, as many of the poor lacked the minimum level of education, skill, knowledge, assets and social status which acted as a major constraint in their endeavours.

RNAE and Poverty and Income Inequality

Impact of RNAE on poverty

Poverty alleviation and reduction of inequality in the distribution of income, wealth and assets have received a considerable importance in the development paradigm, as elsewhere, in India since the 1970s, especially when diversification of the rural economy in favour of the rural non-agricultural sector begun to occur. The distributional impacts of non-agricultural income in rural areas of developing economies are important because additional increase in rural income and employment is occurring outside agriculture. The proportion of households drawing their major segment of employment and income from rural non-agricultural activities is increasing much faster among the resource poor households. As the daily real wage rates and income levels in rural non-agriculture *vis-à-vis* agriculture are higher, the incidence of poverty in rural non-agriculture is far less as compared to agriculture. The growth of RNAE, particularly wage based/self-employed jobs accessible to the poor, plays a vital role in addressing the problem of poverty as government expenditure for poverty alleviation programmes has been gradually constrained in the context of globalisation.

Evidence from five countries in Asia and Africa shows an inverse association between the size of landholdings and the share of non-agricultural income in total household income, implying that for the smallest landholding categories, the non-agricultural income sources accounted for more than 50 percent of the household income (Liedholm and Kilby 1989). Household level studies by Lanjouw (1999) and Elbers and Lanjouw (2001) in Ecuador have also observed that the expansion of RNAE, particularly more productive ones, significantly raised the income levels of the poor. The growth of RNAE could also contribute indirectly to the reduction of poverty *via* raising agricultural wages and improved agricultural productivity of labour due to shift of workers to the rural non-agricultural sector (Lanjouw and Shariff 2004). Similarly, a micro-level study in rural China by Kung and Lee (2001) showed a decline in the incidence of poverty due to an increase in the share of non-agricultural income. Escobal (2001) in rural Peru has found that for the poor, rural non-agricultural activities were "last resort" activities in nature. Although they were not associated with higher levels of labour productivity, they, to some extent, contributed to a decline in the incidence of poverty. Thus, evidence from other developing countries shows that the growth of RNAE certainly contributed to the reduction of poverty in varying degrees (Samuel 1979; Liedholm and Kilby 1989; Lanjouw 1999; Elbers and Lanjouw 2001; Kung and Lee 2001; Escobal 2001).

In India, at the aggregate level, Dev (1990), Unni (1991), Verma and Verma (1995) and Parthasarathy, Shameem and Reddy (1998) have considered the incidence of rural poverty as one of the determinants of RNAE and found a negative association between these two variables. Most of the studies interpreted this association that regions with a higher incidence of poverty had a lower share of workers in RNAE due to lack of demand for non-farm goods and services. The reduction of poverty would rather promote RNAE. Dev (1990) in his study, however, has found that at the macro level, the incidence of poverty was less in rural non-agriculture and more in agriculture, implying that the growth of RNAE had a relatively better poverty reduction effect. Another study by Biradar and Bagalkoti (2004), covering 16 major states in India, has found that the growth of male RNAE was significantly and negatively associated with the incidence of poverty, whereas the growth of female RNAE was positively associated, but statistically insignificant.

Most of the studies based on primary data have focused on exploring the socio-economic characteristics of individuals or households under which different types of non-farm jobs are more likely to be taken up. Some studies (Unni 1997a; Rajasekhar and Biradar 1998; Lanjouw and Shariff 2004), however, have attempted to examine the contribution of RNAE to the reduction of poverty in rural areas. Unni (1997a) has pointed out that the expansion of RNAE not only minimized year-to-year fluctuations in the income levels of the agricultural households, but it also strengthened the base of livelihood of the poor, hence contributing to a sustained decline in the incidence of rural poverty. A study by Rajasekhar and Biradar

(1998) has observed that the income levels of the poor non-farm households had gone up significantly, though per capita income of the poverty-induced RNAE was lower than that of the growth-induced RNAE.

Lanjouw and Shariff (2004) have found that non-agricultural incomes accounted for roughly one-third of the household income, which varied considerably across the states and income quintiles. While in aggregate terms, total non-agricultural incomes were not particularly important for the richer quintiles than the poorer quintiles; the poorer households earned significant shares of total income from casual rural non-agricultural wage employment[22]. But, the growth of casual wage employment had a lesser poverty reduction effect, as compared to that of self-employment in the rural non-agricultural sector. The rich mainly earned non-agricultural income from salaried employment. The study has also pointed out that the indirect impact of RNAE on poverty reduction was observed through an increase in agricultural wage rates due to shifting of workers to rural non-agriculture[23]. The poor were not particularly well placed to benefit from the expansion of the rural non-agricultural sector. Low educational attainments, scant wealth and lower social status appeared to have restricted their access to the relatively more attractive, lucrative non-agricultural jobs that were more likely to lift them out of poverty. The expansion of RNAE certainly improved the income levels of the rural non-agricultural households and contributed to a decline in the incidence of poverty.

Impact of RNAE on income inequality

Another important effect of RNAE, is related to the distributional pattern of income and assets across rural households. Whether the growth of RNAE or income reduces inequality or not depends on, *inter alia*, the rate of benefits accruing to the poor and non-poor from RNAE in which they are involved. If the poor are able to get more income from RNAE than the non-poor, the inequality in the distribution of income tends to fall. It has been, therefore, hypothesised that the growth of RNAE reduces inequality in the distribution of income. A few empirical studies, however, have presented divergent results relating to the distribution of non-agricultural income among rural households.

Although non-agricultural income levels of households with a smaller size of landholding exceeded the incomes of those with a larger size of landholding, a "vertical J" relationship between the rural household income and the size of landholding seemed to be evident only in some parts of the world. Since the data on non-agricultural income at different income quintiles were not available, many studies considered the size of landholding as a proxy for rural household income and attempted to examine the impact of RNAE on the distribution of income. Different types of non-agricultural activities were found at the lower and higher ends of income distribution. For the poor households, wages from construction, brick-kiln and service type activities are the predominant sources of livelihood. On the other hand, in the case of rich households, income from manufacturing activities tends to be predominant. Liedholm and Kilby hypothesise that these latter activities tend to have higher entry barriers and yield higher returns than agriculture or other types of non-farm jobs (Liedholm and Kilby 1989:346-7). Evidence from some Asian and African countries suggests that the Gini coefficient tended to decline when income from non-agricultural activities was included (Liedholm and Kilby 1989:347). Chinn (1979) and Samuel (1979) in Taiwan point out that the income from non-farm sector reduced income inequality in rural areas. According to Chinn, the non-farm income benefited the poor because the share of non-farm income varied inversely with farm size.

A micro level study by Adams (1994) in rural Pakistan has shown that while the income from agriculture was the single most contributing factor for increasing income inequality, the income from rural non-farm jobs and livestock reduced the income inequality. Additional increments of income from rural non-farm jobs or livestock contributed to a decline in overall income inequality. Within the rural non-agricultural sector, only non-agricultural unskilled labour represented an inequality-reducing source of income; and non-agricultural government employment represented an inequality-raising source of income[24]. Leones and Feldman (1998) in their household level study in northern Leyte, Philippines, have found that income from non-agriculture was more equally distributed, as compared to income from agriculture across all households. An increase in the non-agricultural income tended to increase income inequality to a much lesser extent than an increase in remittance income, though it was more equally distributed than the income from all other sources.

Another category of studies, however, has found that the expansion of RNAE raised the income levels of the poor but it also raised income inequality across rural households. Reardon, Delgado and Matlon (1992) in Burkina Faso, Collier, Radwan and Wangwe (1986) in Tanzania and Matlon (1979) in Nigeria have observed that income from the rural non-farm sources contributed further inequality in the distribution of total income. Such inconsistency was found perhaps due to difference in the distribution of land and educational facilities. In land-scarce, labour-rich settings, like Taiwan and much of Asia, small and inadequate landholdings might push the poorer households out of agriculture into the rural non-agricultural sector. Elbers and Lanjouw (2001) in Ecuador have observed that high productive RNAE sharply raised the average income and reduced poverty, but it raised income inequality as compared to low productive RNAE. Entry into high productive RNAE was confined to workers with increasing levels of education and wealth.

Escobal (2001) in rural Peru has found that with increasing levels of consumption quintiles, the share of income from non-agricultural occupations increased and that from agricultural occupations declined. An increase in income from non-agricultural occupations, particularly from self-employment, followed by wage employment, contributed to an increase in income inequality. Janvry and Sadoulet (2001) in Mexico have found that the income from agriculture, followed by non-agricultural wage, was more unevenly distributed as compared to income from agricultural wage, followed by self-employment and remittance. An increase in the income from agricultural and non-agricultural wage employment tended to have an adverse impact on income distribution, as compared to an increase in the income from agricultural wage, self-employment and remittance. In rural China, a household level study by Kung and Lee (2001) has also observed that an increase in the share of non-agricultural income reduced the incidence of poverty, but aggravated the degree of income inequality.

In India, there are not many studies examining the impact of RNAE on the distribution of income across rural households. A household level study by Unni (1997a) in six villages of Mehasana district in Gujarat has found that returns from economic activity or labour were relatively higher in non-agriculture than in agriculture. But an increase in non-agricultural income raised the degree of inequality in the distribution of rural income and ownership of land. It has been argued that if macroeconomic policies push the workers into low-productive non-agricultural jobs, the level of inequality in rural income is more likely to increase. A household survey conducted in three villages of Moradabad district of Uttar Pradesh (Birthal 1997) has found that incomes from livestock, followed by agriculture, was the most equally distributed, whereas the distribution of income from salaried employment was the most inegalitarian. Income from non-agricultural labour and all other non-agricultural jobs was more unequally distributed, as compared to livestock and agriculture.

An additional increase in agricultural income, followed by livestock and salaried jobs, contributed most to income inequality. An increase in income from business and art crafts contributed the least to income inequality. An increase in income from wage employment, both in non-agriculture and agriculture, reduced income inequality. Whether the income inequality tended to decline or not, depended on the availability and accessibility of non-farm jobs and levels of income accruing to households with different income quintiles. Depending upon the level of entry barriers to RNAE experienced by different income quintiles, the impact of RNAE on income distribution was found to be quite divergent across different parts of the world.

Issues for Research

From the foregoing review of literature, the following important research issues have been identified. As evident from the review of literature that no satisfactory reasons have been discovered for the observed decline in WPRs by many studies. There are few studies available analysing the changing facets of employment both in rural and urban areas during the pre-reform and reform periods. It is evident that, going by sectoral patterns, most of the studies have focused on delineating the sectoral diversification of employment away from the agricultural to the non-agricultural sector, especially in rural areas. From the policy viewpoint at the micro level, it is imperative to explore the diversification of economic activities within and between agricultural and non-agricultural sectors. Inter and intra-activity diversification of workers may help the policy makers to chalk out an appropriate strategy for balanced development of each activity, which has a greater poverty reduction and income equalizing effect.

As regards spatial patterns, the regional dimension of RNAE measured by agricultural prosperity (measured in terms of higher per capita income), rate of literacy, unemployment and incidence of poverty in the years of pre-reform and reform periods may help to decide the type of policy intervention needed to reduce regional disparity in the growth of RNAE. There is hardly any study examining whether the extent of regional variation in RNAE has been on the rise or not, during the reform period *vis-à-vis* the pre-reform period. Examining whether the degree of marginalisation or subsidiarisation of RNAE among agricultural or non-agricultural households is increasing or not, in the era of market-driven economy, is another important issue for research.

As regards the determinants of RNAE, there needs to be conceptual clarity about commercialisation of agriculture. Most of the studies under review have defined commercialisation of agriculture as the proportion of area under non-foodgrains to total cropped area. However, a few crops, which are commercial-oriented, have not been generally included in the list of commercial crops. For instance, paddy, wheat, maize, etc., are some of the important crops cultivated in a larger proportion of area, especially in the irrigated regions, for commercial purposes. As a result, the impact of production linkages of agricultural growth on RNAE may be a misnomer. It is necessary to revise the concept of commercialisation of agriculture by including some other crops like paddy, wheat or maize and examine the impact of production linkages of agricultural prosperity on the growth of RNAE.

There has been an intensive debate on distress diversification of RNAE. As there has been a greater incidence of poverty and unemployment in the economy, some part of the growth of RNAE is ascribed to distress-induced factors, but not necessarily in the residual sector. Most of the studies have considered the unemployment rate as an index of distress in agriculture and attempted to test the residual sector hypothesis. A strong positive association between the proportion of workers in RNAE and the unemployment rate is taken to mean that the growth of RNAE was distress induced and residual in nature. As educated persons constitute a major segment of general unemployment, the unemployment rate cannot be an appropriate indicator capturing distress in agriculture. The factors capturing distress in agriculture may not necessarily be the factors capturing the residual nature of RNAE. The factors, which satisfactorily capture distress in agriculture and quality of RNAE, therefore, are required to be identified. Of course, the incidence of poverty and nature of its association with RNAE may be considered as one of the factors measuring the quality of RNAE, but it is not adequate as it is one of the determinants of it.

The review results of the growth process of RNAE at the state, regional/district and taluk levels show that the growth of RNAE whether it is determined by growth or distress-related factors, depends on, *inter alia,* the level of agricultural prosperity of the state, region/district or village. A study analysing the determinants, nature and quality of rural non-agricultural activities and the extent of their impact on reduction of poverty and income inequality is required to be made in areas with different levels of agricultural prosperity. Such a type of study will help to chalk out well-conceived programmes to tackle the problems of unemployment, poverty and income inequality simultaneously. To provide policy emphasis on the specific determinants of RNAE in the development strategy, there is a need to estimate the extent of contribution of each determinant in areas with different levels of agricultural prosperity. The growth process of RNAE for males is not the same as that of females in areas with different levels of agricultural prosperity. Exploring the growth process and structure of RNAE for males and females separately would be helpful to work out the policy package for promotion of high productive RNAE. Human capital measured in terms of literacy plays an important role in determining the extent and quality of RNAE. Hence, estimating the returns from education at different levels to different types of non-agricultural activities is also another area of research. In the rural non-agricultural sector, "social capital" is one of the key factors in the successful management of non-farm enterprises. Therefore, the role of social capital in the rural non-farm sector needs to be explored.

Depending upon the availability of data and resources, the following important research issues have been identified for analysis in the present study. It has been found that most of the studies have attempted to identify factors contributing to the growth of only male RNAE as a whole. As the factors determining the male RNAE are different from those of female RNAE, exploring the factors that determine the level of female participation in RNAE is also quite essential for policy intervention. The extent of male and female participation in RNAE by age groups, education and size of landholdings is also felt increasingly important. There are few studies conducted : to explore the factors contributing to the growth of sub-activities for males and females separately; to estimate the elasticity of important determinants of male and female RNAE as a whole and also sub-activities; and to identify the male and female rural non-agricultural activities which are mainly growth or distress-induced. It is essential to identify the important activities in the rural non-agricultural sector which are mainly growth induced, more productive and better paid; as also whether such activities have a greater impact on poverty reduction and income equalisation, as compared to that of distress induced, lowly paid activities.

The extent of poverty reduction depends on, *inter alia*, the type of activity status of RNAE into which the poor are involved. The participation of male and female workers in different types of RNAE (self-employment, regular and casual) is, therefore, another area of research. A study exploring the qualitative dimensions of RNAE during the pre-reform and reform periods may help policy makers to promote qualitative employment in the rural non-agricultural sector to be accessed by the poor. Hence, issues relating to the nexus between RNAE and rural poverty, the extent of impact of RNAE on poverty reduction during the pre-reform and reform periods, and poverty alleviation approach in the context of globalisation are required to be examined. More importantly, at the macro level, the impact of RNAE on income inequality has been little studied. The need of the hour is not only to explore the cause and nature of RNAE, but also its impact on poverty, inequality and unemployment. The present study has modestly attempted to cover some of these research gaps identified above under the given data constraint.

NOTES

1. It is rather difficult exercise to work out the levels of employment and income shares from RNAE due to differences in the concepts adopted in the national census and nationwide labour surveys in different countries (Anderson and Leiserson 1980).
2. The data on RNAE in each country of Latin America and the Caribbean are not made available for comparable years. The years for which the data are referred to differ from country to country ranging over the period between 1980-1990. A comparison across the country over time, therefore, becomes difficult.
3. In Western Nigeria, the share of male workers in RNAE accounted for 20 percent and female RNAE for 97 percent in 1966, whereas in the case of Rwanda, males accounted for 9 percent and females for 1 percent in 1978 (Reardon, Berdegue and Escobar 2001).
4. The concept of economic activity according to NSSO is broader than that of the Census. The data are being collected by trained enumerators of the NSSO (Deshpande 2003).
5. The divergent observation, in part, could be due to the change in the period selected for analysis. Krishnamurty (1984) used the census data for 1971 and 1981 and the NSS data for 1972-73 and 1977-78, whereas Basant and Kumar (1990) used the census data for 1961 and 1981and NSS data for 1972-73 and 1983.
6. While the contribution of rural areas to employment generation grew in absolute terms in all the three decades from 1961 to 1991, there was a

sharp contraction in the absolute number of non-agricultural workers in rural areas in the decade 1961-1971.

7. A modified version of Rodger's index was computed at the state and district levels for 1961, 1971 and 1981. This index is designed to measure the changing composition of the non-agricultural segment as well as the relative shares of workers in agriculture and non-agriculture in the total rural workforce (Bhalla 1997).
8. The deceleration in the proportion of female workers in RNAE can be attributed to their own weak educational and skill background. Whatever new jobs which are coming up in this sector are qualitatively superior and skill selective, for which the big mass of unskilled or uneducated female workers is not qualified (Chadha 2001b; Chadha 2001a; Biradar and Bagalkoti 2001).
9. Although the values were relatively higher, the 1961 Census estimates were likely to be underestimated because those cultivators, agricultural labourers and household industry workers who were in secondary activities other than the three, i.e., household industry, cultivation and agricultural labour were not tabulated. In fact, the estimates of both the Censuses can be considered underestimates of secondary work in so far as they do not cover more than the two activities undertaken by a worker (Basant and Kumar 1989:36).
10. A rapid increase in the proportion of workers in casual RNAE was attributed to the severe drought of 1987-88. A larger proportion of rural workers, including female agricultural labourers, were forced to participate in drought relief works as casual labourers; as a result of which the proportion of casual workers in the rural non-agricultural sector increased rapidly.
11. Incidence of person-days unemployment is defined as the ratio of unemployed person-days to labour force person-days (NSSO 2001).
12. Haggblade, Hazell and Brown (1989) identified five different linkages through which the growth of agriculture and non-agriculture tends to mutually reinforce, as economic growth picks up over time. The linkages include capital and labour flows (input market) and backward and forward production linkages and consumption linkages that are important for sectoral diversification.
13. The agricultural growth multipliers in Africa were lower than those measured in Asia for a variety of reasons. In Africa, the rainfall patterns and the geology of its river basins tended to preclude cost-effective irrigation on a large scale as in Asia. Hence, backward linkages into pump supply, canal construction and maintenance, which are currently

important in Asian countries were simply unavailable in Africa (Haggblade, Hazell and Brown 1989).

14. A strong negative correlation coefficient existed between the growth rates of foodgrains production and differences in the proportion of male and female workers in RNAE in 1987-88 over 1977-78 for both females (-0.84) and males (-0.72) (Parthasarathy, Shameem and Reddy 1998).
15. Harriss (1987) attempted to identify regional growth linkages from agriculture to non-agriculture in terms of: (1) nature of linkages from agriculture, i.e., forward, backward production or consumption linkages; and (2) flows of resources, i.e., commodity flow, flows of finance and labour (Harriss 1987: 32).
16. The declining land-man ratio, coupled with low labour use in commercial crops, acted as powerful 'push' out of agriculture, making the population mobile both across space and occupations. Hence, the proletarianisation of the workforce not totally dispossessed of land started much earlier in Kerala, pushing a larger proportion of workers into the rural non-agricultural sector (Eapen 1994: 1287).
17. The inferences drawn from the 1971 Census data at the village level, however, may not reflect the present process of linkage effects of agricultural growth on RNAE in Kerala.
18. The significant positive association between agricultural productivity and rural unemployment could be due to: (1) labour displacing technology in high agricultural productivity regions; (2) in-migration of labour force from neighbouring areas to agriculturally high productivity regions; and (3) NSS unemployment mainly captures visible and open or reported unemployment, which would be higher in agriculturally developed regions (Dev 1990:1531).
19. Non-agricultural employment share refers to the ratio between non-agricultural workers and total rural main workers; incidence of RNAE refers to the proportion of main workers in RNAE to rural district population; and density refers to non-agricultural workers per square km. of district rural geographical area (Shukla 1991).
20. The industry like HMT factory at Pinjore; cement factory at Surajpur; sugar factory, paper, utensil making and railway workshop at Jagadhari-Yamuna Nagar; chemical, scientific instruments manufacturing and military cantonment at the twin (Ambala city and Cantonment) and state capital cities (Chandigarh); and heavy industries like tractor manufacturing, Maruti Udyog, motor cycles like Hero Honda and Rajdhoot manufacturing, chemical and paper industries and their military units at Faridabad, Gurgaon, Panipat, Sonepat and National capital Delhi itself (Singh 1993: 43).

21. The diverse development regions include a backward region (Bihar), an agriculturally developed region with little non-agricultural activities (Andhra Pradesh) and a highly agriculturally developed region with large non-agricultural activities (Punjab) (Chadha 1994).
22. Incomes from non-agricultural employment are classified into three categories: income from causal non-agricultural wage labour; own enterprise activities; and regular/salaried non-agricultural employment. This ordering of the three categories reflects their relative importance as sources of income to the poor (Lanjouw and Shariff 2004).
23. Indirect reduction of poverty can be observed through an expansion of construction activities, which do not require any specific skills and knowledge, yet are generally preferred over agricultural wage employment. Expansion of employment opportunities in construction is likely to siphon labour out of agriculture, thereby leading to an increase in agricultural wage rates. A significant increase in agricultural wage rates can have a greater impact on poverty reduction in rural areas (Lanjouw and Shariff 2004).
24. Government employment which results in income inequality has higher entry costs in the form of education and bribes which make this source of non-farm income more accessible to rich households (Adams 1994).

3 Dynamics of Employment and Unemployment in India

Emerging Issues and Challenges

Introduction

The growth of sustained, qualitative and equitable employment opportunities has been regarded as one of the important policy instruments to tackle the long-standing problems of unemployment, poverty and income inequality since the inception of the planning era. It received the serious attention of planners when they started confronting the problems of growing educated youth unemployment, shrinking employment opportunities in agriculture and the urban organised manufacturing sector, deteriorating quality of employment, persistence of poverty and income inequality even after five decades of developmental struggle. It has further received special attention since the 1990s, at a time when India started bringing about major policy breakthrough at the macro level by adopting the neo-liberal development paradigm (i.e., liberalisation, privatisation and globalisation) and also the commitment of developmental planning to generate "ten million employment opportunities per year" (GoI 2002c)[1] during the tenth plan period and halve the incidence of poverty by 2015, as per the Millennium Development Goals set forth at the UN Assembly in 2000 (WB 2001).

The import-substituting development strategy, followed in the mid-1980s, was given up in favour of the export-linked development strategy in the 1990s for sustained growth of national income as well as aggregate employment thereby achieving a sustained reduction of poverty. With the signing of WTO accord in 1995, India further integrated into the global economic system. The entire policy framework appeared to be gearing up to foster a competitive environment and improve the resource-use efficiency. As a result of such changes at policy levels, every sector of the Indian economy was expected to undergo far-reaching changes. One such explicit change is on the employment front in terms of its growth rate and quality. One of the important channels through which the on-going reform measures may impact the economy is through the labour market and the structure of employment growth, both in terms of quantity and quality, in India. In view of such changes at the macro-level policy environment, on the one hand, and slow-down in the growth rates of national output and employment, on the other, it has become important to examine the dynamics of employment and unemployment in India, especially in the era of globalisation.

Growth of Employment

With the introduction of economic reforms since July 1991, there has been an overwhelming debate on their impact on employment, poverty and the well-being of the poor in India, especially in rural areas. In the 1990s, with a massive use of labour saving techniques and labour displacement mechanisms in the urban organised sector, the share of the unorganised and informal sector, both in rural and urban areas, increased and also became more complex and competitive[2] (Tulpule 1997; Papola 1999). A setback in the employment growth rate and WPRs, particularly in rural areas, on the one hand, and a rapid decline in the quality of existing employment in terms of casualisation and feminisation, on the other, and their implications for sustained economic growth and generation of productive employment opportunities in the rural economy became central issues for research and discussion at different levels of academic debate. The basic questions that have been addressed in this chapter are: Has the growth rate of employment declined during the reform period? If so, to what extent and among whom, male or female or both; among educated persons or persons with lower or higher levels of education or illiterates; among youth or older-age working population; has the decline in employment growth rate correspondingly resulted in an increase in unemployment rates across gender, age groups and levels of education? If not, has the decline in WPRs also been accompanied by an increase in SPR and/or the ratio of population neither in labour market nor in schools/colleges? Answers to these typical questions have been, in what follows, attempted through an analysis of the growth rate of employment and WPRs.

Compound annual growth rate of employment

The most daunting challenge of the present development paradigm is to provide sustained and guaranteed employment not only to the new entrants into the labour market but also to reduce the backlog of unemployment of the past. In the context of the growing labour force resulting from high growth rate of population in the 1970s and 1980s, on the one hand, and a continuous decline in employment elasticity with respect to GDP[3], on the other, generation of guaranteed employment became the central theme of the entire policy-restructuring exercise of development planning since the 1990s. The economic reforms were expected to create more employment opportunities primarily because the growth of output was likely to pick up after the economic reforms. Even after more than a decade of economic reforms, however, the growth of output and employment has not gone up as much as was expected by the protagonists of the economic reforms.

Table 3.1 presents the data on the compound annual growth rate of workers, according to UPS and UP+SS, in India during the period between 1977-78 to 1983 and 1993-94 to 1999-2000. The data, *prima facie*, reveal that the annual growth rate of workers was higher in urban than in rural areas throughout the period under study, implying that a good deal of employment opportunities were largely generated in urban areas. The growth rate of employment declined considerably during the reform period (1993-94 to 1999-2000) as compared to the pre-reform period (1987-88 to 1983 or 1983 to 1993-94), according to both UPS and UP+SS, at the all India level and in rural as well as urban areas. For instance, at the all India level, the annual growth rate of employment, as per UPS, declined significantly from 2.03 percent during 1983 to 1993-94 (2.23 percent during 1987-88 to 1993-94) to as low as 0.83 percent during

1993-94 to 1999-2000. Going by gender, the annual growth rate of employment for males declined from 2.25 percent (2.59 percent during 1987-88 to 1993-94) to 1.63 percent, whereas in respect of females, it declined from 1.45 percent (1.27 percent during 1987-88 to 1993-94) to 0.82 percent during the same period. Similar trends, more or less, existed in rural and urban areas. In rural areas, the growth rate of employment for males, according to UPS estimate, declined from 1.9 percent during 1983 to 1993-94 (2.25 percent during 1987-88 to 1993-94) to 0.61 percent during 1993-94 to 1999-2000. In respect of female workers, it also declined rapidly from 1.15 percent (0.87 percent) to (–) 0.16 percent during the same period.

Similarly, in urban areas, the growth rate of employment for male workers declined from 3.31 percent (3.57 percent) to 2.64 per cent and for female workers from 3.35 percent (3.64 percent) to 1.87 percent, which was found to be quite higher as compared to their

Table 3.1 : Compound Annual Growth Rate of Workers in India, 1977-78 to 1999-2000

(percent)

Period/Areas	*UPS*			*UP+SS*		
	Males	*Females*	*Persons*	*Males*	*Females*	*Persons*
Rural						
1977-78 to 1983	1.51	1.77	1.59	1.82	2.52	2.08
1983 to 1987-88	1.43	1.52	1.46	1.44	0.55	1.12
1987-88 to 1993-94	2.25	0.87	1.84	2.20	1.88	2.09
1983 to 1993-94	1.90	1.15	1.67	2.05	1.40	1.81
1993-94 to 1999-2000	0.61	-0.16	0.34	0.53	-0.32	0.23
Urban						
1977-78 to 1983	4.23	4.18	4.22	3.89	3.44	3.80
1983 to 1987-88	2.97	2.95	2.97	2.57	2.79	2.61
1987-88 to 1993-94	3.57	3.64	5.59	4.40	3.69	4.25
1983 to 1993-94	3.31	3.35	3.32	3.93	3.61	3.86
1993-94 to 1999-2000	2.64	1.87	2.48	1.43	0.84	1.31
All India						
1977-78 to 1983	2.11	2.06	2.10	2.28	2.63	2.40
1983 to 1987-88	1.80	1.71	1.77	1.71	0.82	1.41
1987-88 to 1993-94	2.59	1.27	2.23	2.76	2.13	2.55
1983 to 1993-94	2.25	1.45	2.03	2.51	1.69	2.24
1993-94 to 1999-2000	1.63	0.82	0.83	0.78	-0.15	0.48

Source: NSSO (1981, 1988, 1990, 1996 & 2001).

rural counterparts during the same period. While there had been a negative growth rate of employment for females in rural areas, a significant decline, but still a positive growth, was observed for females in urban areas in the reform regime. The decline for male and female workers was found to be faster according to UP+SS estimates as compared to that from the UPS perspective, especially during the reform period. This clearly indicates that the number of persons who were employed under subsidiary status declined much faster than under the UPS during the reform period as against the pre-reform period. It is, thus, evident from the above analysis that there was a considerable decline in the growth rate of employment in the 1990s; a greater decline took place in rural areas as compared to urban areas. A serious setback occurred in respect of rural female workers who experienced a negative growth rate, whereas in respect of urban females, it registered a positive growth rate, but it declined rapidly during the reform period. A major segment of decline came from the subsidiary status of occupation. Several other studies (Dev 2000a; Sundaram 2001b; Chadha 2001a; Chadha and Sahu 2002; Hirway 2002; Bhalla and Hazell 2003) have also pointed to similar trends of employment growth rates both in rural and urban areas in the 1990s.

Work participation rates

The 55th round of NSS (1999-2000) showed a rapid decline in WPRs[4], especially in rural areas, over most of the earlier rounds. The decline in WPRs has raised several issues relating to the growth of employment and structure of the labour market in India. A widely debated issue is: Why have the WPRs sharply declined in the 1990s? A few studies (Biradar and Bagalkoti 2001; Sundaram 2001b; Chadha and Sahu 2002; Hirway 2002) have attempted to explore the possible reasons for the observed decline in WPRs, especially in the 1990s. They have found that the observed decline in WPRs was due to an increase in SPR and a decline in the share of workers under subsidiary status of occupation. Further, Sundaram (2001b) has argued that the decline in WPRs in the 1990s was exaggerated by the fact that the 1993-94 levels were outliers when compared with earlier quinquennial surveys (Figs. 3.1 and 3.2). This implies that the long-term decline in employment levels was not very significant. Though the explanation provided by these studies is valid to some extent, it does not still explain the entire decline in WPRs.

Hirway (2002)[5], however, went further and, by employing the time-use survey data for one point in time, formulated the hypothesis that the decline in the NSS-based WPRs in the 1990s was a result of the increased share of the unpaid component or informal sector of the SNA (system of national accounting) in the 1990s. The decline in WPRs was also due to a shift of some activities from SNA to the extended SNA category[6]. She has argued that in the process of globalisation some of the persons seemed to have withdrawn from the labour market in favour of schools/ colleges and some have shifted from their market-based activities to the non-market, non-SNA spheres. As a result of which, the WPRs seemed to have declined, as reported by NSS in its fifth quinquennial survey (1999-2000). Since there were no time-series data on the time-use pattern of the people, Hirway could not arrive at any firm conclusion about the above hypothesis (Hirway 2002).

The proportion of people involved in the unpaid category of activities, mostly domestic, such as free collection of vegetables, roots, fire wood and the like, including non-SNA activities, can better explain the hypothesis that the decline in WPRs was partly due to a rise in the ratio of people neither in the labour market nor in schools/colleges. The NSS also presents the data on such category of population with the codes from 1992-99, the population neither in labour market nor in schools/colleges in all its quinquennial survey rounds. Information on this aspect can be used to seek an appropriate explanation for the observed decline in WPRs in the 1990s.

The pre-1987-88 years may be considered to capture the changes in WPRs during the pre-reform period, whereas the post-1987-88 years may be considered to capture the changes in WPRs during the reform period. As the year 1987-88 was not a normal year, it was not seriously considered by many scholars when the data for 1983 were available. In the absence of data for 1983, the data for 1987-88 were used to examine the changes in WPRs before and after the introduction of economic reforms with some amount of risk and reliability. It is

Table 3. 2 : Work Participation Rates in India, 1972-73 to 1999-2000

Year/Status	Rural			Urban			All India		
	Males	*Females*	*Persons*	*Males*	*Females*	*Persons*	*Males*	*Females*	*Persons*
UPS									
1977-78	53.7	24.8	39.8	49.7	12.3	31.9	52.9	22.4	37.1
1983	52.8	24.8	39.1	50.0	12.0	32.0	52.1	21.8	37.4
1987-88	51.7	24.5	38.5	49.6	11.8	31.5	51.2	21.7	36.9
1993-94	53.8	23.4	39.0	51.3	12.1	32.7	53.2	20.6	37.5
1999-2000	52.2	23.1	38.0	51.3	11.7	32.4	52.0	20.3	36.5
UP+SS									
1972-73	54.5	31.8	42.8	50.1	13.4	–	–	–	–
1977-78	55.2	33.1	44.4	50.8	15.6	34.1	54.3	29.7	42.3
1983	54.7	34.0	44.5	51.2	15.1	34.0	53.8	21.6	42.0
1987-88	53.9	32.3	43.4	50.6	15.2	33.7	53.1	28.5	41.2
1993-94	55.3	32.8	44.4	52.1	15.5	34.7	54.5	28.6	42.0
1999-2000	53.1	29.9	41.7	51.8	13.9	33.7	52.7	25.9	39.7
CWS									
1983	51.1	22.7	37.2	49.2	11.8	31.5	–	–	–
1987-88	50.3	22.0	36.4	49.3	11.9	31.4	–	–	–
1993-94	53.1	26.7	40.3	51.1	13.9	33.4	–	–	–
1999-2000	51.0	25.3	38.4	50.9	12.8	32.7	–	–	–
CDS									
1983	48.2	19.8	34.2	47.3	10.6	29.9	–	–	–
1987-88	50.1	20.7	35.8	47.7	11.0	30.2	–	–	–
1993-94	50.4	21.9	36.6	49.6	12.0	31.7	–	–	–
1999-2000	47.8	20.4	34.4	49.0	11.1	30.9	–	–	–
Census (main workers)									
1971	53.5	13.1	33.8	48.8	6.6	29.3	52.5	11.9	32.9
1981	52.6	16.0	34.8	48.5	7.3	29.2	51.6	14.0	33.4
1991	51.8	18.6	35.7	48.6	8.1	29.5	50.9	15.9	34.1
2001	44.5	16.8	31.0	47.5	9.1	29.3	45.3	14.7	30.5
Census (main+ marginal workers)									
1981	53.8	23.2	38.9	49.1	8.3	30.0	52.6	19.8	36.8
1991	52.5	26.7	40.0	48.9	9.2	30.2	51.6	22.3	37.5
2001	52.4	31.0	42.0	50.8	11.5	32.2	51.9	25.7	39.3

Source: GoI (1981,1991 & 2001); NSSO (1981, 1988, 1990, 1996 & 2001).

evident from the data presented in Table 3.2 that at the all India level, WPRs, according to UPS, began to show a marginal decline in the late 1980s; even before the introduction of economic reforms, and sharply increased in the early 1990s; and again, steeply declined in the late 1990s both in rural and urban areas (Figs. 3.1, 3.2, 3.3 and 3.4).

Figure 3.1: Trends in Male WPRs (UP+SS) in Rural India, 1972-73 to 1999-2000

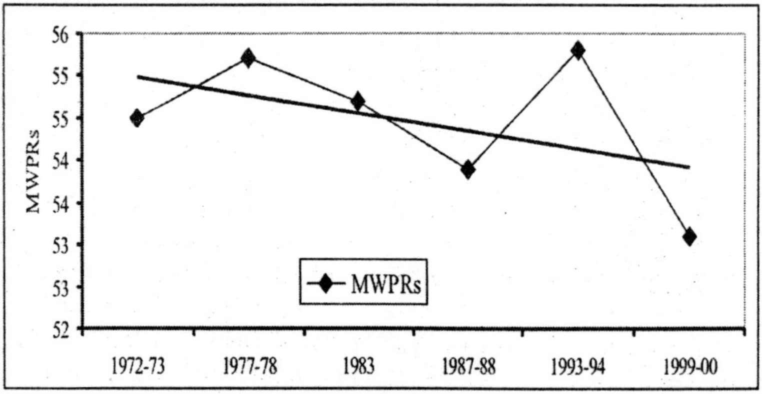

Note : MWPRs: Male work participation rates.
Source : Table 3.2.

Figure 3.2 : Trends in Female WPRs (UP+SS) in Rural India, 1972-73 to 1999-2000

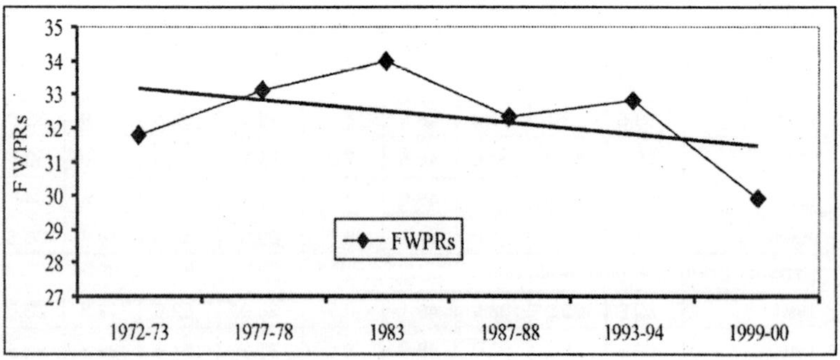

Note: FWPRs: Female work participation rates.
Source: Table 3.2.

Figure 3.3 : Trends in Male WPRs (UP+SS) in Urban India, 1972-73 to 1999-2000

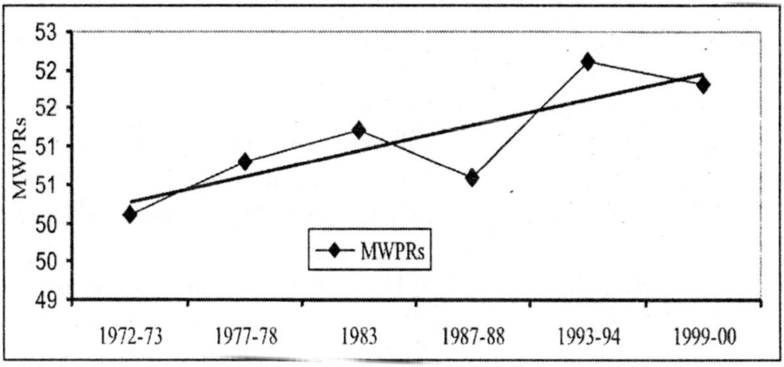

Note : MWPRs: Male work participation rates.
Source: Table 3.2.

Figure 3.4 : Trends in Female WPRs (UP+SS) in Urban India, 1972-73 to 1999-2000

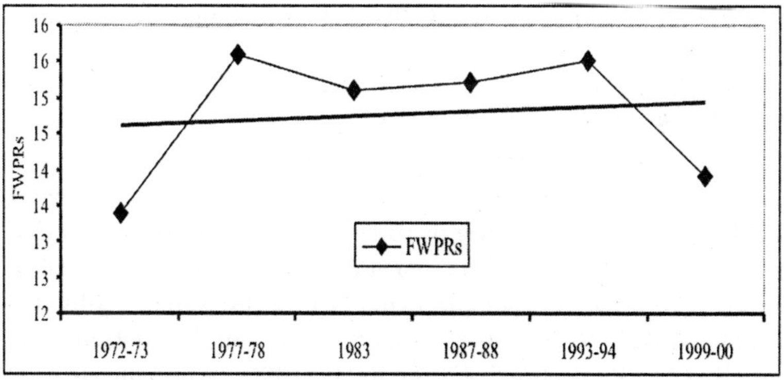

Note: FWPRs : Female work participation rates.
Source: Table 3.2.

The decline in WPRs, which started even before the reform measures, was not only higher but also accelerated in rural as against urban areas during the reform period. As regards female WPRs, the data show that although the share of female participation was smaller, a significant decline took place in rural *vis-à-vis* urban areas in the 1990s. According to UP+SS, although more or less similar trends existed, the decline in WPRs was much faster, as compared to UPS estimation during the same period, particularly in the late 1990s. In other words, the proportion of workers seeking jobs in subsidiary status declined sharply during the reform period. This implies that the employment market started becoming more complex and competitive, creating limited job opportunities only for those who had qualifications, skills and knowledge. The WPRs as per CWS and CDS also, by and large, revealed similar trends.

The WPRs as per CWS and CDS, however, reported to be lower as compared to those of UPS and UP+SS estimates both in rural and urban segments of the labour market in India. The WPRs, according to CDS, showed a greater decline as compared to that of the UPS estimate in 1999-2000 over 1993-94 in almost all the segments of the labour market. It has also been observed that the difference between the estimation of WPRs by NSS and that by the Census was reported to be far less in respect of male workers, but it was too large in respect of female workers both in rural and urban areas. The Census appeared to have underestimated most of the activities being carried out by women.

The Census data, by and large, present similar trends in WPRs as reported by NSS in the 1990s (Figure 3.5). If we take into consideration the WPRs with marginal workers, they did not show any decline for most of the persons in 2001 over 1991, excepting for a negligible decline for males in rural areas. This clearly shows that the share of marginal workers increased significantly due to lack of employment opportunities and inability to secure suitable jobs that could reasonably absorb them throughout the year. In brief, the data show that WPRs declined sharply in the 1990s, especially in rural areas. It is also important to identify who experienced a greater decline in WPRs in the 1990s; whether the decline in WPRs was quite significant among the children, youth, adults or older-age working population and among the illiterate or literate workers, both in rural and urban segments of the labour market in India.

Figure 3.5 : Trends in Male and Female WPRs in India, 1971 to 2001 (Census)

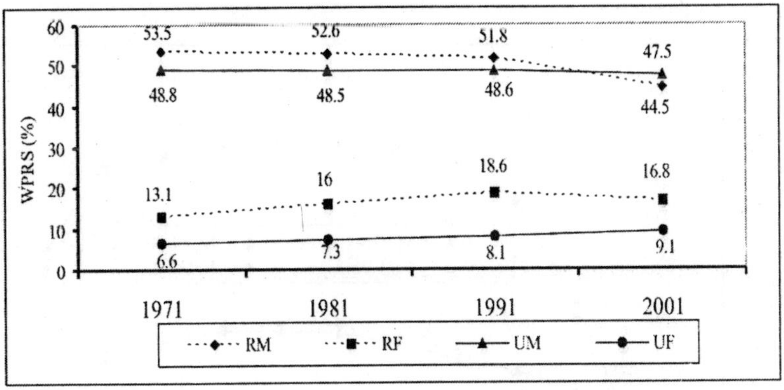

Note: RM: Rural males; RF: Rural females; UM: Urban males; UF: Urban females
Source: Table 3.2.

Work participation rates by age groups

The age of the workers is also one of the important factors that determine the level and pattern of WPRs in different occupations, particularly in the unorganised rural non-agricultural sector. The WPRs by age groups estimated for 1983 were not directly comparable with the subsequent quinquennial NSS surveys[7]. The data presented in Figure 3.6 show that the WPRs tended to increase with rising levels of age groups up to a certain point and declined thereafter with further rising levels of age groups in varying degrees across gender and areas; the increase was quite sharp from the age groups of 5-9 to 25-29 years and became a little flat from the age-groups of 30-34 to 50-54 years, and slowly declined thereafter. Such age-profile WPRs constituted what could be described as the "inverted U-shaped curve" with varying contours across gender and areas (Figure 3.6). At both extremes, the WPRs tended to be lower and higher at intermediate level. At the initial stage of life span (age groups 5-9 to 20-24 years), most of the persons would be in schools and colleges for acquisition of education and skills so that they could have better job options at a

Figure 3.6 : Pattern of Male and Female WPRs (UPS) by Age Groups in Rural India, 1993-94 to 1999-2000

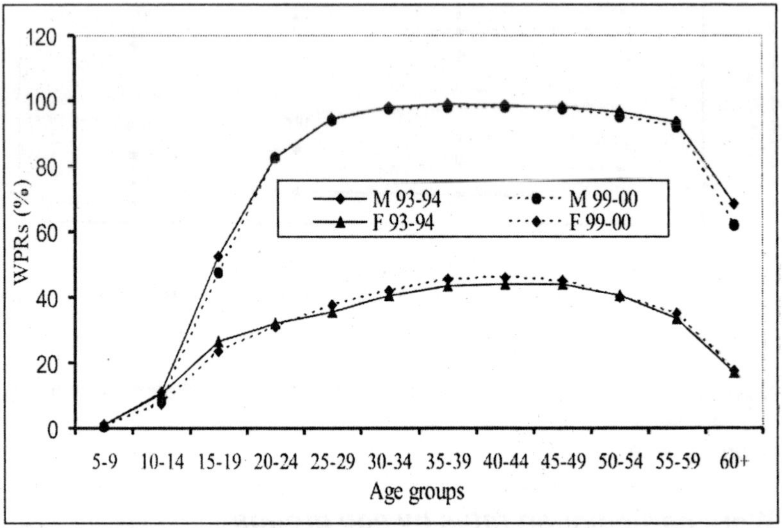

Note : M: Males; F: Females
Source : NSSO (1997, 2001).

later stage of life. At the intermediate level (age groups 20-25 to 50-54 years) most of the people were found to be engaged in one or the other economic activities. When they reached the age group of 50-54 years, most of them started withdrawing from the labour market due to ageing and/or retirement.

The data presented in Table 3.3 reveal that a big chunk of children in the age-group 5-14 years were found to be engaged in work instead of being in schools/colleges. In view of India's commitment that all children in the age-group 5-14 years should get compulsory education, the elimination of child labour deserves special attention. It is, however, not true that they miss out schooling because they are put to work or that all children who do not go to schools work. Only a small percentage of those children who are not attending schools/colleges actually work[8]. The choice is not between schooling and work; it is between idleness and work. The reasons why many children do not get education are not the same as the reasons why some children go

Table 3.3 : Male and Female WPRs (UPS) by Age Groups in India, 1983 to 1999-2000

Age group (Years)	*Rural*				*Urban*			
	1983	*1987-88*	*1993-94*	*1999-2000*	*1983*	*1987-88*	*1993-94*	*1999-2000*
Males								
5-9	2.0	0.7	0.9	0.5	0.5	0.5	0.4	0.3
10-14	21.3	14.5	11.2	8.2	9.3	7.3	5.9	4.6
15-29	78.9	74.7	74.0	72.1	64.0	60.2	60.1	58.3
30-44	98.1	97.8	98.4	97.8	97.2	97.4	97.4	96.8
45-59	95.0	95.2	96.3	95.3	92.1	92.5	93.1	91.8
60+	64.1	64.6	68.3	62.2	48.5	46.1	42.9	38.6
All	52.8	51.7	53.8	52.2	50.0	49.6	51.3	51.3
Females								
5-9	1.8	0.6	1.1	0.6	0.5	0.3	0.3	0.1
10-14	17.0	12.6	10.4	7.4	5.4	4.5	3.5	2.8
15-29	36.1	35.0	31.1	30.7	14.5	13.9	13.3	12.4
30-44	45.7	46.3	42.5	44.3	23.4	23.1	23.0	22.3
45-59	40.7	41.4	40.1	40.7	22.8	22.1	23.1	21.8
60+	14.5	16.0	17.2	17.4	11.6	9.3	9.1	8.2
All	24.8	24.5	23.4	23.1	12.0	11.8	12.1	11.7

Source: NSSO (1988, 1990, 1996 & 2001).

for work (Ghose 2004). Lack of public provided educational facilities as also the inability of many households to meet even the minimum expenses involved seem to be the important reasons, among others, why many children are not in schools/colleges. The higher incidence of child labour is due to extreme poverty and not due to unwise or misguided parental choice (Basu and Van 1998; Basu 1999; Ghose 2004).

The data also indicate that the decline in male WPRs at the younger age groups from 5-9 to 15-29 years, started to occur even before the economic reforms. As a matter of fact, the decline was found to be too large during the pre-reform period itself (1983 to 1987-88) as compared to the reform period (1993-94 to 1999-2000). The decline in male WPRs among adults and aged-workers (age groups 30-44 to 60+ years) began to occur during the reform period. A greater decline in male WPRs came from the younger age group of 10-14 years during the pre-reform period followed by the reform period. A significant decline in male WPRs took place in the case of age group 60+ years. The decline in male WPRs could also be partly attributed to the ageing of the male working population. But, no such decline appeared to have occurred during the pre-reform period. The data also indicate that in most of the age groups, the decline in male WPRs as per UP+SS was found to be higher than the decline as per UPS, especially in the reform regime (NSSO 2001)[9]. This lends some credence to the hypothesis that the decline in WPRs was also due to a decline of workers employed under subsidiary status of occupation (Sundaram 2001b).

In respect of females, although a similar pattern of "inverted U-shaped" WPRs existed, the level of female participation at different age groups was much lower than that for males (Figure 3.6). The incidence of female child labour also continued to be persistent, but it was lower as compared to that of males. A greater decline took place in female WPRs, only among the younger age groups from 5-9 to 15-29 years during the pre-reform and reform periods; a greater decline occurred in the former than in the latter period. Unlike the male WPRs, a moderate rise took place in the age groups 30-44, 45-59 and 60+ years during the reform period as compared to the pre-reform period.

The female WPRs as per UP+SS were quite higher than those as per UPS estimate (NSSO 2001). This appears to reflect the fact that the females, who also held a major portion of the responsibility of housework, were involved in gainful employment for a shorter period of time that could be classified under subsidiary status of occupation. As in the case of male WPRs, the decline in female WPRs as per UP+SS was greater than the decline as per UPS (NSSO 2001). This follows that, as mentioned in respect of male workers earlier, the female WPRs under subsidiary status of occupation declined much faster than female WPRs as per UPS estimate during the reform *vis-à-vis* the pre-reform period. The squeezing of the employment market is reflected in terms of the decline in the number of workers involved in subsidiary status of occupation.

In urban areas too, the data presented in Figure 3.7 show that the pattern of WPRs across different age groups, by and large, constituted a similar "inverted U-shaped curve", but it slightly differed from that of rural workers. In urban areas, the level of male WPRs in all the age groups was much lower than that of rural areas. More importantly, the level of WPRs among male children (age groups 5-9 and 10-14 years) was significantly lower in urban *vis-à-vis* rural areas. The gap in male-female WPRs was much higher in urban as compared to rural areas, implying that a larger number of females tended to participate in rural as compared to urban segment of the labour market in India.

Figure 3.7 : Pattern of Male and Female WPRs (UPS) by Age Groups in Urban India, 1993-94 to 1999-2000

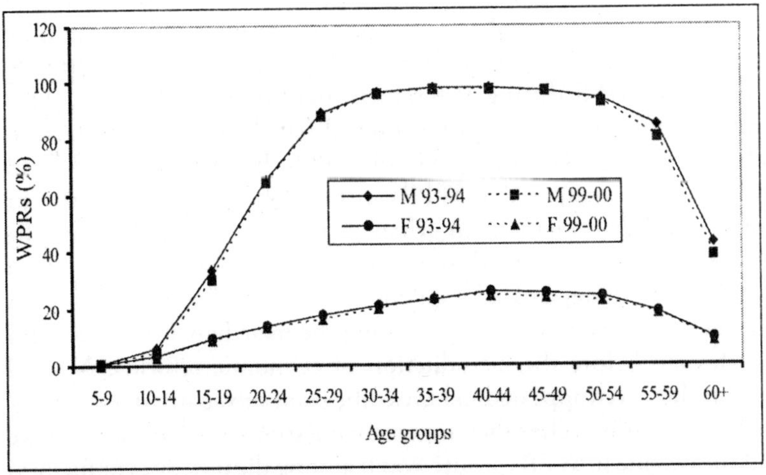

Note : M : Males; F : Females.
Source : NSSO (1997, 2001).

The data presented in Table 3.3 also indicate that in urban areas, the WPRs started to decline among the male youth followed by the children during the pre-reform as well as the reform period; a greater decline occurred during the pre-reform period itself. A greater decline in male WPRs in the age-group of 60+ years took place during the pre-reform period, as compared to the reform period. During the reform period, the male WPRs across different age groups showed a decline, but a greater decline took place in the age groups of 60+ followed by 15-29 years. The decline, in most cases, was found to be relatively much greater according to UP+SS estimate, as compared to that of UPS.

In respect of females in urban areas (Table 3.3), the rate of their participation was much lower than that of males in urban areas and males and females in rural areas. The data also indicated that in urban areas, a decline in female WPRs, as per UPS, occurred during the pre-reform and reform periods; a greater decline took place across almost all the age groups during the pre-reform period itself. The reform years brought about relatively more decline in WPRs among the female adults as compared to that among female youth in urban areas. The decline in female WPRs, according to UPS, in urban areas was found to be lower than the decline in male WPRs in urban areas and female WPRs in rural areas during the reform period. The decline in female WPRs was found to be higher according to UP+SS estimate than as per UPS during the reform period (NSSO 2001). This indicates a decline in the proportion of female workers involved in subsidiary status of occupation in the 1990s. It is, thus, obvious from the foregoing analysis that WPRs in the aggregate, particularly for males in rural and females in urban areas, declined more during the reform period as compared to the pre-reform period. The analysis of WPRs at the disaggregated level does not present a similar picture. A greater decline started to occur among the male female children and youth even before the introduction of economic reforms. A rapid decline, however, appeared to have come from subsidiary status during the reform period. The other important questions, however, are: Where did the decline in WPRs occur? Was it among the illiterates or literates at lower or higher levels of education?

Work participation rates by level of education

The level of WPRs will also be determined by the level of education of the workers. Education is regarded as one of the important components of "human capital" embodied in a person. It has been argued that workers with higher educational attainments are more likely to secure better jobs and more remuneration, especially in the non-agricultural sector (Biradar 2006b; Biradar and Jayasheela 2008). The data presented in Figure 3.8, *prima facie*, reveal that with increasing levels of educational attainments, the WPRs for males in rural areas, aged 15 and above, according to both UPS and UP+SS, tended to decline up to a certain level (secondary) and increased thereafter (higher secondary, graduate and above). This implies that at both extremes of educational attainments, the WPRs for males were found to be quite high and at the intermediate level, it was lower (Biradar 2004). Such pattern of male WPRs at different levels of education constituted a "saucer-shaped curve" with varying degrees across gender (see, Figure 3.8).

Figure 3.8 : Pattern of Male and Female WPRs (UPS) by Level of Education in Rural India, 1993-94 to 1999-2000

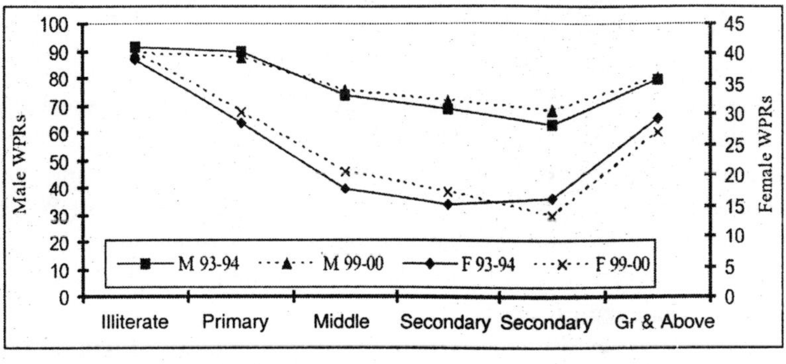

Note : M : Males; F : Females.
Source: Table 3.4.

The data also indicate that in rural areas, there had been a decline in WPRs for male persons, aged 15 and above, in the years of pre-reform and reform period, excepting for 1987-88 which happened to be an abnormal year, the year in which, as mentioned earlier, a severe drought took place across different parts of the country. The decline in rural male WPRs was much higher during the reform (1993-94 to 1999-2000) period as compared to the pre-reform period (1983 to 1993-94). This decline was not uniform across different levels of educational attainments of the workers.

A greater decline in WPRs occurred among the rural male workers at illiterate and literate up to primary level of education during the reform period. On the other hand, it increased at relatively higher levels of educational attainments. This implies that in rural areas, the male workers at illiterate or literate at lower levels of education were not able to secure employment in the competitive job market of globalisation. A greater emphasis has been, therefore, laid not only on educating and skilling the workers but also on a continuous process of skilling, re-skilling, multi-skilling and skill modulation[10] (GoI 2002a; GoI 2002c; Mathur and Mamgain 2002; Singh 2003).

In the case of female workers in rural areas, the average skill level expectedly was much lower than that for the male workers. What appears somewhat surprising is that the average educational background of females involved in economic activities was much lower than that of the females involved in household non-economic activities (Ghose 2004; Biradar 2004). The pattern of female WPRs appeared to be slightly different from that of male WPRs at different levels of educational attainments in rural areas. Although the saucer-shaped curve was evident for female WPRs in rural areas, the slope of its initial part was quite steeper, and then became a little flatter and then tended to rise sharply (see, Figure 3.8). This indicates that the WPRs for rural females tended to fall sharply at lower levels and increased swiftly at higher levels of educational attainments much faster than that for male counterparts.

The data on WPRs for females by different levels of educational attainments show that a greater decline took place among those female workers who had higher secondary and graduate and above levels of education; at lower levels, particularly at middle and secondary levels of education, it increased during the reform period, as compared to the pre-reform period. In the 1990s, as the employment market was tightening and becoming skill-selective, female workers with higher educational attainments experienced a sharp decline in WPRs mostly due to non-availability of employment that suited their educational qualifications and/or due to their ageing. According to UP+SS, the female WPRs in rural area showed that jobs taken up in subsidiary status of occupation declined significantly during the reform period. Such decline was quite significant in respect of female workers with graduate and above levels of education, followed by workers with illiterate and literate up to primary level of education during the same period.

In urban areas, the WPRs at different levels of education were found to be lower, and more so in respect of females, as compared to those of rural workers (Table 3.4). As observed in respect of male WPRs in rural areas, male WPRs in urban areas also constituted the

Table 3.4 : WPRs for Male and Female Persons (15+Years) by Level of Education in India, 1983 to 1999-2000

Level of education	*Rural*				*Urban*			
	1983	*1987-88*	*1993-94*	*1999-00*	*1983*	*1987-88*	*1993-94*	*1999-00*
UPS				**Males**				
Illiterate	91.0	90.2	91.3	89.1	86.3	85.5	86.6	83.6
Primary	88.5	87.0	89.8	87.2	84.1	82.7	84.4	82.4
Middle	70.2	68.8	73.5	75.2	68.8	66.2	71.3	72.5
Secondary	66.5	66.4	68.3	71.3	65.2	64.5	66.3	66.1
Hr. Secondary	NA	NA	62.9	67.9	NA	NA	58.9	59.9
Gr.& Above	77.4	76.6	79.2	80.2	81.0	79.9	80.7	79.7
All	85.6	83.9	84.6	82.9	76.9	75.4	75.8	74.5
UP+SS								
Illiterate	91.8	91.1	91.8	89.5	86.8	86.1	87.0	83.9
Primary	90.1	89.0	90.9	88.0	85.2	83.9	85.0	83.0
Middle	75.3	73.9	77.0	76.8	71.0	68.0	72.3	73.2
Secondary	73.3	72.9	72.8	73.7	67.6	66.4	67.7	66.8
Hr. Secondary	NA	NA	68.6	71.3	NA	NA	60.7	60.8
Gr.& Above	83.9	82.7	83.4	83.6	82.5	81.1	81.8	80.6
All	87.7	86.3	86.4	84.1	78.4	76.8	76.8	76.8
UPS				**Females**				
Illiterate	40.7	41.4	39.1	40.4	24.1	23.0	23.3	22.9
Primary	27.7	27.4	28.6	30.3	13.4	13.2	15.0	14.6
Middle	15.8	17.6	17.7	20.5	7.1	8.1	9.1	9.9
Secondary	15.8	17.5	15.3	17.3	13.5	12.5	10.8	10.4
Hr. Secondary	NA	NA	16.1	13.4	NA	NA	12.6	11.1
Gr.& Above	22.0	25.0	29.3	26.9	27.9	29.7	28.2	25.2
All	37.3	37.3	34.6	35.0	18.0	17.5	17.5	16.6
UP+SS								
Illiterate	54.3	52.6	54.0	51.3	29.6	29.2	30.0	27.1
Primary	41.8	39.1	41.6	40.3	18.1	17.5	20.3	17.7
Middle	29.3	28.6	29.0	29.0	11.1	11.3	13.1	12.9
Secondary	26.9	26.2	25.8	25.7	15.6	15.1	13.4	12.4
Hr. Secondary	NA	NA	23.4	20.6	NA	NA	14.7	12.4
Gr.& Above	29.5	30.7	36.6	31.0	29.7	31.5	30.1	27.3
All	50.9	48.5	48.6	45.2	22.5	22.0	22.3	19.7

Note : NA: Not available.
Source : NSSO (1988, 1990, 1996 & 2001).

"saucer-shaped curve" in varying degrees across gender (Figure 3.9). The saucer-shaped distribution of male WPRs at different levels of education in urban areas showed much better position than that of rural workers in the sense that at higher levels of education, the WPRs were found to be higher in urban areas *vis-à-vis* rural areas, but at the categories of illiterate and literate up to primary level of education, the WPRs were found to be higher in rural areas *vis-à-vis* urban areas.

The data also indicate that urban male WPRs, according to both UPS and UP+SS, showed a greater decline during the reform period, as compared to the pre-reform period. At different levels of education, although the decline in male WPRs in urban areas was quite perceptible in almost all the categories of education, as per UPS and UP+SS, a rapid decline was observed in respect of male workers with illiterate and literate up to primary level of education, followed by a smaller decline at graduate and above levels of education during the reform period.

In respect of females in urban areas, the pattern of WPRs at different levels of education also formed the "U-shaped curve" (Figure 3.9). This is a mirror image of the "inverted U-shaped curve" along which the ratio of females engaged in full-time housework tends to increase as the level of education rises. The end-points on the "inverted U-shaped curve" are above those on the U-shaped curve. This means that the percentage of females involved in housework exceeded the percentage of females participating in the labour market at different levels of education. The U-shaped curve for females shows that the WPRs tended to decline much faster at lower levels of education and increased sharply at higher levels of education. The decline in WPRs for females in urban areas was also noticed even before the economic reforms, but a greater decline took place after the economic reforms. The extent of decline, however, was not uniform at different levels of educational attainments. A greater decline in WPRs for females was observed at higher levels of education (higher secondary and graduate and above) and a marginal decline occurred at illiterate and literate up to primary level of education during the reform period.

The female WPRs showed a greater decline according to UP+SS *vis-à-vis* UPS estimate, particularly at illiterate, literate up to primary

Figure 3.9 : Pattern of Male and Female WPRs (UPS) by Level of Education in Urban India, 1993-94 to 1999-2000

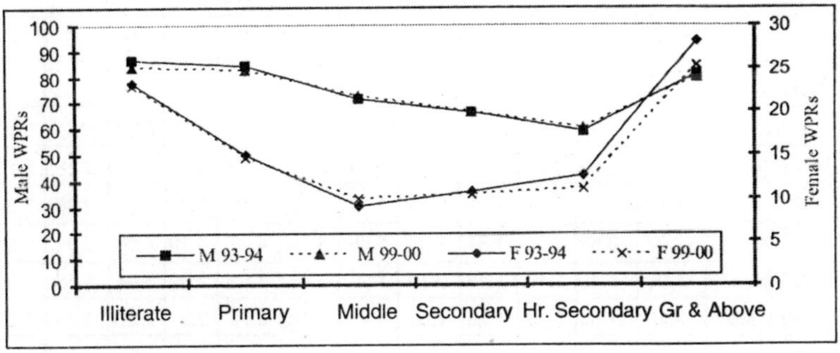

Note : *M* : Males; F : Females.
Source : Table 3.4.

and at graduate and above levels of education. Such trends appear to indicate that in the highly competitive employment market of urban areas, the proportion of females working under subsidiary status of occupation declined significantly, as compared to that of females working under usual status of occupation. Such decline seemed to have occurred among those female workers who did not have higher education.

Unemployment Rates

The decline in the growth rate of employment and WPRs are, *ceteris paribus*, expected to result in a corresponding one-to-one rise in unemployment rates according to any method of measurement. In case the employment growth rates and WPRs have not resulted in a corresponding rise in unemployment rates, the plausible explanations are explored in what follows. An effort to analyse the changes in unemployment rates, in general, and unemployment rates by age groups and levels of education, in particular, has been made to explore alterative explanations/reasons for the decline in WPRs in the 1990s.

The data presented in Table 3.5 reveal that unemployment rates continued to be traditionally higher in urban than in rural areas. This is partly due to the fact that the greater dominance of organised sector

Table 3. 5 : Unemployment Rates in India, 1973-73 to 1999-2000

Year/Area	*UPS*			*UP+SS*			*CWS*			*CDS*		
	M	*F*	*P*	*M*	*F*	*P*	*M*	*F*	*P*	*M*	*F*	*P*
Rural												
1972-73	–	–	–	1.16	0.48	0.92	3.02	5.51	3.86	6.83	11.22	8.21
1977-78	2.22	5.52	3.26	1.28	2.01	1.54	3.37	4.13	3.74	7.12	9.18	7.70
1983	2.12	1.41	1.91	1.41	0.66	1.13	3.72	4.26	3.88	7.52	8.98	7.94
1987-88	2.87	3.52	3.07	1.81	2.27	1.98	4.16	4.27	4.19	4.58	6.91	5.25
1993-94	1.96	1.40	1.79	1.43	0.79	1.20	2.98	3.01	2.99	5.64	5.55	5.61
1999-2000	2.10	1.50	1.96	1.70	1.00	1.43	3.10	3.70	3.91	7.20	6.80	7.21
Urban												
1972-73	–	–	–	4.79	6.05	5.03	5.96	9.09	6.53	8.02	13.67	8.99
1977-78	6.48	7.26	8.77	5.40	12.40	7.01	7.12	10.93	7.86	9.40	14.55	10.34
1983	5.86	6.90	6.04	5.08	4.85	5.02	6.69	7.46	6.81	9.23	10.99	9.52
1987-88	6.07	8.77	6.56	5.16	5.93	5.32	6.71	8.93	7.12	8.79	12.00	9.36
1993-94	4.54	8.21	5.21	4.05	6.24	4.52	5.17	8.42	5.83	6.72	10.52	7.43
1999-2000	4.80	7.10	5.23	4.50	5.70	4.63	5.60	7.30	5.89	7.30	9.40	7.65
All India												
1972-73	–	–	–	1.90	1.02	1.61	3.61	5.87	4.32	7.07	11.46	8.35
1977-78	3.07	7.01	4.23	2.09	3.19	2.47	4.29	4.97	4.48	7.59	9.86	8.18
1983	3.02	2.14	2.77	2.28	1.16	1.90	4.44	4.70	4.51	7.93	9.26	8.28
1987-88	6.30	4.19	3.77	2.56	2.72	2.62	4.75	4.92	4.80	5.54	7.61	6.09
1993-94	2.60	2.44	2.56	2.07	1.54	1.89	3.53	3.83	3.62	5.91	6.33	6.03
1999-2000	NA	NA	2.81	NA	NA	2.20	NA	NA	4.41	NA	NA	7.32

Note : NA: Not available; M: Males; F: Females; P: Persons.
Source : NSSO (1981, 1988, 1990, 1996 & 2001).

in urban areas forces people to be either employed in better paid jobs or even remain unemployed for some more time in expectation of getting better jobs instead of working in any low paid subsidiary occupation, which can disguise an effective state of unemployment, and partly due to a faster rise in the rate of literacy among the youth who are not able to get jobs of their choice immediately after completion of their degrees (swelling the ranks of educated youth unemployment).

In rural areas, on the other hand, the people cannot afford to remain unemployed for a longer period of time and they are more likely to involve themselves in one or the other economic activity however less remunerative or lowly paid it may be to them. It is, however, interesting to note that the differences in unemployment rates, particularly by CWS and CDS, between rural and urban areas narrowed down considerably in the 1990s. While CDS and CWS are expected to measure the incidence of unemployment for a shorter duration, UPS and UP+SS measure it for a longer period of time. Since people cannot remain unemployed for a longer period of time, the number of persons reporting unemployed as per UPS and UP+SS estimate is likely to be smaller as compared to CDS and CWS.

The unemployment rates for both males and females by all measurements, excepting for CDS, used to be quite higher in the years of the pre-reform period as compared to those in the reform period. The unemployment rates increased in 1999-2000 over 1993-94 in both rural and urban segments of the labour market. The increase in unemployment rates was found to be relatively greater according to CDS as compared to the unemployment rates according to UPS, UP+SS and CWS in both the rural and urban segments of the labour market during the same period.

Going by gender, it has been found that the unemployment rates were significantly higher for females as against males in the 1970s and 1980s in rural areas and in the 1990s in urban areas. But the difference between the male-female unemployment rates decreased significantly over time. This could be partly due to rising SPRs and also the ratio of females neither in labour market nor in schools/ colleges to the total female population. The data show that in rural areas, the unemployment rates modestly declined for females *vis-à-vis* males during the pre-reform period (1983 to 1993-94 or even 1987-88 to 1993-94), whereas in the case of the reform period (1993-94 to 1999-2000), it showed a slightly greater increase in unemployment rates for males *vis-à-vis* females as per any method of unemployment measurement, excepting for CWS. A striking feature in rural areas is that during the reform period, the unemployment rates, as per any method of estimate showed a greater increase in respect of males as against females. Of course, the unemployment rates as per CDS increased sharply in 1999-2000 over 1993-94, but still they were quite lower than those in the years of the pre-reform period.

In the urban segment of the labour market, on the other hand, the data, *prima facie*, reveal that the unemployment rates, as per most of the estimates, for both males and females, tended to be more in the years of the pre-reform period than in the reform period. It has been, however, observed that while the unemployment rates for males experienced a marginal increase during the reform period; the unemployment rates for females declined sharply during the same period. A similar observation has also been made by other studies (Chadha and Sahu 2002; Ghose 2004; Biradar 2004). The rise in unemployment rates for males was much lower during the reform period than the decline during the pre-reform period in both rural and urban segments of the labour market in India. This indicates that in a long-term perspective, the unemployment rates appeared to be declining rather than rising rapidly. Nobody, however, can absolutely deny the fact that there has been an increase in unemployment rates in the 1990s. The important questions, however, are: Among whom did it increase rapidly? Was it among the children, youth or adults and illiterates or literates? To what extent did it increase? These typical questions are sought to be addressed in what follows.

Unemployment rates by age groups

As the incidence of unemployment varied across different categories of age groups, an attempt has been made to examine who experienced a greater incidence and decline in unemployment rates, whether among the children, youth or adults or aged workers in the years of the reform period, as compared to those in the pre-reform period, whether in rural or urban areas of India. The data presented in Figure 3.10 indicate that the unemployment rates for males and females tended to rise sharply with increasing levels of age groups up to a certain level (age groups from 5-9 to 20-24 years) and thereafter declined rapidly. Such a pattern of unemployment by age groups constituted the "inverted V- shaped curve", showing that the unemployment rates were quite/

Figure 3.10 : Pattern of Unemployment Rates (UPS) for Males and Females by Age Groups in Rural India, 1993-94 to 1999-2000

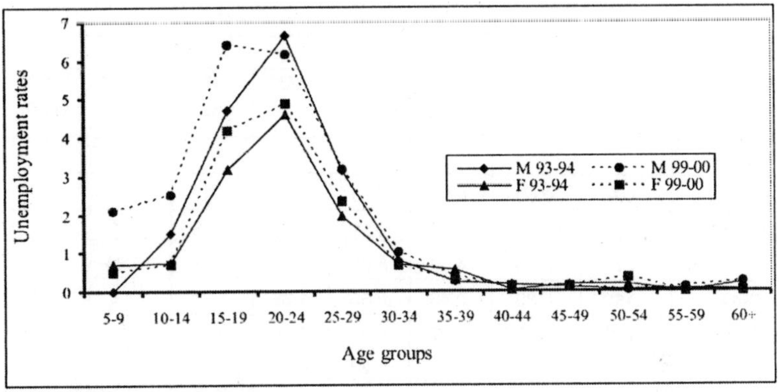

Note : M: Males; F: Females.
Source : NSSO (1997, 2001).

significant among the youth, followed by children and adults[11]. The problem of unemployment was ostensibly serious among the youth in the 1990s (Visaria 1998; Gumber 2000; Chadha 2000; Chaubey 2000; Parthasarathy and Nirmal 2000). In the age groups from 30-34 to 60+ years, the incidence of unemployment for males was observed to be far less in the rural segment of the labour market.

The data presented in Table 3.6 indicate that unemployment rates for rural males in most of the age groups increased during the reform period, as compared to the pre-reform period. The increase was quite significant, particularly for teenagers and male youth (age groups 5-9, 10-14 and 15-29 years), excepting for the youth in the age category of 20-24 years, who experienced a negligible decline. The above decline may be among those male children and youth who had no, or lower, levels of education. A sharp increase in unemployment rates among the children can be partly attributed to the ban on child labour in several production and service delivery systems and partly to the incidence of poverty that prevented the children of poor households from attending schools/colleges.

Among the adults (age groups 30-44 years), a negligible increase in unemployment rates took place during the reform period *vis-à-vis* the pre-reform period. The unemployment rates showed a marginal

Table 3.6 : Unemployment Rates for Males and Females (UPS) by Age Groups in India, 1983 to 1999-2000

Age group (Years)	Rural				Urban			
	1983	*1987-88*	*1993-94*	*1999-2000*	*1983*	*1987-88*	*1993-94*	*1999-2000*
Males								
5-9	0.5	0.00	0.00	2.09	5.3	0.0	1.51	1.01
10-14	3.00	3.33	1.49	2.51	11.0	9.88	5.24	6.13
15-29	4.70	6.24	4.83	5.09	12.0	13.6	10.84	11.51
30-44	0.50	0.83	0.43	0.55	1.38	1.21	1.11	1.41
45-59	0.20	0.50	0.07	0.07	0.70	0.69	0.28	0.45
60+	0.20	0.46	0.02	0.23	0.64	1.07	0.31	0.11
All	2.12	2.87	1.96	2.10	5.86	6.07	4.54	4.80
Females								
5-9	1.11	0.00	0.70	0.50	0.00	0.00	0.00	2.35
10-14	1.22	3.08	0.74	0.69	2.52	4.35	2.38	3.53
15-29	2.80	5.36	3.23	3.70	15.54	18.79	19.39	16.70
30-44	0.54	2.62	0.46	0.38	2.05	3.49	2.71	2.85
45-59	0.37	1.92	0.13	0.17	0.65	1.13	0.39	0.65
60+	0.64	1.84	0.22	0.00	0.17	1.08	0.37	0.15
All	1.41	3.52	1.40	1.50	6.90	8.77	8.21	7.10

Note : NA: Not available.
Source : NSSO (1998, 1990, 1996 & 2001).

increase for males in the age group of 60+ years during the same period. It has been observed that the increase in male WPRs, in most cases, was higher than the decline in unemployment rates during the pre-reform period, whereas during the reform period, the extent of decline in male WPRs was higher than a corresponding increase in unemployment rates.

In respect of females, the data presented in Figure 3.10 indicate that the unemployment rate was slightly lower than that for their male counterparts. The unemployment rate, as per UPS perspective, tended to rise up to a certain level (age groups 20-24 years) and sharply declined thereafter. Such a pattern of unemployment rate for females across different age groups also formed the "inverted V-shaped curve".

This simply implies that the problem of unemployment was quite significant among female youth, especially in the age group of 20-24 years, as compared to any other age group. Unlike in the case of male children, the unemployment rate for female children (age groups 5-9 and 10-14 years) declined both in the pre-reform and reform years: a greater decline took place during the pre-reform period, indicating that a rapid rise in SPR started even well before the introduction of economic reforms. Among the female youth (age groups 15 to 29 years), the unemployment rates declined rapidly during the pre-reform period and increased moderately during the reform period.

The unemployment rates for females in the age group of 30-44 and 60+ years registered a decline during both the periods; a greater decline took place during the pre-reform period *vis-à-vis* the reform period. A rapid decline in female unemployment rates in 1993-94 over 1987-88 was attributed to drought in the year 1987-88, due to which a larger proportion of females seemed to have been forced to seek employment, but some of them appeared to have obtained work and some of them, it seems, did not, but were available for work and reported as unemployed. During 1987-88, the WPRs as well as the unemployment rates were, therefore, reported to be quite phenomenal, as compared to the position which prevailed in 1983 and 1993-94. It is an established fact that, in most cases, females tend to go for work only in times of distress in agriculture or if there is no male breadwinner in the household or otherwise if they are highly paid.

A striking trend witnessed during the reform period was that, the decline in female WPRs, particularly among the youth (age groups 15-29 years), was found to be higher than the rise in unemployment rates among them. This indicates that a larger proportion of female youth had withdrawn from the labour market in favour of schools/ colleges to acquire higher educational qualifications. In other age groups, particularly 30-44 and 60+ years, the unemployment rates for females seemed to have shown gradually a declining tendency.

In urban areas, the data presented in Figure 3.11 indicate that the distributional pattern of male and female unemployment rates also constituted the "inverted V-shaped curve" (Figure 3.11). The unemployment rates were found to be quite significant among male

Figure 3.11 : Pattern of Unemployment Rates (UPS) for Males and Females by Age Groups in Urban India, 1993-94 to 1999-2000

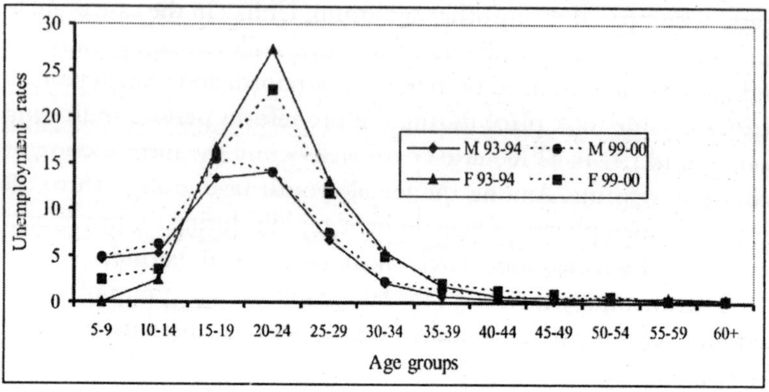

Note : M: Males; F: Females.
Source : NSSO (1997, 2001).

youth and children and that in urban areas, as compared to female youth and children and that in rural areas. The unemployment rates among children (age groups 5-9 and 10-14 years), contrary to the expectation in view of the back-to-school programme, increased marginally in the reform period, as compared to a significant decline in the pre-reform period.

The data presented in Table 3.6 show that the unemployment rates among male children and youth showed a greater increase during the reform period, as compared to the rapid decline during the pre-reform period. In the age groups of 30-44 and 45-59 years, a marginal increase took place in male unemployment rates during the same period. Unemployment rates for males in the age-group 60+ years witnessed a marginal decline during both the periods; a greater decline appeared to have occurred during the pre-reform period.

As far as females are concerned, the data indicate that the unemployment rates at different age groups also formed the "inverted V-shaped curve" but slightly different from that for their male counterparts. It has been observed that the incidence of unemployment for females, particularly among the youth, was found to be significantly higher than that for males in urban areas and females in rural areas. This implies that although the WPRs for females were found to be lower, there was a greater representation in the labour market among those females who had completed graduate and post-graduate degree courses. The graduate and post-graduate degree holders generally wish to secure better jobs which would make them wait for some more time. This results in a greater magnitude of unemployment among the female youth in urban areas. The unemployment rates among children (age groups 5-9 and 10-14 years) increased during the reform period, as compared to the no-unemployment incidence in the age-group 5-9 years and decline in the age-group 10-14 years during the pre-reform period. Even in urban areas, the incidence of female child labour increased during the reform period as compared to the pre-reform period.

Although the unemployment rates among female youth were found to be too high, a rapid decline took place during the reform period, as compared to a rise during the pre-reform period. This shows that, in the urban labour market, the economic reforms seemed to have brought new employment opportunities to those female workers who had completed graduate and post-graduate courses and possessed better skills and knowledge. While for female adults, especially in the age categories of 30-44 and 45-59 years, the economic reform measures seemed to have resulted in a marginal rise in the unemployment rates. The female labour force in the age category of 60+ years showed a marginal decline in the unemployment rate in the years of reform period, as compared to the pre-reform period. This could be partly due to the ageing of the female working population in urban areas. It is, thus, evident that in the urban segment of the labour market, unemployment rates among male youth and children declined rapidly during the pre-reform period, compared to an increase during the reform period. As far as females in urban areas were concerned, the incidence of unemployment among female children registered a rising tendency and that of female youth a declining tendency during the reform period, as compared to the pre-reform period.

Unemployment rates by level of education

There has been an extensive discussion on the issue of unemployment rates among educated persons, particularly in respect of persons with higher educational attainments in the era of globalisation. An attempt, therefore, has been made here to examine the unemployment rates by genders, levels of education and are as rural and urban during the pre-reform and reform periods. The data presented in Figure 3.12 illustrate that with increasing levels of educational attainments, the unemployment rates tended to increase rapidly among the male and female labour force in rural and urban segments of the labour market; the increase was much faster among persons in the categories of higher secondary, graduate and above levels of education. The incidence of educated unemployment was found to be quite higher in respect of females as compared to males, particularly at higher levels of education (Figure 3.12).

Figure 3.12 : Pattern of Unemployment Rates (UPS) for Males and Females by Level of Education in Rural India, 1993-94 to 1999-2000

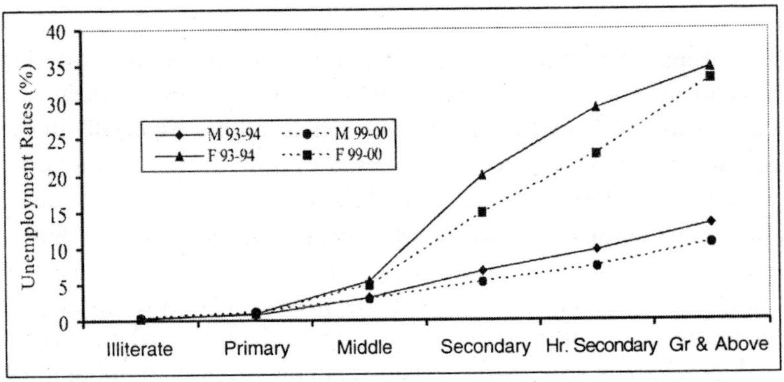

Note : M: Males; F: Females.
Source : Table 3.7.

The data presented in Table 3.7 indicate that the unemployment rates among relatively more educated persons (secondary, higher secondary and graduate and above) were found to be significantly higher in rural than in urban areas. This implies that employment opportunities for educated persons were largely created in urban areas, as compared to rural areas, during the reform period. That is why a large chunk of educated persons in rural areas, soon after getting their educational qualifications, seemed to have migrated to urban areas in

Table 3.7 : Unemployment Rates for Males and Females (15 + years) by Level of Education in Rural India, 1983 to 1999-2000

Level of education	*Rural*				*Urban*			
	1983	*1987-88*	*1993-94*	*1999-00*	*1983*	*1987-88*	*1993-94*	*1999-00*
UPS				**Males**				
Illiterate	0.24	0.90	0.30	0.40	1.42	1.80	1.10	1.40
Primary	0.90	2.10	0.80	1.10	3.10	4.60	2.50	3.00
Middle	2.83	5.40	3.00	2.80	5.61	8.80	5.70	5.60
Secondary	5.14	10.50	6.70	5.20	5.77	8.80	6.30	5.50
Hr. Secondary	NA	NA	9.60	7.30	NA	NA	8.50	8.30
Gr.& Above	7.46	15.00	13.20	10.60	5.76	7.40	6.40	6.60
UP+SS								
Illiterate	0.42	0.50	0.20	0.30	1.66	1.50	0.90	1.30
Primary	1.31	1.30	0.40	0.90	3.62	3.80	2.30	2.80
Middle	3.87	3.50	2.20	2.30	6.35	7.60	5.40	5.20
Secondary	7.76	6.80	5.00	4.30	6.51	7.40	5.60	5.10
Hr. Secondary	NA	NA	7.30	6.10	NA	NA	7.50	7.70
Gr.& Above	11.35	10.20	9.80	8.20	6.39	6.70	5.60	6.30
UPS				**Females**				
Illiterate	0.07	2.40	0.20	0.20	0.12	2.10	0.40	0.60
Primary	0.27	3.50	1.00	0.90	0.58	6.40	4.50	2.50
Middle	2.02	13.70	5.30	4.70	1.84	21.60	15.70	11.10
Secondary	6.68	33.50	19.90	14.70	3.74	22.80	20.00	14.40
Hr. Secondary	NA	NA	29.10	22.70	NA	NA	22.20	18.90
Gr.& Above	13.31	37.30	34.60	33.10	6.30	21.00	20.60	16.30
UP+SS								
Illiterate	0.20	1.70	0.20	0.10	0.22	1.00	0.30	0.50
Primary	0.57	1.80	0.50	0.60	0.74	4.00	2.90	1.90
Middle	2.53	7.70	3.00	3.20	2.06	14.30	10.90	7.90
Secondary	7.96	23.00	11.30	9.80	4.00	17.00	15.70	12.00
Hr. Secondary	NA	NA	19.90	14.30	NA	NA	19.20	16.70
Gr.& Above	15.60	30.40	27.40	29.80	7.44	18.70	18.90	14.70

Note : NA: Not available.
Source : NSSO (1988, 1990, 1996 & 2001).

search of better employment options and hence a marginal rise in unemploy-ment rates among highly educated male persons in urban areas and decline in rural areas in the reform years.

In the rural segment of the labour market, the unemployment rates, as per both UPS and UP+SS, increased marginally for male persons in the categories of illiterate and literate up to primary level of education and a rapid decline in the categories of middle, secondary, higher secondary and graduate and above levels of education during the reform period, as compared to the pre-reform period. A rapid decline took place among those who had higher educational attainments, particularly those who were graduates and post-graduates in the reform regime, as compared to a significant rise in the pre-reform regime. This implies that economic reforms have created substantial employment opportunities for persons with higher educational attainments than for those having no education or education at lower levels.

In respect of females, it has been found that the unemployment rates for females with more educational attainments, especially graduation and above, were found to be greater as compared to those for males at the same level of education. A rapid decline, however, took place among all the educated females, more so in respect of higher secondary and secondary levels of education during the reform period as compared to a rapid rise during the pre-reform period. In rural areas, the decline in unemployment rates for females with higher educational attainments was greater than the decline in male unemployment rates in the 1990s (see, Figure 3.12).

In the urban segment of the labour market (Table 3.7), the data illustrate that the unemployment rates for males as well as females tended to increase with increasing levels of education up to a certain level (higher secondary) and declined sharply at graduate and above levels of education (Figure 3.13), implying that the persons with higher educational attainments were more likely to be represented in the labour market by securing jobs suited to their qualifications. It has been found that the unemployment rates for males at higher levels of educational attainments were observed to be more in urban than in rural areas. The male-female difference in unemployment rates was higher in urban areas as compared to rural areas. The problem of

Figure 3.13 : Pattern of Unemployment Rates (UPS) for Males and Females by Level of Education in Urban India, 1993-94 to 1999-2000

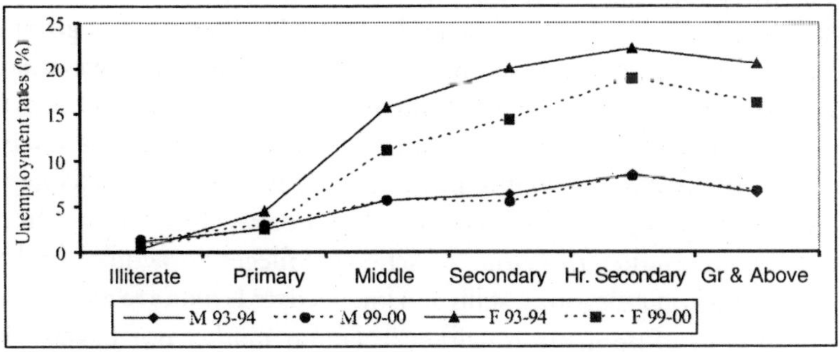

Source : Table 3.7.

educated unemployment for females, particularly at graduate and post-graduate levels, was reported to be more acute in the rural than in urban segment of the labour market; but a rapid decline took place in the urban segment of labour market in the 1990s.

In respect of females in urban areas, the data indicate that the unemployment rates tended to increase with increasing levels of education up to a certain level (higher secondary) and declined thereafter. The unemployment rates for females, particularly at higher levels of education, was found to be quite higher, as compared to that for males in urban areas. The data also indicate that the unemployment rate for the illiterate category increased marginally during the reform period, as compared to the pre-reform period, and in the case of all other categories of education, it declined significantly during both the periods; a greater decline took place during the reform period, particularly at middle and secondary levels of education. The decline in female unemployment rates was higher than that of a decline in male unemployment, especially in the categories of middle, secondary and higher secondary levels of education in the urban segment of the labour market. This indicates that in the urban segment of the labour market, huge employment opportunities were generated for those females who had higher educational attainments, as compared to that of males with the same level of education during the reform period.

Factors Contributing to the Decline in WPRs in the 1990s

The foregoing analysis reveals that WPRs, in aggregate, had sharply declined in the 1990s in varying degrees across gender, age groups, level of education and areas. The extent of decline in WPRs was much higher than the increase in unemployment rates during the reform period. This has generated a nation-wide debate on what could have caused the decline in WPRs in the 1990s. Is it due to a decline in employment opportunities? Is it due to withdrawal of workforce from the labour market in favour of schools/colleges to acquire higher educational qualifications, skills and knowledge? Is it due to an increase in the ratio of population neither in labour market nor in schools/colleges? Or is it due to a combination of all these factors? An attempt, in what follows, has been made to explore the possible reasons and satisfactory explanations for the observed decline in WPRs across, gender, age groups, level of education and areas, especially in the 1990s.

Increase in unemployment rates

The decline in employment opportunities in the 1990s was expected to result in a corresponding one-to-one rise in unemployment rates by any method of estimates. If the entire decline in WPRs was not due to the decline in employment opportunities, the unemployment rates were not expected to increase correspondingly. The data on WPRs and unemployment rates, as analysed in the previous sections, have clearly shown that the decline in WPRs was much higher than the corresponding rise in unemployment rates in the 1990s. Some part of the decline in WPRs, however, has been attributed to a rise in unemployment rates. Table 3.8 provides the data on point-to-point changes in WPRs and unemployment rates for both males and females in rural and urban areas in India during 1977-78 to 1999-2000. The data, as per UPS, show that the increase or decrease in WPRs did not match with the unemployment rates during the pre-reform and reform periods. For instance, in rural areas, the male WPRs increased by 2.1 percent in 1993-94 over 1987-88 or by 1 percent over 1983, the years of the pre-reform period, whereas the unemployment rates

Table 3.8 : Changes in WPRs and Unemployment Rates in India,1983 over 1987-88 to 1999-2000 over 1993-94

Period/ Status	*Change in WPRs*						*Change in unemployment rates*					
	Rural			*Urban*			*Rural*			*Urban*		
UPS	M	F	P	M	F	P	M	F	P	M	F	P
1983/77	-0.90	0.00	-0.70	0.30	-0.30	0.10	-0.10	-4.11	-1.35	-0.62	-0.36	-2.73
1987/83	-1.10	-0.30	-0.60	-0.40	-0.20	-0.50	0.75	2.11	1.16	0.21	1.87	0.52
1993/87	2.10	-1.10	0.50	1.70	0.30	1.20	-0.91	-2.12	-1.28	-1.53	-0.56	-1.35
1993/83	1.00	-1.40	-0.10	1.30	0.10	0.70	-0.16	-0.01	-0.12	-1.32	1.31	-0.83
1999/93	-1.60	-0.30	-1.00	0.00	-0.40	-0.30	0.14	0.10	0.17	0.26	-1.11	0.02
UP+SS												
1983/77	-0.50	0.90	0.10	0.40	-0.50	-0.10	0.13	-1.35	-0.41	-0.32	-7.55	-1.99
1987/83	-0.80	-1.70	-1.10	-0.60	0.10	-0.30	0.40	1.61	0.85	0.08	1.08	0.30
1993/87	1.40	0.50	1.00	1.50	0.30	1.00	-0.38	-1.48	-0.78	-1.11	0.31	-0.80
1993/83	0.60	-1.20	-0.10	0.90	0.40	0.70	0.02	0.13	0.07	-1.03	1.39	-0.50
1999/93	-2.20	-2.90	-2.70	-0.30	-1.60	-1.00	0.27	0.21	0.23	0.45	-0.54	0.11
CWS												
1983/77	-	-	-	-	-	-	0.35	0.13	0.14	-0.43	-3.47	-1.05
1987/83	-0.80	-0.70	-0.80	0.10	0.10	-0.10	0.44	0.01	0.31	0.02	1.47	0.31
1993/87	2.80	4.70	3.90	1.80	2.00	2.00	-1.18	-1.26	-1.20	-1.54	-0.51	-1.29
1993/83	2.00	4.00	3.10	1.90	2.10	1.90	-0.74	-1.25	-0.89	-1.52	0.96	-0.98
1999/93	-2.10	-1.40	-1.90	-0.20	-1.10	-0.70	0.12	0.69	0.92	0.43	-1.12	0.06
CDS												
1983/77	-	-	-	-	-	-	0.40	-0.20	0.24	-0.17	-3.56	-0.82
1987/83	1.90	0.90	1.60	0.40	0.40	0.30	-2.94	-2.07	-2.69	-0.44	1.01	-0.16
1993/87	0.30	1.20	0.80	1.90	1.00	1.50	1.06	-1.36	0.36	-2.07	-1.48	-1.93
1993/83	2.20	2.10	2.40	2.30	1.40	1.80	-1.88	-3.43	-2.33	-2.51	-0.47	-2.09
1999/93	-2.60	-1.50	-2.20	-0.60	-0.90	-0.80	1.56	1.25	1.60	0.58	-1.12	0.22

Note : M: Males; F: Females; P: Persons.
Source : Computed from Tables 3.2 & 3.5.

declined by -0.16 or -0.91 during the same period. During the reform period, the male WPRs declined by –1.60 percent, whereas the unemployment rate increased by 0.14 percent in 1999-2000 over 1993-94 in rural areas. A similar trend, by and large, was observed, even as per UP+SS, CWS and CDS with some minor exceptions.

In respect of females in rural areas, however, it has been observed that the percent point decline in WPRs, as per UPS, was much higher than the rise in unemployment rates during the reform period. The decline in female WPRs did not result in a corresponding rise in female unemployment rates to the extent that occurred in respect of males during the same period. The above analysis implies that as the rise in unemployment rate is not equal to the decline in WPRs, the SPRs and/or the ratio of population neither in labour market nor in schools/colleges are expected to rise in varying degrees across gender, age groups and level of education both in rural and urban areas.

In urban areas, as observed in the case of rural male workers, the data in aggregate, as per UPS, indicate that the increase in male WPRs resulted in a corresponding decline in unemployment rates during the pre-reform period, whereas during the reform period, no change in male WPRs was accompanied by a marginal rise in unemployment rate. The rise in unemployment rate for males in urban areas might be due to migration of the labour force with higher educational attainments from rural areas in search of better employment options during the reform period.

In respect of females in urban areas, it was observed that the WPRs witnessed a marginal decline, and the unemployment rates by all methods of estimation also showed a moderate decline in the reform period. But the decline in WPRs was found to be much higher than that in unemployment rates. This provides some evidence to believe that in urban areas, the female SPRs or the ratio of females neither in labour market nor in schools/colleges seemed to have increased to the extent that the WPRs and unemployment rates declined during the same period. A greater increase in female SPRs, therefore, seemed to have come mainly from a decline in WPRs in the age groups of 5-9 and 10-14 years. It is, thus, evident that the decline in WPRs in rural and urban areas, in most cases, could be attributed to a rise in unemployment rates as well as in the ratio of population neither in labour market nor in schools/colleges.

Increase in student population ratios

The data presented in Table 3.9 reveal that the SPRs increased rapidly for males and females in rural areas during the pre-reform and reform periods, but a rapid rise had begun to take place even before the economic reform measures. The rise in SPRs was observed to be more in the age groups 5-9 and 10-14 years, as compared to the age groups of 15-29 years both in respect of males and females. Further, the female SPRs increased more than the male SPRs during both the periods. The increase in male and female SPRs appeared to be higher than the decline in male and female WPRs at the relevant age groups in rural areas. A greater decline in male and female WPRs, especially in the age groups of 10-14 and 15-29 years was accompanied by a greater rise in male and female SPRs in the same age groups during 1993-94 to 1999-2000. This implies that the workers in the age groups

Table 3.9 : Age-Specific Student Population Ratios (UPS) for Males and Females in Rural India, 1983 to 1999-2000 (Percent)

Age group	*Student population ratio*				*Change in Student population ratio*			
	1983	*1987-88*	*1993-94*	*1999-2000*	*1987-88/ 1983*	*1993-94/ 1987-88*	*1993-94/ 1983*	*1999-00/ 1993-94*
Males				**Rural**				
5-9	48.4	27.6	67.4	70.5	-20.8	39.8	19.0	3.1
10-14	60.6	66.7	76.3	78.4	6.1	9.7	15.7	2.1
15-29	15.1	14.6	19.1	19.8	-0.5	4.5	4.0	0.7
All	19.0	16.7	26.3	27.9	-2.3	9.5	7.3	1.6
Females								
5-9	34.4	23.5	56.2	63.2	-10.9	32.7	21.8	7.0
10-14	33.1	42.1	55.5	64.2	9.0	13.3	22.4	8.7
15-29	4.0	5.0	7.4	10.1	1.0	2.5	3.4	2.7
All	12.0	9.3	17.2	20.7	-2.8	7.8	17.2	3.6
Males				**Urban**				
5-9	71.0	72.6	84.0	83.1	1.8	11.4	13.2	-0.9
10-14	79.0	80.7	86.2	87.3	1.9	6.6	7.2	1.1
15-29	25.0	24.8	27.9	28.5	-0.3	3.1	2.8	0.7
All	26.0	25.6	27.6	27.6	-0.4	2.0	1.7	0.0
Females								
5-9	65.0	67.7	80.3	80.4	2.3	12.6	14.9	0.1
10-14	67.0	72.3	81.3	82.6	5.5	9.0	14.5	1.3
15-29	15.0	16.7	21.4	23.6	1.6	4.7	6.2	2.2
All	20.0	21.6	24.4	25.0	1.2	2.8	4.0	0.6

Note : NA: Not available.

Source : NSSO (1988, 1990, 1996 & 2001).

of 10-14 and 15-29 years had withdrawn themselves from the labour market in favour of schools/colleges to acquire more qualifications, skills and knowledge in the era of globalisation.

In the case of urban areas (Table 3.9), it has been found that a greater rise in male SPRs, in aggregate, took place during the pre-reform period, as compared to no change during the reform period. A rapid rise in male SPRs took place, especially in the age-groups of 5-9 and 10-14 years during the pre-reform period itself as compared to a marginal rise during the reform period. In respect of females, the SPRs increased during both the periods; a greater increase, however, occurred during the pre-reform period. Although the female SPRs were lower than those for males, a greater rise took place in respect of females, as compared to males, during both the periods. The decline in male and female WPRs, especially in the age groups of 10-14 and 15-29 years, was accompanied by a rise in male and female SPRs in the same age groups during 1993-94 to 1999-2000. The decline in WPRs among the youth in urban areas, thus, could be partly attributed to a rise in SPRs among them. In the knowledge-based economy, there has been a clear-cut indication that human capital formation, particularly among the youth, seemed to have occupied a greater importance both in rural and urban areas in the 1990s.

Rise in the ratio of population neither in labour market nor schools/colleges

Another plausible and a more satisfactory explanation for the decline in WPRs is the increase in the ratio of population neither in the labour market nor in schools/colleges to total population. The data presented in Figure 3.14 show that the ratio of male persons neither in the labour market nor in schools/colleges tended to decline rapidly up to the age-group of 25-29 years and then became very marginal and flat up to the age group of 40-44 years, and thereafter tended to rise significantly (Figure 3.14). The distributional pattern of such population share by age groups constituted what could be described as the "bathtub-shaped curve". At both extremes of age groups, the ratio of male persons neither in labour market nor schools/colleges was found to be quite significant and at the intermediate level from the age-groups of 30-34 to 44-49 years, it was very minimal.

Figure 3.14 : Age-Profile of Male Persons neither in Labour Market nor in Schools/Colleges in Rural India, 1993-94 to 1999-2000

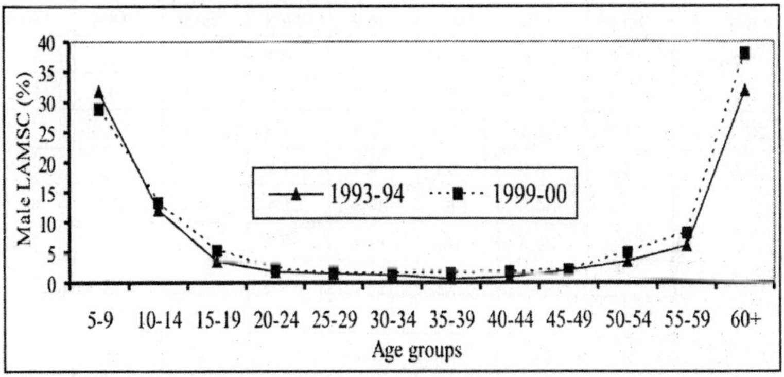

Note : LAMSC: Proportion of male persons neither in labour market nor in schools/colleges.
Source: NSSO (1997, 2001).

The data presented in Table 3.10 also indicate that, in rural areas, the ratio of male persons neither in the labour market nor in schools/ colleges, in aggregate, appeared to have declined significantly during the pre-reform period, as compared to a marginal increase during the reform period. Such segment of male persons among the teenagers (age-groups 5-9 and 10-14 years) declined significantly during the pre-reform period; a smaller decline took place in the age-group 5-9 years and a marginal rise in the age-group 10-14 years during the reform period. In the age-groups 15-29 and 30-44 years, the ratio seemed to have increased during both the periods; a greater increase appeared to have taken place during the reform period. For the older age groups from 45-59 to 60+ years, the ratio of such male persons declined during the pre-reform period, as compared to a rapid increase during the reform period.

In respect of females in rural areas, the distributional pattern of female persons neither in the labour market nor in schools/colleges by age groups was not the same as that for male persons. The ratio of female *vis-à-vis* male persons neither in the labour market nor in

Table 3.10 : Ratio of Age-Specific Male and Female Population (UPS) neither in Labour Market nor in Schools/Colleges in India, 1983 to 1999-2000

Age group	*Rural*				*Urban*			
	1983	*1987-88*	*1993-94*	*1999-2000*	*1983*	*1987-88*	*1993-94*	*1999/2000*
Males								
5-9	49.7	71.7	31.8	28.7	28.7	26.9	15.5	15.9
10-14	17.5	17.9	12.1	13.1	10.8	10.5	6.4	7.7
15-29	2.1	9.5	6.0	7.0	2.0	11.5	9.3	10.2
30-44	1.3	2.1	1.4	2.1	1.3	2.4	2.5	3.1
45-59	4.8	4.8	3.7	4.7	7.2	7.4	7.7	8.2
60+	35.7	34.8	31.7	37.8	51.1	52.7	57.0	61.1
All	13.3	29.4	10.2	11.1	9.1	20.2	17.4	17.8
Females								
5-9	63.8	75.9	42.7	36.3	34.1	32.0	19.4	18.9
10-14	49.7	44.5	34.1	28.3	27.7	23.0	15.1	14.5
15-29	58.9	58.9	60.7	58.6	67.7	66.7	62.7	61.8
30-44	54.0	53.0	57.5	55.6	76.0	76.2	76.4	76.9
45-59	59.1	58.1	60.1	59.0	77.0	77.6	76.8	78.0
60+	84.3	83.3	82.3	82.2	88.2	90.2	90.6	91.4
All	51.4	64.6	55.4	52.2	53.9	64.5	61.5	61.7

Note : NA: Not available.
Source : NSSO (1988, 1990, 1996 & 2001).

schools/colleges was found to be very high in almost all the age groups. The ratio of females in such category of population in different age groups appeared to have declined during both the periods; a greater decline, however, took place during the pre-reform period. The ratio of such female persons declined significantly in the age groups of 5-9 and 10-14 years during both the periods; a greater decline occurred during the pre-reform period. A smaller decline took place in the ratio of females in the age-group 60+ years during both the periods; a greater decline was observed in the pre-reform period. It is evident that the decline in WPRs was accompanied by a marginal rise in the unemployment rate and a rapid rise in SPRs in the 1990s.

In urban areas, the data provided in Figure 3.15 show that the ratio of male persons neither in the labour market nor in schools/ colleges in different age groups seemed to have formed what could be described as the "boat-shaped curve", which is slightly different from that observed for rural male persons (Figure 3.15). Such segment of population was found to be very small in the age groups of 5-9 to 20-24 years and very large in the age group of 50-54 years, as compared to that for males in rural areas.

Figure 3.15 : Age-Profile of Male Persons (UPS) neither in Labour Market nor in Schools/ Colleges in Urban India, 1993-94 to 1999-2000

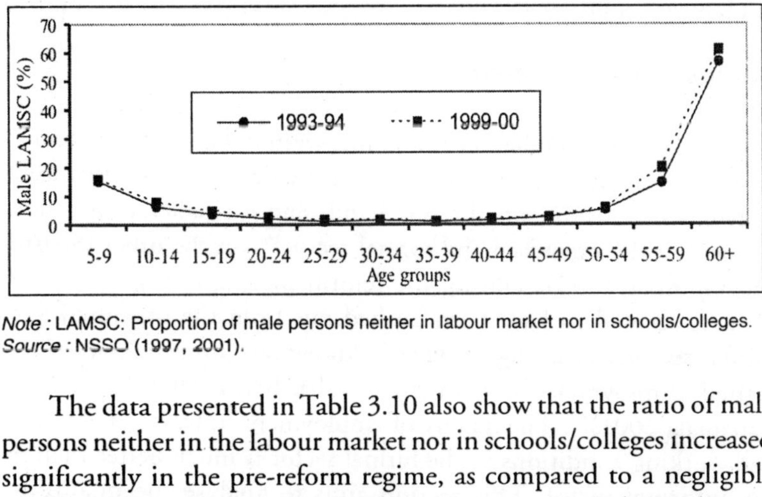

Note : LAMSC: Proportion of male persons neither in labour market nor in schools/colleges.
Source : NSSO (1997, 2001).

The data presented in Table 3.10 also show that the ratio of male persons neither in the labour market nor in schools/colleges increased significantly in the pre-reform regime, as compared to a negligible decline in the reform regime. In the age-groups 5-9 and 10-14 years, however, a rapid decline took place in the pre-reform regime and a moderate rise occurred in the reform regime. In the rest of the age groups, including 60+ years, a greater rise took place in the ratio of such segment of male persons during the pre-reform period, as compared to a smaller rise during the reform period. As regards females in urban areas, it has been found that the ratio of females neither in the labour market nor in schools/colleges did follow any meaningful pattern as observed in the case of males in urban areas, but appeared to have increased with rising levels of age groups. In urban areas, such segment of female persons was observed to be very large, as compared to males and even males and females in rural areas.

Although it is difficult to dovetail the decline in WPRs with the rise in unemployment rates, SPRs and the ratio of population neither in the labour market nor in schools/colleges; it is evident that the decline in WPRs is attributed partly to a rise in unemployment rates, mostly among the illiterates and literates up to primary level of education and among the youth and old-aged persons; partly to a withdrawal of workers, in the age groups of 5-9, 10-14 and 15-29 years from the labour market in favour of schools /colleges; and partly to a rise in the ratio of population neither in the labour market nor in schools/colleges, especially in respect of males in rural areas and males and females in urban areas.

Quality of Employment

It is a well known fact that the labour market in India is heterogeneous, comprising a large segment of workforce involved in the unorganised and informal sector, on the one hand, and a smaller proportion engaged in the organised and formal sector[12], on the other (Shariff and Gumber 1999). During the reform period, a larger share of workers seemed to have been pushed out from the organised and formal sectors, resulting in what is known as "informalisation" of formal employment both in rural and urban areas (Unni 2001; Vanamala 2001)[13]. The quality of employment in terms of earnings and working conditions in the formal sector is much better than in the informal sector. This section aims to analyse the quality of employment in rural and urban areas during the pre-reform and reform periods.

Casualisation of employment

The hypothesis that the higher the rate of unemployment, the higher will be the rate of casualisation of employment has been advanced (Minhas and Majumdar 1987; Jayadevan 1996). If there is a surplus of labour, from the demand side, the employer can demand the service of labour at lower wages on contract/temporary basis. The surplus labour paves the way for greater casualisation of employment. Similarly, another hypothesis that the higher the degree of casualisation, the higher the proportion of people living below the poverty line has also been formulated (Jayadevan 1996). This is due to the fact that the labourers from the supply side, especially in the case of resource-poor households, instead of waiting for long to secure a better job, accept any job at any wage rate and under any kind of work situations, thereby leading to lowering of their level of earnings and persistence of chronic poverty among such households.

It has been argued that the reform measures have caused a rapid growth of casual wage paid employment, which constitutes a major share of the informal sector. In the market-driven economy, with a view to earning more profits and becoming a world-class company, the entrepreneurs started to bring about structural changes in their organisational, managerial, technical and operational setup leading to cost-effectiveness and profit maximisation (Vanamala 2001). As a result of these deliberate changes, some of the jobs generated in the informal sector during the reform period were casual, low-end, lowly paid contractual jobs (Vanamala 2001). Informalisation of employment, pushing workforce out of formal work conditions led to a rapid rise in the share of wage paid employment in casual and/or contractual work situations.

The data presented in Table 3.11 show that in rural areas, a larger proportion of employment was composed of self-employed and casual wage labourers both for males and females. The regular salaried/wage employment accounted for a smaller share of total employment in rural areas as compared to urban areas. The share of casual wage employment for males, as per UP+SS, increased rapidly from 22 percent in 1972-73 to 36.2 percent in 1999-2000. In respect of females, it also rapidly increased from 31.4 percent to 40 percent during the same period. A major segment of increase in casual wage employment took place at the cost of self-employment.

The process of casualisation and informalisation of formal employment has accelerated due to a simultaneous operation of supply and demand side of employment. From the supply-side of employment, the casualisation reflects the non-availability of adequate and regular self-employment opportunities. From the demand-side, an increase in the proportion of casualisation is the result of the

Table 3.11 : Distribution of Workforce by Status of Employment in Rural India, 1972-73 to 1999-2000

(*Percent*)

Status of employment	*Rural*						*Urban*					
	1972-73	*1977-78*	*1983*	*1987-88*	*1993-94*	*1999-2000*	*1972-73*	*1977-78*	*1983*	*1987-88*	*1993-94*	*1999-2000*
Males	**UPS**											
SE	NA	NA	59.5	57.5	56.4	54.4	NA	NA	40.2	41.0	41.1	41.2
RSE	NA	NA	10.6	10.4	8.5	9.0	NA	NA	44.5	44.4	42.7	41.9
CWE	NA	NA	29.9	32.1	34.6	36.6	NA	NA	15.3	14.6	16.2	16.9
Females												
SE	NA	NA	54.1	54.9	51.0	50.0	NA	NA	45.8	47.1	45.4	45.3
RSE	NA	NA	3.7	4.9	3.4	3.9	NA	NA	25.8	27.5	28.6	33.3
CWE	NA	NA	42.2	40.2	45.3	46.1	NA	NA	28.4	25.4	26.2	21.4
Males	**UP+SS**											
SE	65.9	62.8	60.4	58.6	57.9	55.0	39.3	40.4	40.7	41.7	41.7	41.5
RSE	12.1	10.6	10.8	10.0	8.3	8.8	50.7	46.4	44.6	43.7	42.1	41.7
CWE	22.0	26.7	28.8	31.4	33.8	36.2	10.1	13.2	14.8	14.6	16.2	16.8
Females												
SE	64.5	62.1	62.2	60.8	58.5	57.0	48.4	49.5	46.5	47.1	45.2	45.3
RSE	4.1	2.8	3.1	3.7	2.9	3.0	27.9	24.9	26.2	27.5	28.6	33.3
CWE	31.4	35.1	34.7	35.5	37.6	40.0	23.7	25.6	27.3	25.4	26	21.4

Note : SE: Self-employment; RSE: Regular salaried employment; CWE: Casual wage employment

Source : NSSO (1981, 1988, 1990, 1996 & 2001).

employer's greedy attempt to reduce labour costs by employing labour on "hiring and firing" basis to mainly escape from the labour laws (Jayadevan 1996; Vanamala 2001).

In urban areas, the share of workers, both male and female, in casual wage employment was found to be far less as compared to rural areas. In urban areas, the share of females in casual wage employment was quite significant as compared to that of males. The data also indicate that the rate of increase in casual wage employment was quite lower in urban areas than in rural areas. While the share of male casual wage employment marginally increased, the share of female casual wage employment declined significantly during the reform period. The share of males in causal wage employment, as per UP+SS, constantly increased from 10.1 percent to 16.8 percent during 1972-

73 to 1999-2000. In respect of females, it increased from 23.7 percent in 1972-73 to 27.3 percent in 1983, and thereafter declined to 21.4 percent during 1999-2000. The growth of male casual wage employment took place at the cost of regular salaried/wage employment, whereas in respect of females, it took place at the cost of self-employment. A similar observation was also made other studies (Jayadevan 1996; Sahu 2003; Biradar and Bugalkoti 2001).

The foregoing analysis reveals that the economic reforms appeared to have resulted in the growth of casual wage of employment, particularly in respect of females in rural areas and males in urban areas. An important and widely debated question is how the incidence of poverty, particularly in rural areas, had declined even when there was a sharp increase in the rate of unemployment, casualisation and informalisation of formal employment in the 1990s. Because, the income levels of the workers involved in such casual, lowly paid employment would be lower than the income levels of those workers involved in self-employment or regular salaried/wage employment in agricultural or non-agricultural sectors and expected to decline with growing casualisation over time. In this context, it has been hypothesised that a major decline in the incidence of poverty was due to the shift of workers away from agriculture to the rural non-agricultural sector where they could get regular employment and higher wage rates.

Rural Non-Agricultural Employment in India: A Synoptic View

In view of the increasing population pressure on agriculture, diminishing land frontiers, declining employment elasticities in agriculture and urban organised manufacturing sector and excess supply of labour force, the growth of rural non-agricultural activities has become an important strategy in combating the problems of unemployment and poverty in rural areas. Since the 1970s, the rural economy has begun to diversify its occupational structure away from agriculture in favour of the non-agricultural sector, absorbing a large chunk of labour force in rural areas.

Table 3.12 : Distribution of Workforce (UP+SS) in Non-agricultural Employment in India, 1972-73 to 1999 2000 (Percent)

Year	*Rural*			*Urban*			*All India*		
	Males	*Females*	*Persons*	*Males*	*Females*	*Persons*	*Males*	*Females*	*Persons*
1972-73	16.7	10.4	14.4	89.3	67.1	85.2	31.2	15.7	26.1
1977-78	19.3	11.8	16.6	89.4	68.0	84.8	34.4	18.2	29.0
1983	22.1	12.2	18.4	89.7	68.0	85.2	37.8	18.8	31.4
1987-88	25.4	15.2	21.7	90.9	70.0	86.5	41.3	22.3	35.1
1993-94	25.9	13.6	21.5	91.4	75.3	88.1	43.4	22.4	36.7
1999-2000	28.8	14.7	23.9	93.4	82.5	91.3	46.8	25.0	40.0

Source: NSSO (1981, 1988, 1990, 1996 & 2001).

The data furnished in Table 3.12, *prima facie*, show that there was a moderate shift of workers from agricultural to the rural non-agricultural sector since the 1970s and virtual stagnation in the early 1990s, and started gaining momentum in the late 1990s. In rural areas, the share of workers, both males and females, in the non-agricultural sector was lower in the years of the pre-reform period, as compared to that of the reform period. The shift of workers in favour of the rural non-agricultural sector had inter-regional, gender and activity-wise variations. An analysis of occupational diversification in favour of the rural non-agricultural sector at disaggregated level and its underlying growth process and impact on poverty and income inequality provides relevant policy prescriptions to address the long-standing problems of unemployment, poverty and income inequality in rural areas, especially in the context of globalisation.

Conclusions

There has been an apprehension that the economic reforms have had adverse impact on employment and poverty, particularly in rural India. It is too early to assess the impact of economic reforms on structural transformation of the emplayment market and thereby its impact on poverty reduction at the aggregate level by using the data for only two time periods (i.e., 1993-94 and 1999-2000). The 55th round of NSS on employment and unemployment situation, however, presents mixed results: withdrawal of some segment of workforce from the labour market in favour of schools/colleges; and a marginal rise in unemployment rates and the ratio of population neither in the labour market nor in schools/colleges in 1990s. The dynamics of employment and unemployment in India during the pre-reform and reform periods is succinctly summarised as follows.

The annual growth rate of employment was higher in urban than in rural areas throughout the period under study. The annual growth rate of employment declined considerably during the reform period as compared to the pre-reform period at the all India level; the decline was greater in respect of rural workers as compared to their urban counterparts and also for females as compared to males. In rural areas, there was a considerable decline in the growth rate of employment for male workers, whereas in respect of female workers, a negative growth rate was observed during the reform period *vis-à-vis* the pre-reform period. In urban areas, the annual growth rate of employment witnessed a rapid decline during the same period; the decline was greater in respect of females as compared to males. A major decline in the growth of employment appeared to have come from subsidiary status of occupation both in rural and urban areas.

The WPRs began to show a marginal decline in the late 1980s, even before the reform measures and sharply increased in the early 1990s; and again steeply declined in the late 1990s both in rural and urban areas. The decline in WPRs, which started even before the reform measures, was not only higher but it also accelerated in rural as against urban areas during the reform period. Although the female WPRs were smaller, a significant decline took place as compared to their male counterparts in rural areas during the reform period. The decline in WPRs during the reform period was not uniform across the age groups and levels of education.

Going by the age groups of workers, the WPRs tended to increase with rising levels of age groups up to a certain level and thereafter declined. The increase was quite sharp from the age groups of 5-9 to 25-29 years and became a little flat from 30-34 to 50-54 years, and slowly declined thereafter. Such a pattern of WPRs across different age groups constituted the "inverted U-shaped" curve with varying contours across males and females, and rural and urban areas. The WPRs tended to be lower at both extremes, and higher at intermediate levels. A greater decline in WPRs started to occur among children and youth, as compared to adults and aged workers even before the economic reforms, though the overall WPRs declined during the reform period, especially in rural areas.

The WPRs by levels of education reveal that with increasing levels of educational attainments, the WPRs tended to decline up to a certain level and increased thereafter in varying degrees across males and females, rural and urban areas. This implies that the WPRs were higher at both extremes of educational attainments and lower at the intermediate levels. Such a pattern of WPRs constituted the "saucer-shaped curve" with varying contours across gender and areas. In respect of urban females, the pattern of WPRs at different levels of education formed the "slanting U-shaped curve". This was a mirror image of the "inverted U-shaped curve" along which the ratio of females in full-time housework tended to increase as the level of education rose.

The WPRs for male persons, aged 15 and above declined during both the periods; a greater decline taking place during the reform period. Going by different levels of education, the decline in male WPRs in rural areas was higher among workers who were illiterate or educated at primary level. In the case of rural females, the decline in WPRs was greater in the years of the pre-reform period, as compared to that of the reform period; it rather marginally increased. Such decline was rapid among the female workers in the categories of graduates and above and illiterate and literate up to primary levels education during the same period.

The male WPRs in urban areas showed a greater decline during the reform period as compared to the pre-reform period. Although the decline in urban male WPRs was quite perceptible at each level of educational attainment, a rapid decline, however, was observed among workers who were illiterate and literate up to primary level of education; and even at graduate and above levels of education, a smaller decline took place during the reform period. The decline in urban female WPRs started to occur even before the economic reforms, but a greater decline took place after the economic reforms. The female WPRs showed a greater decline at higher levels of educational attainments and a marginal one at the level of illiterates and literates up to primary level of education. It has been found that the WPRs sharply declined in the 1990s in varying degrees across gender, age groups and levels of education both in rural and urban areas.

The pattern of unemployment rates by age groups for males and females, in rural and urban segments of the labour market, constituted the "inverted V-shaped curve" with varying contours across males and females. The data on unemployment rates by all methods of estimation were traditionally higher in urban than in rural areas. The incidence of unemployment as per UPS and UP+SS was found to be more among the male labour force *vis-à-vis* female labour force in rural areas, among the female labour force *vis-à-vis* male labour force in urban areas, and in urban as compared to rural segment of the labour market in India. The incidence of unemployment, for both males and females, used to be quite higher in the years of the pre-reform period than the reform period. The unemployment rates, as per both UPS and UP+SS, were found to be more among the male and female youth. A greater increase took place in the youth unemployment rates, particularly in rural areas, in 1999-2000 over 1993-94. But, the incidence of youth unemployment for males in urban areas declined during the same period.

The incidence of unemployment was quite higher among highly educated persons and lower among the labour force with illiterate and literate at lower levels. However, a marginal rise took place in the incidence of unemployment for male and female persons with illiterate or with education up to primary level, whereas in the case of persons with higher educational attainments, a greater decline occurred during 1993-94 to 1999-2000; the decline was quite perceptible in respect of females. The decline in WPRs was attributed partly to a rise in unemployment rates among the youth and old-aged persons and among illiterates and educated up to primary level and partly to withdrawal of workers from the labour market in favour of schools/colleges to acquire higher educational qualifications and partly to a rise in the ratio of population neither in the labour market nor in schools/colleges.

The quality of rural employment especially in the case of females seemed to have significantly deteriorated during the reform period. The pace of casualisation and informalisation of formal employment has been increasing in recent years. The incidence of casualisation was not uniform across gender, different levels of education and activities both in rural and urban areas. It was found to be quite higher among illiterate women as against men, particularly in rural areas. As the level of education tended to rise, the proportion of workers experiencing casualisation of employment declined in the 1990s.

The synoptic view of the data shows that there had been a moderate shift of rural workers from agriculture to the non-agricultural sector since the 1970s and virtual stagnation in the early 1990s, and started gaining momentum in the late 1990s. The share of workers, both males and females, in the rural non-agricultural sector was lower during the pre-reform period as compared to the reform period. The shift of workers in favour of the rural non-agricultural sector had inter-regional, gender and activity-wise variations. An analysis of occupational diversification in favour of the rural non-agricultural sector at disaggregated level and its growth process and impact on poverty and income inequality provides relevant policy prescriptions to address the long-standing problems of unemployment, poverty and income inequality in rural areas, especially in the context of globalisation. An attempt, therefore, has been made to analyse the different aspects of employment in the rural non-agricultural sector in subsequent chapters.

NOTES

1. The Planning Commission constituted a Special Group on "Targeting Ten Million Employment Opportunities per Year" under the Chairmanship of Dr. S. P. Gupta. The special group, after examining all the relevant studies and the situation of employment and unemployment, suggested a future strategy to improve the quality of employment and remove unemployment by the turn of this decade (GoI 2002b).

2. The labour market in India is heterogeneous because it encompasses a vast segment of unorganised sector that accounted for 91.7 percent in 1999-2000, which was 91.4 percent in 1993-94. Any change in the macroeconomic policies of the government affects the working class people of this sector, the sector that is not provided adequate social security measures.
3. The employment elasticity based on CDS for the economy as a whole declined from 0.68 to 0.16 over the period between 1983 to 1987-88 and 1993-94 to 1999-2000 (GoI 2002b). Moreover, while employment elasticity in the agricultural sector as a proxy for rural employment has declined from 0.87 to as low as 0.01, the employment elasticity in the manufacturing sector as an index of urban employment also declined from 0.59 to 0.33 during the same period (GoI 2002b: 525, Table 22).
4. Work participation rates refer to the proportion of people in work to total population.
5. Hirway conducted a pilot survey in six states, namely, Haryana, Madhya Pradesh, Gujarat, Orissa, Tamil Nadu and Meghalaya in 1998-99 to collect data on the time-use pattern of people. She has pointed out that data from the time use survey better explained the pattern of work distribution and provided satisfactory reasons for the observed decline in WPRs in the 1990s. The conventional surveys are not adequate to explain the economic work of people. She has questioned the validity of the available workforce data (Hirway 2002).
6. SNA activities are those activities covered in the System of National Accounting and falling within the SNA production boundary. Extended SNA activities are those activities falling outside the SNA production boundary but within the general production boundary, consisting mainly of unpaid services, which can be classified in three categories: household unpaid, management, shopping, etc.; care of children, the aged or disabled in the family; and community services. Non-SNA personal services are those activities which need to be performed by people themselves and cannot be delegated to others (for example rest, sleeping, etc.) (Hirway 2002).
7. The classification of the age groups for 1983 was not similar to the classification of age groups for 1987-88, 1993-94 and 1999-2000. Hence, the temporally comparable age-wise WPRs were estimated for 1987-88, 1993-94 and 1999-2000 to compare with those of 1983, but not at more disaggregated levels of age groups.
8. In 2000, of the 248.4 million children in the age group of 5-14 years in the country as a whole, 181.2 million or 73 percent were full-time students. Of the 67.6 million who were not attending educational institutions,

only 8.6 million or 12.3 percent were in the labour force and another 8.2 million or 12.2 percent involved in housework (Ghose 2004: 5108).

9. The data on male WPRs, as per UP+SS, have not been presented due to lack of space (see, NSSO 2001).
10. There are two important policy documents, namely, the Study Group Report on 'Skill Development, Training and Workers' Education of the National Commission on Labour, Government of India (GoI 2002a) and the report on "Targeting Ten Million Employment Opportunities Per Year" of the Planning Commission, Government of India (GoI 2002c) that have accorded due importance for skill formation/up-gradation by providing education and training.
11. Persons in the age groups 15-19 and 20-24 years are considered as youth in many countries and by international development agencies like United Nations. A further distinction has been made by the International Labour Organisation (ILO) as teenagers in 15-19 and young adults in 20-24 age groups to identify different risk factors associated with their schooling, vocational training and employment (ILO 1985). For the present study, based on the availability of temporally comparable data, the persons in the age category of 15-29 years are considered as the youth.
12. In the recent past, the growth of employment in the unorganised (excluding agriculture) sector was much higher than that of the organised sector. This might be due to disinvestments in public sector undertakings.
13. Although the concept of informalisation is difficult to define, it is used to denote an act of entrepreneurs in which the employer seeks to escape from responsibilities in providing the entitled benefits to the employee even in the formal sector where the employment is conditioned by labour laws. The employer who is lured by the "profit bug" tries to shirk from the entitled responsibilities by hiring workers on a daily basis as casual labour. The nature of employment in this category is more often contractual, casual, insecure, low profile and low earning. The working conditions in the informal system are exploitative and oppressive (Vanamala 2001).

4 Growth of Non-Agricultural Employment in India

A Spatio-Temporal Analysis

Introduction

An analysis of RNAE by gender, age groups, level of education and size of landholdings, *ipso facto*, helps policy makers and development practitioners in deciding about the type of policy intervention needed to tackle the long-standing problems of unemployment, poverty and income inequality simultaneously. Diversification of occupational structure in favour of the non-agricultural sector in rural areas has been regarded as one of the important strategies. The economic growth process of the industrialised countries shows a particular pattern of sectoral diversification in employment as well as income over a period of time. Several studies (Fisher 1939; Clark 1940; Kuznets 1966; Chenery and Syrquin 1975) on the growth process of the industrialised economies have revealed that as economic growth takes place, the share of the primary sector largely consisting of agriculture shows a decline in its relative contribution to national output even though its absolute contribution to national output increases. The growth of the primary sector contributes to the growth of the secondary or industrial sector through its multiple linkages.

Once the economy becomes industrialised, the stage at which the labour force released from the primary sector would be productively utilised, the share of the secondary sector starts declining in its relative importance, and that of the tertiary sector consisting largely of services of all kinds begins to rise. There would also be corresponding changes in the occupational distribution of workforce in each sector of the economy. A monotonic relationship in sectoral diversification will be an outcome of sustained growth dynamism and interaction between the growth-impulses generated in each sector of the economy[1]. The importance of the secondary sector-manufacturing— and its ability to absorb resources released from the primary sector-agriculture— has been visualised in economic thinking as in the Lewis (1954) and Fei and Ranis (1961) models of economic growth.

This pattern of economic growth observed in the industrialised economies has not been observed in most of the developing economies including India. In India, there has been a rapid transition from agriculture to the service sector bypassing the manufacturing sector. Inter-sectoral growth movements since 1950s reveal that while the share of agriculture in national income has been steadily declining, the share of the service sector has witnessed a rapid increase and that of the secondary sector has increased at a much slower pace than what was expected during the same period (Nagaraj 1990; Bhalla 1997; Madheswaran and Dharmadhikary 2000; GoI 2003a). Before the country reached a stage at which it could be called an "Industrialised Economy" the share of the service sector in national income has begun to increase significantly[2]. This is something which is contrary to the experience of the industrialised economies like the USA or UK.

As regards occupational structure, it has been found that the change in the occupational structure of workers has not matched a corresponding shift in the production structure in each sector of the economy (Nagaraj 1990; Bhalla 1997; Parthasarathy, Shameem and Reddy 1998; Bhatt 2003). While there was a continuous decline in the share of agriculture in national output since the 1950s, there was no decline in the share of employment until 1970. The beginning of the 1970s started to show a sustained decline in the share of the workers in agriculture and a corresponding increase in non-agricultural activities in rural areas. At the all-India level, for instance, while the share of workforce in agriculture declined from 74 percent in 1972-73 to 60 percent in 1999-2000 (NSSO 1981, 2001), the share of income from agriculture consistently declined from about two-thirds to a little over one-fourth of the national income during the same period (GoI 2003a). As evident from the review of literature (see,

Chapter 2), the pattern of occupational shift was more pronounced in respect of males than that of females. Within the rural non-agricultural sector, the increase in the share of employment in the service sector was much higher than that of the manufacturing sector; the bulk of the increase in RNAE was explained by an increase in casual wage employment.

Since the 1970s, especially when the sustained decline in poverty was accompanied by a rapid increase in the proportion of rural non-agricultural workers, the growth of RNAE has come to occupy top priority in the development paradigm of rural economy. A considerable decline in the share of agricultural income and marginal decline in the share of agricultural workers resulted in a decline in the growth rate of productivity per worker in agriculture and a substantial increase in the rural non-agricultural sector. This has created ample scope for investment and thereby an increase in production and employment opportunities in the rural non-agricultural sector in India. Consequently, the growth of RNAE has been accelerated in varying degrees across gender, age groups, level of education and size of landholdings.

Analysing the growth of RNAE at more disaggregated levels becomes pertinent in view of the commitment of development planning to generate 'ten million employment opportunities per year over the 10th Plan period' (GoI 2002c) and to 'halve poverty by 2015'— the Millennium Development Goal as set at the UN General Assembly in 2000 (WB 2001). As agriculture and the urban organised manufacturing sector have virtually ceased to employ more labour force, the burden of providing additional employment to the growing labour force has fallen upon the unorganised non-agricultural sector, particularly the rural non-agricultural sector. The important question is whether the rural non-agricultural sector can bear the burden of providing employment to the growing educated labour force in the context of globalisation.

In view of the above analytical framework, it is essential to address the following research issues: Have the growth rates and the size of RNAE continued to increase in the 1990s at the same pace as observed in the 1980s? What are the activities that were growing rapidly in the years of the reform period? What are the states experiencing sustained increase in male and female RNAE? Who are all those getting involved more in RNAE, whether men or women; youth or adults or aged; educated or uneducated; the landless, small and marginal farmers or large farmers? Is the degree of casualisation in RNAE increasing in the years of the reform period? If so, is it among men or women; youth or adults or aged workers; educated or uneducated; the resource-poor or resource-rich households?

Growth of Rural Non-Agricultural Employment

The data presented in Appendix 4.1 reveal that the growth rate of workers in the non-agricultural sector was found to be quite high as compared to the growth rates of workers in agriculture as well as total workers during the period between 1972-73 to 1977-78 and 1993-94 to 1999-2000. The growth rate of workers in agriculture was negative in the 1990s. This implies that a larger part of the increased labour force shifted to RNAE, particularly during the reform period, *albeit* at decreasing rate. Table 4.1 presents the data on growth rates of workers in RNAE, GDP and foodgrains at the aggregate level in India during 1972-73 to 1999-2000. The annual growth rates of male workers in the non-agricultural sector were higher in the years of the pre-reform period, as compared to those of the reform period. This implies that the growth rate of employment in the rural non-agricultural sector had slowed down in the 1990s. The growth of RNAE was found to be greater for males, as compared to females, especially during the reform period. A greater decline in the growth rate of female RNAE took place in the 1990s. It is evident that the rate at which the growth of the rural non-agricultural sector was increasing in the 1980s had not sustained itself in the 1990s.

The data also reveal an interesting trend: the growth rates of workers in the non-agricultural sector, especially in rural areas for females, fluctuated in accordance with the growth rates of agricultural income (at 1993-94 prices) and foodgrains production. It has been found that high growth rates of agricultural income and foodgrains production were accompanied by low growth rates of workers in

Table 4.1 : Growth Rates of Workers (UP+SS) in RNAE, GDP and Foodgrains in India, 1972-73 to 1999-2000

(Per cent/per annum)

Particular	*1972-73 to 1977-78*	*1977-78 to 1983*	*1983 to 1987-88*	*1987-88 to 1993-94*	*1993-94 to 1999-00*	*1972-73 to 1999-00*
Growth rate of NAWs			**Rural**			
Persons	5.07	4.23	4.48	1.93	2.05	3.30
Males	5.03	4.58	4.28	2.54	2.33	3.52
Females	5.19	3.19	5.09	0.36	1.04	2.61
			Rural & Urban			
Persons	4.74	4.05	3.68	3.32	1.97	3.35
Males	4.39	4.24	3.50	3.61	2.04	3.37
Females	6.10	3.31	4.37	2.18	1.70	3.28
Growth rate of GDP*	4.53	3.13	4.76	5.31	6.60	4.81
Agriculture	3.91	1.75	1.70	4.02	3.13	3.03
Manufacturing	5.13	3.41	5.91	5.50	6.92	5.63
Services	5.01	4.80	6.89	6.29	8.84	6.08
Growth rate of foodgrains	4.26	0.92	0.73	3.30	2.03	2.62

Note : * The calculation is based on the data (at 1993-94 prices) as supplied by CSO 2003, New Delhi; NAWs: Non-agricultural workers.

Source : NSSO (1981, 1988, 1990, 1996 & 2001); GoI (2003a).

RNAE, especially in respect of females during the pre-reform and reform periods and more so in the latter. The growth rates of workers, particularly for females, in the rural non-agricultural sector tended to move in the opposite direction to the growth rates of agricultural income and foodgrains production (Figure 4.1). This illustrates that the bulk of the growth of employment in the rural non-agricultural sector was determined by the growth performance of agriculture, mostly through its consumption and production linkages.

Regional pattern of employment

A considerable variation existed in the growth rates of workers, both for males and females, in RNAE across major states in India. Table 4.2 presents the data on annual growth rates of male and female

Figure 4. 1 : Nexus between Growth of Agriculture and Non-Agricultural Employment in Rural India, 1972-73 to 1999-2000

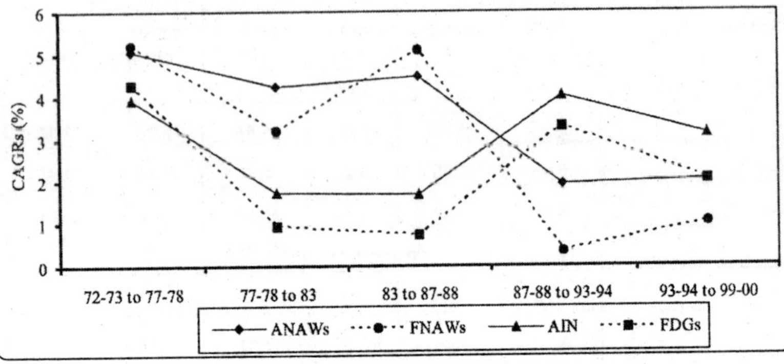

Note : ANAWs: All non-agricultural workers; FNAWs: Female, Non-agricultural workers; AIN: Agricultural Income; FDGs: Food grains; CAGR: Compound Annual growth rate.
Source: Table 4.1.

Table 4.2 : Growth Rate of Workers (UP+SS) in RNAE in India, 1977-78 to 1983 and 1993-94 to 1999-2000 *(Percent/per annum)*

State	*Males*				*Females*			
	1977-78 to 1983	*1983 to 1987-88*	*1987-88 to 1993-94*	*1993-94 to 1999-00*	*1977-78 to 1983*	*1983 to 1987-88*	*1987-88 to 1993-94*	*1993-94 to 1999-00*
Andhra Pradesh	3.99	4.35	1.63	1.28	4.71	3.76	1.55	-0.65
Assam	8.45	4.45	1.38	10.47	0.87	7.43	0.76	4.39
Bihar	3.50	3.34	0.97	4.42	6.24	-7.26	-3.18	13.24
Gujarat	6.73	11.81	0.86	1.73	7.14	30.91	-12.92	-0.46
Haryana	4.91	3.16	6.62	3.30	0.47	2.55	-2.03	-0.74
Karnataka	3.47	3.27	3.12	1.27	1.13	5.70	4.82	-4.70
Kerala	0.75	3.10	2.02	4.76	-2.82	1.90	-1.10	2.56
Madhya Pradesh	4.63	4.79	0.27	4.43	-1.07	17.8	-4.05	6.30
Maharashtra	2.12	5.02	2.37	1.79	-0.41	5.58	2.45	-5.96
Orissa	8.56	4.49	-1.23	2.38	7.43	3.52	-2.41	4.77
Punjab	2.62	7.58	1.85	2.81	-1.64	4.78	-6.80	10.10
Rajasthan	3.21	14.83	1.02	2.38	4.89	26.88	-11.00	1.78
Tamil Nadu	4.41	3.39	1.49	-0.13	3.61	6.84	-0.11	-0.26
Uttar Pradesh	3.17	2.81	4.81	3.59	4.19	-5.23	4.38	4.87
West Bengal	2.26	5.38	3.91	-0.52	-0.61	6.19	6.84	1.04
All India	4.70	4.16	3.06	2.33	3.19	5.09	0.73	0.51

Source: Chadha and Sahu (2002).

workers, as per UP+SS, in RNAE for major states in India during 1977-78 to 1999-2000. It has been found that the growth rate of RNAE for both males and females was higher in most of the states in the years of the pre-reform period (1977-78 to 1983 and 1983 to 1987-88). The higher growth rates of RNAE, however, did not sustain themselves in most of the states in the years of the reform period (1987-88 to 1993-94 and 1993-94 to 1999-2000).

The data also indicate that the growth rates of employment in the rural non-agricultural sector were significantly fluctuated in most of the states without showing a clear-cut directional changes. It has been found that most of the states, namely, Andhra Pradesh, Gujarat, Haryana, Karnataka, Maharashtra, Orissa, Tamil Nadu, Uttar Pradesh and West Bengal, by and large, have a tendency towards gradual decline in the growth rate of male RNAE during the period between 1977-78 to 1983 and 1993-94 to 1999-2000. The decline was sharp during the post-reform period. For instance, Andhra Pradesh showed a decline from 3.99 percent to 1.28 percent during the same period, except for a marginal rise during 1983 to 1987-88. Unexpected or unusual change in the growth rate of male RNAE in most of the states during 1983 to 1987-88 over the previous period was attributed to the total breakdown of agriculture due to severe drought across different parts of the country in the year 1987-88. The distress movement of workers from agricultural to the rural non-agricultural sector for employment was quite evident, and as agriculture picked up in 1993-94, no phenomenal transfer of workers from agricultural to the rural non-agricultural sector took place[3].

In respect of females, a considerable regional variation also existed in the growth rate of RNAE depending on the extent of agricultural prosperity during the pre-reform and reform period. A majority of the states registered positive growth rates of female RNAE in the years of pre-reform period and negative growth rates in the years of the reform period. During the period between 1977-78 and 1983, out of 15 states only five, namely, Kerala, Punjab, Madhya Pradesh, West Bengal and Maharashtra; and during 1983 to 1987-88, only two states, namely, Bihar and Uttar Pradesh had registered negative growth rates in varying degrees. During 1987-88 to 1993-94 as many as nine states, namely, Gujarat, Rajasthan, Punjab, Madhya Pradesh, Bihar, Orissa, Haryana, Kerala and Tamil Nadu; and during 1993-94 to 1999-2000, six states such as Maharashtra, Karnataka, Haryana, Andhra Pradesh,

Gujarat and Tamil Nadu experienced negative growth rates. The number of states experiencing negative growth rates of RNAE for females increased in the years of the reform period, as compared to those of the pre-reform period. None of the states in India revealed a uni-linear pattern of growth rates of workers in RNAE and a majority of the states showed periodic fluctuations (Chadha and Sahu 2002). This shows that the growth of RNAE for females had not been encouraging in most of the states during the reform period, as compared to the pre-reform period. It appears to indicate that male workers were moving out of agriculture in favour of RNAE, whereas female workers were moving out of RNAE in favour of agriculture.

Sectoral pattern of growth rates

It is evident that the growth of RNAE has been on the gradual rise since the 1970s with considerable variations across gender, regions and sectors. Within the rural non-agricultural sector, the growth rate of employment was not uniform across different sub-sectors. The data presented in Table 4.3 indicate that no sector maintained a uniform or consistent rise or decline in the growth rate of employment for males and females; year- to-year fluctuations were quite discernible in most of the sub-sectors. The annual growth rates of workers in most of the rural non-agricultural activities were positive in varying degrees in the years of the pre-reform period and became negative in some activities, namely, utilities (electricity, gas and water); services; and, mining and quarrying in the years of the reform period. The growth rates of male and female workers in rural non-agricultural activities were much higher than in agriculture. The decline in the growth rate of employment for male workers in agriculture during the reform period could partly be attributed to a decline in WPRs among the younger age-groups (5-9, 10-14,15-19 and 20-24 years) and a shift of workers to the rural non-agricultural sector. The decline in RNAE of public utilities and services could be attributed to the changes in government policies in many domains in the backdrop of new economic reforms since 1991. An important trend is that the rural non-agricultural activities, which registered a positive growth since the 1970s, namely, transport, storage and communication; construction; and trade, hotel and restaurant business continued to show an impressive progress, whereas the manufacturing sector started to register a lower growth rate during the reform period *vis-à-vis* the pre-reform period. A large number of small-scale industries became sick and a great majority of them, especially textile based, closed down due to various reasons, among which severe competition from imported goods was one. This seemed to have resulted in a slow growth of employment in small-scale industries during the reform period.

Table 4.3 : Growth Rates of Workers (UP+SS) by Sectors in Rural India, 1972-73 to 1999-2000

(*Percent*)

Sector	*1972-73 to 1977-78*	*1977-78 to 1983*	*1983 to 1987-88*	*1987-88 to 1993-94*	*1993-97 to 1999-00*
Males					
Agriculture	1.31	1.11	0.58	2.51	–0.16
Mining & quarrying	6.96	5.15	4.10	3.40	–1.59
Manufacturing	4.28	3.71	2.47	1.43	1.48
Electricity, gas, water, etc.	24.57	0.00	10.76	3.71	–7.79
Construction	3.71	7.21	12.40	-0.33	7.71
Trade, hotel, restaurant, etc.	7.50	3.65	4.63	4.36	4.88
Transport, storage and communication	5.51	8.87	4.88	4.44	8.62
Services	3.94	4.90	1.65	5.13	–1.74
Non-agriculture	5.03	4.58	4.28	2.54	2.33
All	1.97	1.84	1.42	2.65	0.64
Females					
Agriculture	2.21	2.43	–0.15	2.64	–0.65
Mining & quarrying	0.00	24.57	5.92	0.00	–5.59
Manufacturing	7.33	4.30	1.99	2.95	0.80
Construction	–8.97	3.71	33.03	–18.48	4.10
Trade, hotel, restaurant, etc.	7.78	1.22	2.25	0.00	–2.20
Services	3.71	0.82	2.29	5.15	1.09
Non-agriculture	5.19	3.19	5.09	0.36	1.04
All	2.53	2.52	0.55	2.36	–0.49

Note : Services include finance, insurance and real estate community, social and personal services.

Source : NSSO (1981, 1988, 1990, 1996 & 2001).

In respect of females, the growth of employment in agriculture registered a positive growth rate during the pre-reform period but a negative growth rate during the reform period. Going by different rural non-agricultural activities, it has been found that a negative growth of employment took place in mining and quarrying (-5.59 percent); and trade, hotel and restaurant business (-2.20 percent). A positive growth of employment continued to take place in construction, services and manufacturing; but a significant slow-down occurred in manufacturing and services during the reform period, as compared to the pre-reform period. It is evident that the growth rate of RNAE for males and females slowed down in the 1990s, as compared to the 1980s; a greater setback, however, took place in respect of females. For males, transport, storage and communication; construction; and, trade, hotel and restaurant business were found to be important activities during the reform period. Though the manufacturing sector registered a positive growth rate of employment for males, it was yet lower than that of other activities in the rural non-agricultural sector. In respect of females, construction, manufacturing and services were recorded to be important activities, registering positive growth rates during the same period.

Size and Composition of Workers in RNAE

The sectoral diversification of the rural economy in terms of employment in agriculture was not found to be proportionate to its corresponding changes in the national output. While there had been a constant decline in the proportion of agriculture in the national output since the 1950s, there was no decline in the share of employment until 1970. Since the 1970s, the proportion of workers started to show a sustained decline in the agricultural sector and a corresponding rise in the rural non-agricultural sector. At the all-India level, while the proportion of the workforce in agriculture declined from 74 percent in 1972-73 to 60 percent in 1999-2000 (NSSO 1981, 2001), the proportion of income from agriculture consistently declined from about two-thirds to a little over one-fourth of the national income during the same period (GoI 2003a). This illustrates that the changes in the production structure have not matched the changes in the occupational structure (Bhalla 1997; Parthasarathy, Shameem and Reddy 1998; Bhatt 2003).

A considerable decline in the proportion of income and a marginal decline in the proportion of workers in the agricultural sector created a greater scope for the growth of RNAE. The data presented in Table 4.4 reveal that, at the all India level, the proportion of workers in RNAE had been on the constant rise over time. It increased from 26.11 percent in 1972-73 to 40.03 percent in 1999-2000. The pace of increase was not similar across male and female workers. While the proportion of male workers in RNAE, by and large, witnessed a constant increase, a virtual stagnation occurred in respect of females in the early 1990s and picked up only in the late 1990s. The diversification of occupational structure was much faster in respect of males as compared to females.

Table 4. 4 : Proportion of Workers (UP+SS) in RNAE in India, 1972-73 to 1999-2000

Year	*Rural*			*Urban*			*All India*		
	Males	*Females*	*Persons*	*Males*	*Females*	*Persons*	*Males*	*Females*	*Persons*
1972-73	16.68	10.35	14.42	89.31	67.12	85.17	31.21	15.68	26.11
1977-78	19.34	11.76	16.59	89.35	67.96	84.84	34.43	18.18	28.98
1983	22.11	12.15	18.42	89.70	68.03	85.20	37.83	18.79	31.37
1987-88	25.38	15.16	21.69	90.93	70.00	86.55	41.31	22.34	35.06
1993-94	25.89	13.56	21.49	91.39	75.29	88.13	43.40	22.41	36.65
1999-2000	28.79	14.71	23.94	93.43	82.51	91.28	46.76	25.02	40.03

Source: NSSO (1981, 1988, 1990, 1996 & 2001).

This pattern of changes, observed at the all India level, varied across rural and urban areas. In rural areas, the proportion of male workers in RNAE was much higher than that of females. The share of male workers in non-agricultural occupations increased more than that of females during 1972-73 to 1999-2000, excepting 1987-88. In 1987-88, the country experienced a severe drought due to which employment opportunities in agriculture declined. As a result, rural workers were compelled to take up jobs in the rural non-agricultural sector available within the village or nearby towns. If we drop the year 1987-88, a rapid increase in respect of male and a moderate increase in respect of female RNAE was quite evident in rural areas.

It has also been found that the share of both male and female workers in the urban non-agricultural/informal sector, by and large, increased constantly; a greater increase, however, took place in the 1990s especially in respect of females. Further, the share of females in the urban non-agricultural sector increased much faster than that of females in the rural non-agricultural sector in the 1990s as compared to the 1980s.

In urban areas, female employment in the non-agricultural sector was not affected much due to the drought in 1987-88. It is clear that the size of the non-agricultural sector, particularly for males in rural areas and females in urban areas, had constantly increased during the entire period; the increase was a little faster in the pre-reform than in the reform regime. The share of males in RNAE increased much faster than that of females in rural areas. There has been a widening gap between male and female participation in RNAE over the years (Figure 4.2), implying that female workers were less likely to diversify away from agriculture in favour of RNAE, and their accessibility was, to a greater extent, constrained by several socio-economic factors prevailing in rural areas. All in all, the size of the rural non-agricultural sector was expanding in varying degrees across male and female workers; but the rate at which it increased in the 1980s had not sustained itself in the 1990s.

Figur 4.2 : Size and Composition of Workers in RNAE in India, 1972-73 to 1999-2000

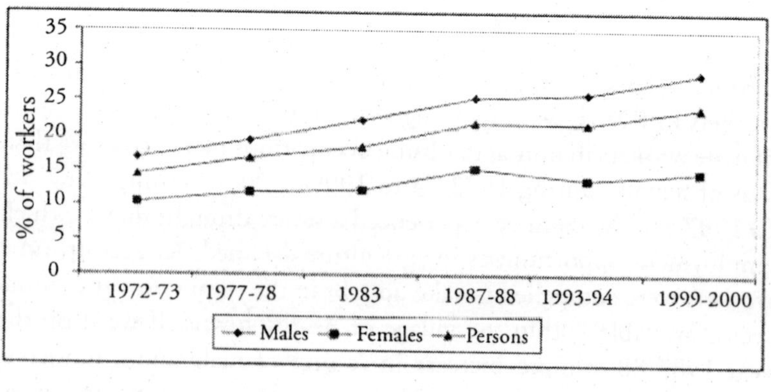

Source : Table 4.4.

Regional pattern of RNAE

The above analysis suggests that the size of the rural non-agricultural sector had been expanding over time in varying degrees across gender. This trend is not similar across states. The data (Table 4.5) reveal that a majority of the states had a higher proportion of male workers in RNAE than that at all-India rural levels, among which Kerala, followed by Himachal Pradesh and Haryana, recorded the highest share in RNAE. A sustained increase in the share of male workers in RNAE was witnessed in states such as Kerala, Himachal Pradesh, Haryana, Tamil Nadu, Punjab, Uttar Pradesh, Maharashtra and Karnataka during 1972-73 to 1999-2000. The growth impulses generated in the agricultural sector seemed to have sustained the growth of RNAE in states like Haryana and Punjab. The remaining states such as Andhra Pradesh, Bihar, Madhya Pradesh, Orissa, Rajasthan and Gujarat, by and large, experienced a rising but inconsistent trend in the proportion of male workers in RNAE, except for West Bengal, which, in fact, showed a steep decline during 1993-94 to 1999-2000. A rapid increase in the proportion of male workers in RNAE was experienced by most of the states in the 1980s as compared to the 1990s. A majority of the states witnessed a steep decline in the proportion of male RNAE in 1993-94 over 1987-88 and but the trend picked up in 1999-2000. The drought of 1987-88 seemed to have disturbed the steady growth of RNAE in a majority of the states, and more so in Rajasthan, Gujarat and Orissa (Parthasarathy, Shameem and Reddy 1998). The values of coefficient of variation reveal that the regional disparity started to show a decline in the years of the pre-reform period, but it again began to rise in the years of the reform period.

An effort has also been made to analyse the regional pattern of male and female RNAE across the states ranked with the agricultural prosperity, rate of literacy and incidence of unemployment in 1983 and 1999-2000. An effort has also been made to explore such insights by selecting a few indicators reflecting the growth as well as distress situation which prevailed in rural areas. The percentage distribution of male workers in RNAE by states ranked according to ascending order of PCIA[4], as a proxy for agricultural prosperity, reveals that

Table 4.5 : Distribution of Male Workers (UP+SS) in RNAE by Major States in India, 1972-73 to 1999-2000

(Percent)

State	Proportion of workers in RANE						Percent Point Change				
	1972-73	1977-78	1983	1987-88	1993-94	1999-00	1977-78 over 1972-73	1983 over 1977-78	1987-88 over 1983	1993-94 over 1987-88	1999-00 over 1993-94
Andhra Pradesh	21.4	19.7	25.6	25.9	24.4	25.6	-1.7	5.9	0.3	-1.5	1.2
Assam	–	–	–	23.5	21.8	35.3	–	–	–	-1.7	13.5
Bihar	17.8	16.9	18.7	20.0	18.0	21.0	-0.9	1.8	1.3	-2.0	3.0
Gujarat	16.1	15.6	21.1	31.4	28.9	28.6	-0.5	5.5	10.3	-2.5	-0.3
Haryana	19.9	22.5	27.8	29.1	39.1	40.4	2.6	5.3	1.3	10.0	1.3
Himachal Pradesh	18.9	22.6	22.9	31.3	34.2	46.2	3.7	0.3	8.4	2.9	12.0
Karnataka	14.8	16.8	18.4	20.4	21.2	21.5	2.0	1.6	2.0	0.8	0.3
Kerala	44.3	40.8	42.2	45.8	46.8	57.2	-3.5	1.4	3.6	1.0	10.4
Madhya Pradesh	9.9	10.8	12.8	14.7	12.8	15.8	0.9	2.0	1.9	-1.9	3.0
Maharashtra	17.6	19.6	20.4	24.2	24.7	26.2	2.0	0.8	3.8	0.5	1.5
Orissa	18.4	15.4	21.8	25.1	21.3	23.0	-3.0	6.4	3.3	-3.8	1.7
Punjab	20.6	22.2	22.5	31.2	31.9	36.3	1.6	0.3	8.7	0.7	4.4
Rajasthan	15.6	17.5	19.0	34.8	30.4	32.7	1.9	1.5	15.8	-4.4	2.3
Tamil Nadu	24.6	26.1	31.1	34.8	36.0	37.8	1.5	5.0	3.7	1.2	1.8
Uttar Pradesh	18.1	19.8	21.3	21.1	23.7	28.2	1.7	1.5	-0.2	2.6	4.5
West Bengal	22.1	22.3	26.9	27.8	35.3	33.6	0.2	4.6	0.9	7.5	-1.7
All India	16.7	19.4	22.1	25.4	25.9	28.8	0.2	2.9	3.1	0.4	2.7
CV (%)	37.8	32.9	29.0	27.0	31.2	34.2	–	–	–	–	–

Note: CV: Coefficient of variation. CV was estimated for the data on proportion of male workers in RNAE for fifteen major states.
Source: NSSO (1981, 1988, 1990, 1996 & 2001).

most of the states with higher PCIA had a major proportion of male workers in RNAE as compared to states with lower PCIA, excepting Tamil Nadu and Bihar (Figure 4.3).

A greater increase took place in the proportion of male workers in RNAE in most of the agriculturally developed states *vis-à-vis* backward states in 1999-2000 over 1983[5]. It follows from this that the growth of agriculture contributed to the growth of male RNAE during the reform period. States with a lower PCIA also, in part, seemed to have experienced a moderate increase in male RNAE, indicating that the distress in agriculture also contributed to the growth of male RNAE during the same period. The distress movements of male workers into the rural non-agricultural sector due to distress in agriculture could not be ruled out, *in toto* (Jayaraj 1989; Parthasarathy, Shameem and Reddy 1998; Rajasekhar and Biradar 1998; Biradar 2009).

Figur 4.3 : Precentage Distribution of Male Workers in RNAE by Major States Ranked According to Ascending Order of PCIA, 1983 and 1999-2000

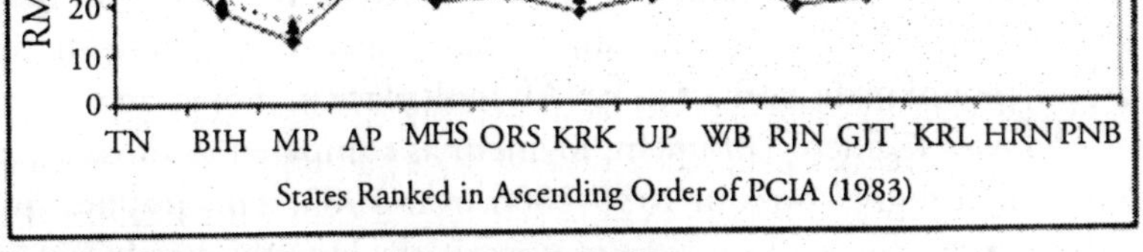

Note : RMNAWs : Rural male non-agricultural workers.
Source : Table 4.5.

A similar pattern of distribution of male workers in RNAE by the states ranked according to ascending order of the rate of male literacy (RML)[6] in 1983 and 1999-2000 shows that states with a higher RML had a considerable proportion of male workers in RNAE, as compared to states with a lower RML (Figure 4.4). A greater increase in the proportion of male RNAE took place in states with a higher

RML than that with lower ones in 1999-2000 over 1983. This indicates that the rate of literacy, as a surrogate to human capital, was also found to be another important factor in determining the growth of male RNAE during the reform period.

Figure 4.4 : Percentage Distribution of Male Workers in RNAE by Major States Ranked According to Ascending Order of RML, 1983 and 1999-2000

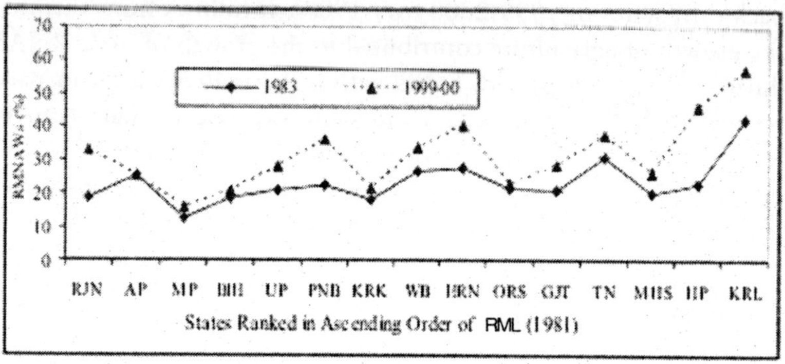

Note: RMNAWs: Rural male non-agricultural workers.
Source: Table 4.5.

The percentage distribution of male workers in RNAE (Figure 4.5) across different states ranked according to ascending order of the unemployment rate, as per CDS[7], during the same period also indicates that states with a higher incidence of unemployment had a larger share of male workers in RNAE in 1983 and 1999-2000. This follows that the distress factors also contributed partly to the growth of male RNAE. The data, however, show that a greater increase in the proportion of male workers in RNAE took place in those states which had a lower incidence of unemployment as compared to those states which had higher ones in 1999-2000 over 1983. This implies that although the distress factors were at work, the growth of male RNAE in states with a higher incidence of unemployment was determined by some other factors in the 1990s. This observation does not, therefore, lend any support to the distress diversification hypothesis in respect of males.

In the case of females, the data (Table 4.6) show that the proportion of female workers in RNAE increased continuously only in two states, viz., Kerala and Himachal Pradesh during 1972-73 to

Figure 4.5 : Percentage Distribution of Male Workers in RNAE by Major States Ranked According to Ascending Order of Unemployment Rate (CDS), 1983 and 1999-2000

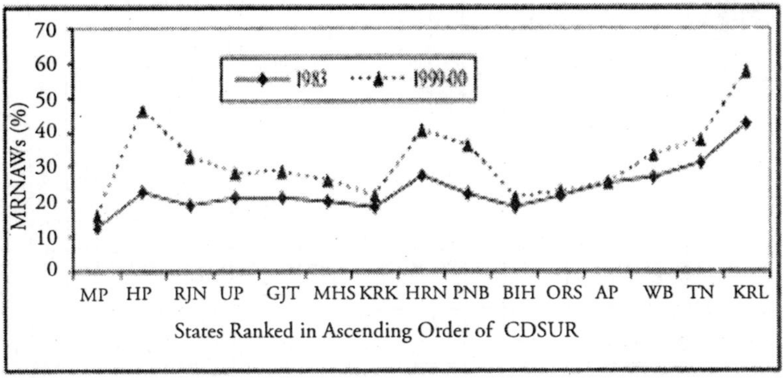

Note: RMNAWs: Rural male non-agricultural workers; CDS UR: Current daily status unemployment rates.
Source: Table 4.5.

1999-2000. An inconsistent rise in the proportion of female workers in RNAE occurred in many states such as West Bengal, Tamil Nadu, Orissa, Bihar, Uttar Pradesh, Punjab, Assam, Madhya Pradesh and Rajasthan, many of which witnessed a sharp rise in 1987-88 and a greater setback in 1993-94 and picked up in 1999-2000. An inconsistent decline was experienced by some other states like Andhra Pradesh, Gujarat, Maharashtra and Karnataka in the 1990s (1993-94 to 1999-2000), as compared to the 1980s (1983 to 1987-88). A significant inconsistent variation was also found in female RNAE in most of the states. The inconsistent variation across these states has been attributed to drought in 1987-88 (Parthasarathy, Shameem and Reddy 1998). It has been found that the coefficient of variation for female RNAE was much higher than that for male RNAE. The values of coefficient of variation for female RNAE started declining in the years of the pre-reform period and began to rise in the years of the reform period; the pace of increase was much greater than the pace of decline.

A distributional pattern of female workers in RNAE by states ranked according to ascending of PCIA suggests that, unlike as in the case of male workers, states with a higher PCIA had a lower proportion of female workers in RNAE *vis-à-vis* the states with a lower

Table 4.6 : Distribution of Female Workers in RNAE (UP+SS) by Major States in India, 1972-73 to 1999-2000

(*Per cent*)

State	*Proportion of Workers in RNAE*						*Percent point change*				
	1972-73	*1977-78*	*1983*	*1987-88*	*1993-94*	*1999-00*	*1977-78/ over 1972-73*	*1983- over 1977-78*	*1987-88- over 1983*	*93-94/ over 87-88*	*99-00/ over 93-94*
Andhra Pradesh	13.8	14.6	18.7	17.9	16.3	15.7	0.8	4.1	-0.8	-1.6	-0.6
Assam	-	-	-	17.7	16.8	20.6	-	-	-	-0.9	3.8
Bihar	14.8	11.0	11.8	9.8	8.1	14.3	-3.8	0.8	-2.0	-1.7	6.2
Gujarat	8.7	5.6	7.7	24.2	9.4	8.0	-3.1	2.1	16.5	-14.8	-1.4
Haryana	15.6	9.5	9.7	7.5	6.8	7.9	-6.1	0.2	-2.2	-0.7	1.1
Himachal Pradesh	1.6	1.8	2.4	3.6	4.5	4.9	0.2	0.6	1.2	0.9	0.4
Karnataka	10.9	12.5	11.8	14.5	15.4	12.2	1.6	-0.7	2.7	0.9	-3.2
Kerala	41.4	27.4	29.1	34.3	37	40.2	-14.0	1.7	5.2	2.7	3.2
Madhya Pradesh	4.4	5.3	6.1	8.9	6.4	8.4	0.9	0.8	2.8	-2.5	2.0
Maharashtra	6.6	7.8	7.1	8.6	8.8	6.1	1.2	-0.7	1.5	0.2	-2.7
Orissa	18.3	14.2	19.0	22.0	15.0	19.6	-4.1	4.8	3.0	-7.0	4.6
Punjab	37.0	10.0	7.4	8.4	7.3	9.4	-27.0	-2.6	1.0	-1.1	2.1
Rajasthan	8.4	4.7	6.1	16.7	7.0	8.1	-3.7	1.4	10.6	-9.7	1.1
Tamil Nadu	15.6	16.4	18.2	22.9	21.5	24.1	0.8	1.8	4.7	-1.4	2.6
Uttar Pradesh	15.0	10.9	10.3	8.7	10.0	12.5	-4.1	-0.6	-1.6	1.3	2.5
West Bengal	43.1	31.3	24.9	29.2	41.1	45.9	-11.8	-6.4	4.3	11.9	4.8
All India	10.4	11.8	12.2	15.2	13.6	14.7	-2.4	0.4	3.0	-1.5	0.8
CV (%)	76.7	66.2	60.2	57.2	77.5	77.5	–	–	–	–	–

Note: CV: Coefficient of variation. CV was estimated for the data on proportion of male workers in RNAE for fifteen major states.

Source: NSSO (1981, 1988, 1990, 1996 & 2001).

PCIA, excepting Kerala and West Bengal (Figure 4.6). An increase in the proportion of female workers in RNAE appeared to have recorded in those states which had a lower PCIA than the states with higher ones in 1999-2000 over 1983. This implies that higher income from agriculture seemed to have prevented the female workers from participating in RNAE unless they were secure and highly paid.

Figure 4.6 : Percentage Distribution of Female Workers in RNAE by Major States Ranked According to Ascending Order of PCIA, 1983 and 1999-2000

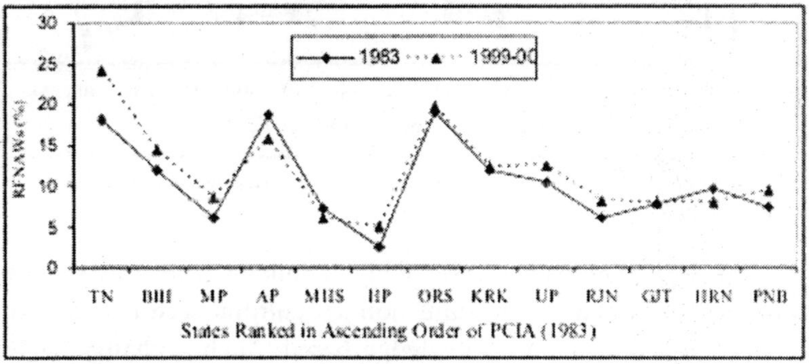

Note : RFNAWs : Rural female non-agricultural workers.
Source : Table 4.6.

In times of distress in agriculture, some share of female labourers, mostly belonging to resource-poor households, would be forced to work outside agriculture so as to stabilise their household income and the living standard. In the absence of agricultural growth, the distress movements of female workers in favour of RNAE, therefore, were quite evident. A similar observation was also made by other studies (Jayaraj 1989; Parthasarathy, Shameem and Reddy 1998; Rajasekhar and Biradar 1998; Biradar 2006a). As in the case of male workers, the percentage distribution of female workers in RNAE by states ranked in ascending order of the rate of female literacy (RFL)[8] reveals that states with a higher RFL had a larger proportion of female workers in RNAE in the years 1983 and 1999-2000 (Figure 4.7).

It has also been found that states with higher RFL experienced a greater increase in the proportion of female workers in RNAE, as compared to states with a lower RFL in 1999-2000 over 1983, except for Bihar, Uttar Pradesh and Rajasthan, which experienced a moderate

Figure 4.7 : Percentage Distribution of Female Workers in RNAE by Major States Ranked According Ascending Order of RFL, 1983 and 1999-2000

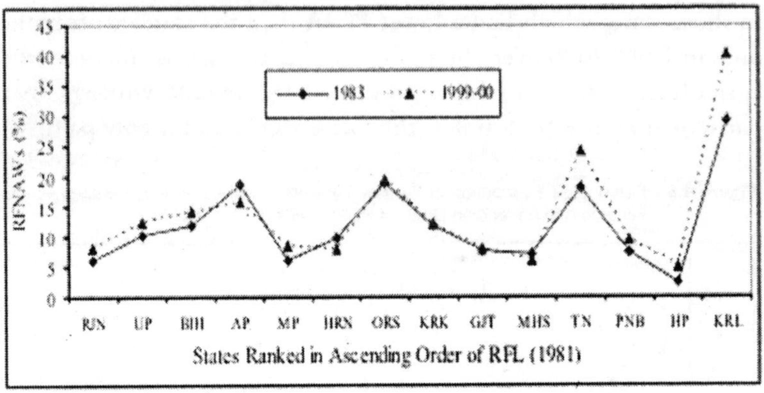

Note: FRNAWs: Female rural non-agricultural workers; RFL: Rate of female literacy.
Source: Table 4.6.

rise although they had a lower RFL. It follows from this that the activities generated in the rural non-agricultural sector were more likely to be skilled and knowledge based. Such a change in the occupational structure appeared to have resulted in the withdrawal of workers (in the age-groups 5-9, 10-14, 15-19 and 20-24 years) from the labour market in favour of educational institutions to acquire better educational qualifications, skills and knowledge.

The distress diversification of female RNAE was explicitly evident from the percentage distribution of female workers in RNAE in states ranked according to ascending order of female unemployment rate in 1983 and 1999-2000 (Figure 4.8). States with a higher incidence of female unemployment *vis-à-vis* those with lower ones showed a significant share of females in RNAE in the years 1983 and 1999-2000. A rapid increase in the proportion of female workers in RNAE also took place in those states with a higher incidence of female unemployment, as compared to states with lower ones in 1999-2000 over 1983. The foregoing regional pattern of analysis reveals that the regional variation in the proportion of male workers in RNAE had been better explained by growth related factors, whereas in the case of females, it had been largely explained by distress related factors.

Figure 4.8 : Percentage Distribution of Female Workers in RNAE by Major States Ranked According to Ascending Order of Unemployment Rate (CDS), 1983 and 1999-2000

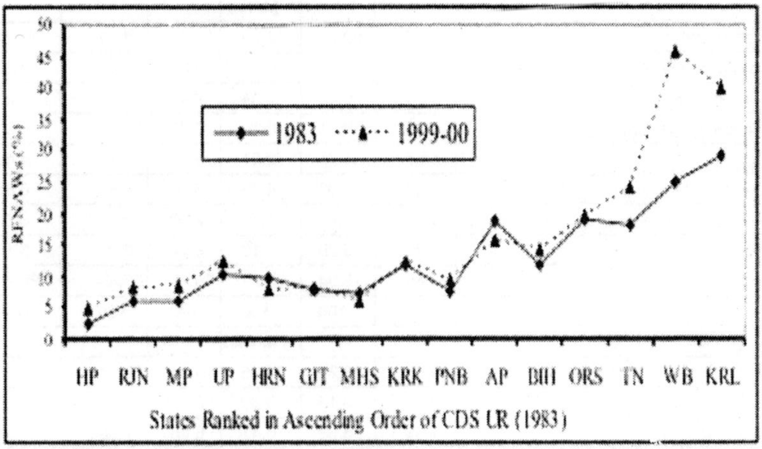

Note: RFNAWs: Rural female non-agricultural workers: CDS UR: Current daily status unemployment rates.
Source: Table 4.6.

Size and composition of workers by activity

It has been found that the proportion of workers, especially in respect of males, in agricultural employment consistently declined and a corresponding increase took place in the rural non-agricultural sector. An increase in the proportion of workers in RNAE was not uniform across different activities in the rural non-agricultural sector (Table 4.7).

The data show that among all the activities, manufacturing; services; trade, hotel and restaurant business; construction; and transport, storage and communication constituted major share of RNAE for males and females. Other activities such as mining and quarrying; electricity, gas and water were found to have a smaller share in RNAE. It is imperative to identify the key activities providing sustained employment opportunities for a large chunk of workers in the days to come.

It has been found that construction; transport, communication and storage; and trade, hotel and restaurant business were emerging as important activities in the rural non-agricultural sector in the era

Table 4.7 : Distribution of Workers (UP+SS) by Sub-Sectors in Rural India, 1972-73 to 1999-2000 *(Percent)*

Sector	*1972-73*	*1977-78*	*1983*	*1987-88*	*1993-94*	*1999-00*
			Males			
Agriculture	83.32	80.66	77.79	74.62	74.11	71.21
Mining & quarrying	0.39	0.50	0.59	0.67	0.69	0.62
Manufacturing	5.74	6.42	7.04	7.41	6.98	7.28
Electricity, gas, water, etc.	0.08	0.21	0.20	0.30	0.32	0.21
Construction	1.57	1.71	2.21	3.70	3.20	4.49
Trade, hotel and restaurants	3.07	4.00	4.36	5.10	5.54	6.81
Transport, storage and communication	1.02	1.21	1.69	2.00	2.18	3.20
Services	4.80	5.28	6.12	6.19	6.98	6.19
Non-agriculture	16.68	19.34	22.21	25.38	25.89	28.79
All	100	100	100	100	100	100
			Females			
Agriculture	89.65	88.24	87.85	84.84	86.01	85.29
Mining & quarrying	0.14	0.13	0.33	0.43	0.38	0.29
Manufacturing	4.68	5.88	6.41	6.88	7.08	7.55
Construction	1.13	0.63	0.66	2.69	0.86	1.08
Trade, hotel and restaurants	1.56	2.00	1.88	2.04	1.82	1.67
Transport, storage and communication	0.00	0.13	0.11	0.11	0.40	0.39
Services	2.84	3.00	2.76	3.01	3.44	3.73
Non-agriculture	10.35	11.76	12.15	15.16	13.60	14.71
All	100	100	100	100	100	100

Source: NSSO (1981, 1988, 1990, 1996 & 2001).

of globalisation. While no significant employment opportunities were generated in manufacturing, probably in the household sector, the share of employment in services showed a marginal decline in the reform period. In the case of females, it has been found that services and manufacturing were growing much faster than any other activities and also than those of the male workers. But in construction; transport, communication and storage; and trade, hotels and restaurants, the employment opportunities were stagnated or not growing significantly during the same period.

Regional pattern of size and composition of workers by activity

A considerable regional variation existed in the proportion of workers, both males and females, in manufacturing; construction; trade, hotels and restaurants; transport, storage and communication; and services across different states and over time.

Manufacturing

The data presented in Table 4.8 indicate that the regional variation in the proportion of male workers in the manufacturing sector was quite perceptible, ranging from as high as 14 percent in Tamil Nadu to as low as 3 percent in Assam in 1999-2000. The value of the co-efficient of variation also increased from 35 percent to 39 per cent during 1987-88 to 1999-2000. This indicates that the regional variation in the proportion of male workers in manufacturing employment was not only large but it also showed an increasing tendency over time.

The data also indicate that as many as eight states, namely, Haryana, Himachal Pradesh, Bihar, Punjab, Uttar Pradesh, Tamil Nadu, Madhya Pradesh and Maharashtra witnessed an inconsistent rise in the proportion of male workers in manufacturing activities during the same period. In most of the above states, the increase was slightly greater than the decline. Only Assam showed a sustained rise during the same period. A continuous decline in the proportion of male workers in manufacturing activities took place in only five states, namely, Andhra Pradesh, Karnataka, Kerala, Orissa and Rajasthan, whereas an inconsistent decline occurred in two states of Gujarat and West Bengal during the same period. In most of the agriculturally developing states, there were some indications of employment generation for males in the manufacturing sector during the reform period.

In respect of females, it was observed that the regional variation in the proportion of female workers in manufacturing sector, according to UPS, was much higher than that of male workers, ranging from as high as 31 percent in West Bengal to as low as 2 percent in Himachal Pradesh during 1999-2000. The co-efficient of variation for females was found to be significantly higher than that for males, indicating a wider regional disparity. A consistent increase in the proportion of female workers in the manufacturing sector was registered only by four states, namely, Bihar, Punjab, Uttar Pradesh and Kerala, whereas an inconsistent increase occurred in six states, viz., Orissa, West Bengal,

Table 4. 8 : Distribution of Workers in Manufacturing by Major States in Rural India, 1987-88 to 1999-2000

(percent)

State	Males						Females					
	UPS			UP+SS			UPS			UP+SS		
	1987-88	1993-94	1999-00	1987-88	1993-94	1999-00	1987-88	1993-94	1999-00	1987-88	1993-94	1999-00
Andhra Pradesh	7.9	6.4	5.3	7.9	6.5	5.3	8.1	7.6	5.9	7.7	7.4	6.0
Assam	2.1	2.2	3.0	2.0	2.2	2.9	4.5	3.1	6.2	9.8	8.7	8.3
Bihar	5.0	3.4	5.3	4.9	3.4	5.3	3.9	4.3	9.2	4.1	3.9	8.5
Gujarat	9.5	12.5	10.1	9.2	12.3	10.1	3.6	5.4	2.7	3.6	4.1	2.2
Haryana	8.7	5.5	9.3	8.4	5.3	9.2	2.6	2.5	3.4	2.8	1.4	2.0
Himachal Pradesh	7.2	5.5	7.1	6.6	4.8	6.7	1.8	1.8	1.6	1.8	1.6	1.1
Karnataka	6.3	5.3	5.2	6.2	5.4	5.2	9.6	9.9	5.7	8.8	8.4	5.7
Kerala	10.7	9.9	9.6	10.4	9.7	9.4	23.7	24.1	25.1	19	19.2	19.3
Madhya Pradesh	4.9	3.3	3.9	4.9	3.2	3.9	5.2	3.6	3.6	5.0	3.2	4.2
Maharashtra	7.4	6.7	6.9	7.1	6.6	7.0	2.8	3.1	1.8	2.7	3.0	2.2
Orissa	6.3	5.9	5.6	6.1	5.7	5.6	13.4	6.7	12.3	11.4	7.5	12.9
Punjab	9.8	6.2	7.7	9.7	6.2	7.7	5.5	6.3	10.8	2.8	1.3	2.3
Rajasthan	7.9	5.5	5.4	7.7	5.3	5.4	4.0	1.4	3.5	4.1	1.4	2.8
Tamil Nadu	13.7	12.8	13.8	13.4	12.8	13.8	14.1	13.7	14.5	12.9	12.9	14.2
Uttar Pradesh	7.3	7.2	8.4	7.2	7.0	8.3	3.9	4.7	8.3	3.8	4.7	6.4
West Bengal	9.6	11.9	11.0	9.1	11.7	10.9	27.3	35.8	30.5	19.6	30.0	36.1
All India	7.6	7.0	7.3	7.4	7.0	7.3	7.5	7.5	7.7	6.9	7.0	7.6
CV (%)	34.6	47.1	39.0	37.4	47.8	39.2	91.2	110.4	91.4	76.1	104.0	107.3

Source: NSSO (1990, 1996 & 2001).

Rajasthan, Tamil Nadu, Assam and Haryana during 1987-88 to 1999-2000. While a continuous decline was experienced by only Andhra Pradesh, an inconsistent decline took place in Karnataka, Gujarat, Maharashtra and Himachal Pradesh during the same period. In most of the states, an increase in the proportion of female workers in the manufacturing sector was found to be relatively more than that for males, and it was quite perceptible during the reform period.

Construction

The data presented in Table 4.9 show that the regional variation in the proportion of male as well as female workers in construction was much higher than that of manufacturing activities; but it was on the decline over time. The proportion of male workers in construction, according to UPS, was found to be as high as 18.3 percent in Himachal Pradesh and as low as 2.3 percent in Assam during 1999-2000. The data also indicate that while seven states experienced a continuous rise in the proportion of male workers in construction, an inconsistent rise took place in seven other states during 1987-88 to 1999-2000; the rise, in most cases, was a little higher than the decline during 1993-94 to 1999-2000. No state, in fact, experienced a constant decline during the same period; but West Bengal witnessed a marginal decline and Maharashtra remained unchanged during 1993-94 to 1999-2000.

In respect of females, it was observed that their proportion in construction, in most cases, was lower than that of male workers. The coefficient of variation in construction was also relatively greater in the case of females as compared to males; but it was, by and large, on the decline over time. During the period between 1987-88 and 1999-2000, while Kerala and Haryana showed a constant rise, Orissa and Tamil Nadu, Madhya Pradesh, Gujarat, Bihar, Andhra Pradesh experienced an inconsistent rise in the proportion of females in construction; a consistent decline took place in Rajasthan, followed by Maharashtra and Assam; and an inconsistent decline in West Bengal and Karnataka. Employment of females in construction activities showed a marginal rise in most of the states during the reform period.

Table 4. 9 : Distribution of Workers in Construction by Major States in Rural India, 1987-88 to 1999-2000

(Percent)

State	Males						Females					
	UPS			UP+SS			UPS			UP+SS		
	1987-88	1993-94	1999-00	1987-88	1993-94	1999-00	1987-88	1993-94	1999-00	1987-88	1993-94	1999-00
Andhra Pradesh	2.8	2.6	3.4	2.9	2.6	3.4	1.4	0.7	0.8	1.2	0.6	0.8
Assam	1.2	0.8	2.3	1.1	0.8	2.2	0.8	0.3	0.0	0.3	0.1	0.2
Bihar	2.4	1.6	2.8	2.3	1.6	2.8	1.0	0.2	0.6	0.8	0.2	0.5
Gujarat	9.6	3.0	3.3	9.5	3.0	3.3	20.3	2.0	2.4	16.9	1.4	1.8
Haryana	3.8	6.6	8.8	3.8	6.3	8.7	0.4	1.3	5.0	0.4	0.4	0.8
Himachal Pradesh	11.9	13.7	18.3	11.3	11.9	17.2	0.4	0.6	0.6	0.3	0.4	0.4
Karnataka	2.2	2.0	2.1	2.2	2.0	2.1	1.2	0.7	0.6	1.0	0.7	0.5
Kerala	6.5	7.5	13.0	6.3	7.5	12.6	1.2	3.2	3.8	0.8	2.1	2.7
Madhya Pradesh	1.4	1.2	2.3	1.4	1.2	2.2	1.9	0.4	1.3	1.7	0.4	1.2
Maharashtra	3.5	3.4	3.4	3.5	3.4	3.4	3.1	1.3	1.0	2.9	1.2	0.9
Orissa	4.0	2.3	3.8	3.9	2.2	3.9	3.1	1.4	2.5	2.5	1.1	1.8
Punjab	4.1	4.8	7.9	4.0	4.7	7.8	0.0	0.0	0.8	0.0	0.0	0.1
Rajasthan	13.3	10.7	12.0	13.4	10.5	11.9	10.2	3.3	2.9	10.2	2.5	2.4
Tamil Nadu	3.2	3.6	5.8	3.2	3.6	5.8	1.3	0.8	1.7	1.2	7.0	1.6
Uttar Pradesh	2.5	2.7	4.5	2.4	2.6	4.4	0.5	0.3	0.4	0.4	0.2	0.5
West Bengal	1.9	2.8	2.7	1.8	2.7	2.7	1.2	2.8	0.6	0.6	1.6	0.4
All India	3.7	3.3	4.5	3.7	3.2	4.5	3.2	1.1	1.2	2.7	0.9	1.1
CV (%)	80.4	83.1	78.9	80.6	78.6	76.7	173.2	89.3	89.4	175.7	136.9	76.8

Source: NSSO (1990, 1996 & 2001).

Trade, hotel and restaurant

In view of the growing rural-urban linkages, a larger proportion of workers with better education find jobs in trade, commerce, hotel and restaurant activities in rural areas. With the introduction of reform measures in major sectors of the economy, the scope for employment generation increased in trade, commerce, hotel and restaurant business. The data presented in Table 4.10 show that the proportion of male workers finding employment in trade, hotel and restaurant business was not uniform across different states. The coefficient of variation was found to be quite lower in the proportion of male workers in trade, hotel and restaurant business than in construction, implying a greater volatility in construction employment. States which are progressive in terms of agriculture, education and infrastructure (Punjab, Haryana Kerala and Tamil Nadu), have had a larger share of male workers in trade, hotel and restaurant business, as compared to the backward states (Bihar, Gujarat and Rajasthan and Madhya Pradesh) in 1999-2000.

The data also indicate that ten states experienced a constant rise, whereas four states witnessed an inconsistent rise in the proportion of male workers in trading, hotel and restaurant business during 1987-88 to 1999-2000; a greater increase, however, took place during 1993-94 to 1999-2000. A great majority of the states experienced a constant rise in the proportion of male workers in trade, hotels and restaurants, but a negligible decline occurred in Bihar during 1987-88 to 1999-2000. Regional variation in the proportion of male workers finding employment in trade, hotels and restaurants, more or less, appeared to have shown a rising trend during 1993-94 to 1999-2000.

In respect of females, it was found that the regional variation was relatively greater than in the case of males during the same period. A few states, viz., Punjab, Haryana and Kerala had a greater share of female workers, as per UPS, ranging from about 7 to 5 percent; quite a few states had a lower share of female workers, ranging from about a little less than 2 to 3 percent in trade, hotel and restaurant business in 1999-2000. According to UPS perspective, four states witnessed a consistent rise in the proportion of female workers, whereas four other states experienced an inconsistent rise, and two states of Rajasthan and Uttar Pradesh showed no change during 1993-94 to 1999-2000.

Table 4. 10: Distribution of Workers in Trade, Hotels and Restaurants by Major States in Rural India, 1987-88 to 1999-2000

State	Males						Females					
	UPS			UP+SS			UPS			UP+SS		
	1987-88	1993-94	1999-00	1987-88	1993-94	1999-00	1987-88	1993-94	1999-00	1987-88	1993-94	1999-00
Andhra Pradesh	5.1	5.4	5.8	5.1	5.4	5.8	4.2	3.3	3.1	3.9	3.4	3.0
Assam	7.0	8.1	10.2	6.6	8.2	9.9	2.5	2.2	1.8	1.7	1.9	1.5
Bihar	5.3	5.4	5.3	5.3	5.3	5.3	2.0	2.2	2.0	1.3	1.9	1.8
Gujarat	3.6	4.2	5.5	3.6	4.2	5.6	0.7	0.8	1.7	0.3	1.0	1.6
Haryana	6.6	7.2	7.4	6.3	7.4	7.3	0.6	2.5	6.1	0.4	1.3	2.6
Himachal Pradesh	2.9	5.7	6.4	2.7	5.2	6.1	0.2	0.7	0.9	0.3	6.0	0.6
Karnataka	4.7	4.8	5.7	4.7	4.8	5.8	3.0	2.0	2.4	2.3	2.2	2.5
Kerala	13.2	11.9	16.3	12.7	11.6	15.8	4.8	4.8	5.1	3.7	3.6	3.7
Madhya Pradesh	2.8	2.2	3.9	2.8	2.2	3.9	1.1	0.8	0.9	1.0	0.7	1.0
Maharashtra	4.8	4.5	5.8	4.7	4.5	5.7	1.3	1.5	1.3	1.2	1.8	1.3
Orissa	5.8	5.3	6.0	5.7	5.2	6.1	3.8	3.4	2.1	3.4	2.9	1.9
Punjab	4.7	6.3	8.1	4.5	6.3	8.1	3.3	2.7	6.7	1.1	1.0	1.1
Rajasthan	4.1	3.8	5.5	4.1	3.7	5.6	0.7	0.8	0.8	0.6	0.7	0.6
Tamil Nadu	6.9	6.3	7.2	6.8	6.4	7.1	4.5	2.6	3.5	4.1	2.8	3.4
Uttar Pradesh	4.2	5.2	6.8	4.1	5.4	6.7	1.7	2.3	2.3	1.5	2.1	1.9
West Bengal	7.6	9.0	10.3	7.1	9.0	10.2	3.9	4.0	3.4	2.6	2.7	2.8
All India	5.2	5.5	6.8	5.1	5.5	6.8	2.4	2.2	2.3	2.1	1.8	1.6
CV (%)	44.5	38.3	40.7	43.4	38.1	39.3	64.9	52.6	65.9	67.5	60.1	49.0

Source: NSSO (1990, 1996 & 2001).

However, a continuous decline took place in only two states of Andhra Pradesh and Assam during 1987-88 to 1993-94, and an inconsistent decline occurred in four states, namely, Orissa, Bihar, West Bengal and Maharashtra during the same period. The pace of increase in employment in trade, hotel and restaurant activities was found to be higher for males than for females.

Transport, Sforage and Communication

Employment in transport, storage and communication was another important segment of the rural non-agricultural sector. In the knowledge-based economy, the importance of transport, storage and communication seemed to have gained momentum and thereby educationally qualified and skilled workers were more likely to secure better jobs. The data provided in Table 4.11 show that an increase in the proportion of male workers in transport, storage and communication took place in most of the states, excepting Haryana during 1987-88 to 1999-2000; an inconsistent rise occurred in nine states and a consistent rise in six other states during the same period. There has been an encouraging sign of employment generation for males in transport, storage and communication during the reform period. In the case of females, their proportion finding jobs in transport, storage and communication was marginal. No discernible trend was found across the states and over time.

Community and personal services

Finance, insurance, community and personal services are important components of the services sector. Table 4.12 provides data on the proportion of workers, both for males and females, in community and personal services during 1987-88 to 1999-2000. The regional variation in the proportion of male workers in community and personal services was also quite perceptible; but it was very low and on the rise over time. The proportion of male workers, according to UPS, in community and personal services ranged from 17.2 percent in Assam to 4.1 percent in Madhya Pradesh in 1999-2000. Although the coefficient of variation in the proportion of females in community and personal services was lower, it increased over time. The data also

Table 4.11 : Distribution of Workers in Transport, Storage and Communication by Major States in Rural India, 1987-88 to 1999-2000 ***(Percent)***

State	*Males*						*Females*					
	UPS			*UP+SS*			*UPS*			*UP+SS*		
	1987-88	*1993-94*	*1999-00*	*1987-88*	*1993-94*	*1999-00*	*1987-88*	*1993-94*	*1999-00*	*1987-88*	*1993-94*	*1999-00*
Andhra Pradesh	1.9	1.8	3.0	1.8	1.9	3.0	0.1	–	–	0.1	–	0.1
Assam	2.2	1.7	3.3	2.0	1.6	3.3	0.2	0.2	0.6	0.1	0.1	0.3
Bihar	1.9	1.3	1.9	1.8	1.3	1.9	0.1	–	0.1	0.0	–	0.1
Gujarat	2.6	2.2	3.9	2.4	2.5	3.9	0.1	–	0.2	0.1	–	0.1
Haryana	2.2	5.3	4.7	2.1	5.1	4.6	0.1	–	–	0.1	0.4	0.0
Himachal Pradesh	1.2	1.9	4.5	1.2	1.6	4.3	0.1	–	0.2	0.1	–	0.1
Karnataka	1.5	1.2	2.4	1.4	1.2	2.4	0.1	–	0.0	0.0	–	0.0
Kerala	5.9	5.9	9.3	5.6	5.8	9.0	1.0	0.7	0.3	0.7	0.8	0.2
Madhya Pradesh	0.6	0.8	1.2	0.6	0.7	1.2	–	–	–	0.0	0.1	–
Maharashtra	2.1	2.0	3.3	2.1	1.9	3.2	–	1.0	–	0.0	1.0	–
Orissa	1.2	1.1	1.7	1.1	1.1	1.8	–	–	–	0.0	–	–
Punjab	3.9	3.6	5.6	3.8	3.6	5.6	0.5	–	–	1.0	–	–
Rajasthan	2.1	1.9	3.0	2.0	1.9	3.0	0.0	–	–	0.0	–	–
Tamil Nadu	2.8	3.7	4.3	2.8	3.6	4.3	0.2	0.1	0.2	0.2	0.1	0.2
Uttar Pradesh	1.5	2.1	3.0	1.5	2.1	2.9	–	–	–	0.0	–	–
West Bengal	3.3	3.6	4.3	3.1	3.5	4.2	0.1	0.6	–	0.1	0.3	–
All India	2.1	2.2	3.2	2.0	2.2	3.2	0.1	1.0	1.0	0.1	0.4	0.4
CV (%)	55	60	51	55	60	50	157	193	167	181	176	138

Source: NSSO (1990, 1996 & 2001).

Table 4.12 : Distribution of Workers in Community and Personal Services by Major States in Rural India, 1987-88 to 1999-2000 **(perunt)**

State	*Males*						*Females*					
	UPS			*UP+SS*			*UPS*			*UP+SS*		
	1987-88	*1993-94*	*1999-00*	*1987-88*	*1993-94*	*1999-00*	*1987-88*	*1993-94*	*1999-00*	*1987-88*	*1993-94*	*1999-00*
Andhra Pradesh	7.1	7.1	7.0	6.9	7.1	7.1	4.9	4.5	5.3	4.5	4.3	5.1
Assam	11.0	8.9	17.2	10.4	8.5	16.6	14.1	9.6	15.8	5.6	5.9	10.4
Bihar	4.6	5.6	5.1	4.5	5.6	5.1	2.2	1.9	3.6	2.1	1.6	3.3
Gujarat	5.9	6.0	5.0	5.6	11.4	5.0	3.0	3.4	2.3	2.6	2.6	1.8
Haryana	7.7	13.9	8.8	7.4	13.4	8.7	7.3	9.0	14.4	3.6	3.4	2.5
Himachal Pradesh	8.9	10.5	10.6	8.1	9.2	9.9	1.5	1.9	4.4	1.1	1.5	2.6
Karnataka	4.5	6.4	5.1	4.5	6.4	5.1	1.5	3.8	2.9	1.3	3.6	2.9
Kerala	9.1	10.1	8.0	8.3	10.0	8.1	14.8	15.6	19.5	9.4	11.1	13.6
Madhya Pradesh	4.1	3.8	4.1	4.0	3.7	4.1	1.0	0.9	1.9	0.9	0.9	1.9
Maharashtra	6.2	7.7	6.2	6.0	7.6	6.3	1.8	2.4	1.7	1.6	3.4	1.6
Orissa	6.4	5.7	4.8	6.3	5.7	5.0	4.2	3.3	2.5	3.5	2.6	2.8
Punjab	7.6	9.6	5.9	7.2	9.6	10.4	16.1	25.3	31.6	4.2	4.7	5.7
Rajasthan	5.1	6.2	4.6	5.0	6.1	4.6	1.4	1.9	1.9	1.3	1.5	1.4
Tamil Nadu	7.6	8.9	5.8	7.5	8.9	5.8	4.7	5.2	4.6	4.2	4.7	4.4
Uttar Pradesh	5.9	6.7	5.6	5.7	6.5	5.5	3.2	3.5	5.3	2.8	2.3	3.6
West Bengal	6.2	8.0	4.9	6.0	8.0	5.0	10.4	11.6	8.4	6.1	6.4	6.7
All India	6.4	7.1	6.1	6.2	7.0	6.1	3.7	4.0	4.3	3.0	3.4	3.7
CV (%)	27.7	31.4	48.1	26.2	30.5	45.6	90.5	100	106.3	66.1	65.8	77.0

Source: NSSO (1990, 1996 & 2001).

reveal that most of the states experienced a sharp rise in the proportion of male workers in community and personal services during 1987-88 to 1993-94 and a decline during 1993-94 to 1999-2000. In the 1990s, community and personal services showed no encouraging signs of employment generation.

In respect of females, there was a moderate increase in the proportion of female workers in community and personal services during the same period. According to UPS, while six states, namely, Punjab Haryana, Himachal Pradesh, Kerala, Uttar Pradesh and Rajasthan experienced a sustained rise in the proportion of female workers in community and personal services, only Orissa showed a constant decline during 1987-88 to 1999-2000. States such as Assam, Andhra Pradesh, Bihar and Madhya Pradesh experienced a moderate decline during 1987-88 to 1993-94 and a sharp rise during 1993-94 to 1999-2000. Similarly, states such as Gujarat, Karnataka, Maharashtra, Tamil Nadu and West Bengal showed a marginal rise during 1987-88 to 1993-94 and a decline during 1993-94 to 1999-2000; but the decline was relatively greater, in most cases, than the rise. In the 1990s, it appears that the female workers were replacing the male workers in some of the service related activities. There have been, however, encouraging signs of employment generation for females in community and personal services during the reform period.

RNAE by Other Characterstics of Workers

The extent of workers' participation in RNAE also depends on, *inter alia*, the age of the workers, their level of education and size of landholdings. An effort has been, therefore, made to analyse the trends and patterns of RNAE by such characteristics of the workers at the all-India rural level.

RNAE by age groups of workers

An analysis of RNAE by age group of the workers identifies whether the younger, children or old-aged persons are more likely to take up jobs in the agricultural and non-agricultural occupations, and also the extent of child labour in both the sectors. The data provided in Table 4.13 indicate that the male WPRs in agriculture tended to rise as the level of age groups rose up to a certain level, i.e., from 5-9 to 45-59 years and it declined thereafter (60+ years).

Table 4.13 : Male WPRs by Age Groups in Agriculture and Non-Agriculture in Rural India, 1983 to 1999-2000

(Percent)

Age group (Years)	*Agriculture*				*Non-agriculture*			
	1983	*1987-88*	*1993-94*	*1999-00*	*1983*	*1987-88*	*1993-94*	*1999-00*
UPS								
5-9	1.82	0.52	0.90	0.54	0.13	0.09	0.15	0.11
10-14	18.31	12.43	9.25	6.38	2.52	2.10	2.04	1.70
15-29	60.17	54.64	53.94	50.36	17.56	20.08	20.03	21.67
30-44	71.64	68.21	68.83	66.46	25.37	29.62	29.57	31.49
45-59	74.10	71.53	72.47	69.98	20.24	23.51	23.72	25.53
60 +	53.11	54.36	58.64	51.82	10.66	10.48	10.14	10.10
All	40.55	38.13	39.72	37.18	11.66	13.53	14.14	15.01
UP+SS								
5-9	2.42	2.21	0.93	0.55	0.13	0.09	0.16	0.11
10-14	22.36	16.49	11.59	7.39	2.52	2.46	2.36	1.89
15-29	63.97	58.42	57.08	51.92	17.56	20.81	20.53	22.07
30-44	72.01	68.61	69..21	66.70	25.37	29.71	29.60	31.54
45-59	74.65	72.28	72.95	69.95	20.24	23.47	23.80	25.58
60 +	55.79	55.93	59.90	53.34	10.56	10.90	10.29	10.33
All	42.42	40.10	41.03	37.85	11.36	13.80	14.33	15.17

Source: NSSO (1988, 1990, 1996 & 2001).

On the other hand, the male WPRs in the rural non-agricultural sector tended to increase with rising age groups (age-groups 5-9 to 30-44 years) and declined thereafter (age-groups 45-59 to 60+ years). This constituted the "inverted U-shaped curve" (Figure 4.9). The male WPRs were much higher in the agricultural sector, as compared to the non-agricultural sector in rural areas. While the former started falling from the age-group 45-59 years, the latter from the age-group 30-34 years. This shows that the younger persons were more likely to take up jobs in the rural non-agricultural sector and relatively older persons in the agricultural sector (Figure 4.10).

Figure 4.9 : Age-Profile of Male in Agriculture and Non-Agriculture in Rural India 1993-94 to 1999-2000

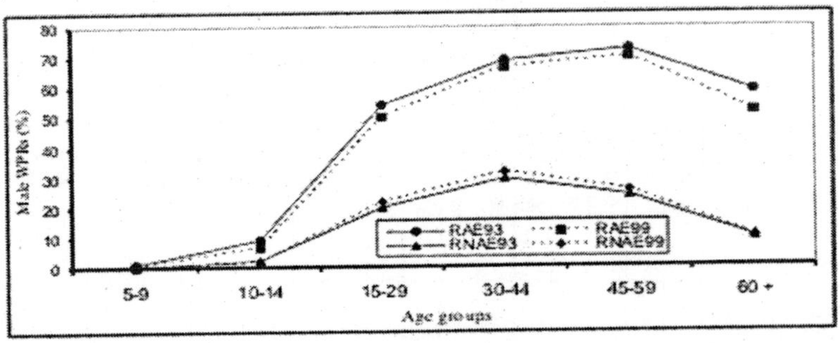

Note: RAE: Rural agricultural employment; RNAE: Rural non-agricultural employment.
Source: Table 4.11.

Figure 4.10 : Proportion of Male Workers by Age Groups in Agriculture and Non-Agriculture in Rural India, 1999-2000

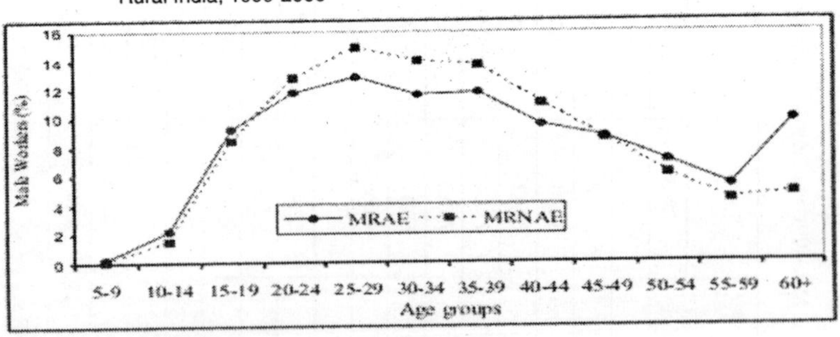

Note: MRAE: Male rural agricultural employment;
MRNAE: Male rural non-agricultural employment.
Source: Prepared from NSS data (NSSO 2001).

The data also indicate that the male WPRs declined in the agricultural sector and increased in the rural non-agricultural sector during the pre-reform as well as the reform periods; while agriculture witnessed a greater decline during the reform period *vis-à-vis* the pre-reform period, the rural non-agricultural sector witnessed a greater increase during the pre-reform *vis-à-vis* the reform period. The decline in male WPRs, *inter alia*, was attributable to a rise in SPRs and persons neither in the labour market nor in schools/colleges. This process of rising SPR seemed to have started well before the economic reforms in India.

Going by different age groups, it was found that the participation rate of child labour was much higher in the agricultural than in the rural non-agricultural sector; but it declined rapidly in both the sectors; a greater decline took place in agriculture during the pre-reform period as compared to the reform period. Further, while agriculture witnessed a greater decline in the age-groups 15-29, 30-44 and 45-59 years, the rural non-agricultural sector showed a rise during both the periods; but the pre-reform years showed a relatively greater decline in respect of agriculture. The proportion of workers in the age group of 60+ years also declined in both the sectors; a greater decline took place in agriculture than in the rural non agricultural sector. This shows that a part of the decline in WPRs was also due to the ageing of the workers.

The data presented in Table 4.14 illustrate that the proportion of youth (15-29 years) and adults (30-44 years) was more in non-agriculture and less in agriculture, on the one hand, and the proportion of child labour (age-groups 5-9 to 10-14 years) and aged (age-groups 45-59 to 60+ years) workers was found to be more in agricultural and less in the rural non-agricultural occupations. This supports our earlier observation that a larger proportion of children and old-aged workers (60+years) seemed to have been involved in the agricultural occupation and the youth and adults with higher education and skills seemed to have been involved in the rural non-agricultural occupations (Figure 4.10).

The proportion of child labour declined continuously both in the agricultural and non-agricultural occupations in rural areas during the pre-reform and reform periods, and a greater decline took place in the agricultural occupations during the pre-reform period. The proportion of aged workers (45-59 to 60+ years) showed, by and large, a rising tendency in the agricultural sector and a continuous

Table 4.14 : Distribution of Male Workers by Age Groups in Agriculture and Non-Agriculture in Rural India, 1983 to 1999-2000

(Percent)

Age group (Years)	*Agriculture*				*Non-agriculture*			
	1983	*1987-88*	*1993-94*	*1999-00*	*1983*	*1987-88*	*1993-94*	*1999-00*
UPS								
5-9	0.66	0.20	0.30	0.20	0.16	0.10	0.14	0.10
10-14	6.13	4.20	2.80	2.20	2.93	2.00	1.73	1.45
15-29	36.14	36.00	34.97	33.90	36.69	37.30	36.47	36.13
30-44	28.69	29.90	31.47	32.90	35.35	36.60	37.97	38.61
45-59	19.72	20.50	20.48	21.00	18.74	19.00	18.83	18.97
60 +	8.68	9.20	9.99	9.80	6.06	5.00	4.85	4.73
All	100	100	100	100	100	100	100	100
UP+SS								
5-9	0.83	0.80	0.30	0.20	0.16	0.10	0.14	0.10
10-14	7.15	5.30	3.39	2.50	2.93	2.30	1.98	1.60
15-29	36.73	36.60	35.83	34.33	36.69	37.90	36.88	36.42
30-44	27.58	28.60	30.64	32.43	35.35	36.00	37.51	38.28
45-59	18.99	19.70	19.96	20.62	18.74	18.60	18.63	18.82
60 +	8.72	9.00	9.88	9.91	6.06	5.10	4.86	4.79
All	100	100	100	100	100	100	100	100

Source: NSSO (1988, 1990, 1996 & 2001).

Table 4.15 : Female WPRs by Age Groups in Agriculture and Non-Agriculture in Rural India, 1983 and 1999-2000

(Percent)

Age group (Years)	*Agriculture*				*Non-agriculture*			
	1983	*1987-88*	*1993-94*	*1999-00*	*1983*	*1987-88*	*1993-94*	*1999-00*
UPS								
5-9	1.64	0.59	0.94	0.60	0.13	0.09	0.18	0.07
10-14	14.77	10.24	8.51	6.01	2.09	2.38	1.81	1.40
15-29	31.17	28.33	25.90	25.25	4.76	6.61	5.26	5.20
30-44	39.51	38.39	36.16	37.32	6.00	7.84	6.42	6.99
45-59	35.20	35.15	34.93	35.03	5.36	6.33	5.04	5.76
60 +	12.65	13.51	14.62	15.04	2.82	2.52	2.42	2.39
All	21.42	20.28	19.77	19.48	3.28	4.28	3.57	3.68
UP+SS								
5-9	2.25	2.21	1.12	0.59	0.13	0.07	0.23	0.10
10-14	21.81	15.51	11.88	7.87	2.09	2.70	2.19	1.74
15-29	44.70	38.47	37.81	33.59	4.76	7.59	6.78	6.37
30-44	55.11	50.28	51.61	48.93	6.00	9.07	8.16	8.20
45-59	49.70	45.76	47.91	45.23	5.36	7.26	6.15	6.61
60 +	19.72	18.66	20.87	19.01	2.82	2.98	3.06	2.70
All	30.55	27.40	28.30	25.51	3.28	4.92	4.53	4.37

Source: NSSO (1988, 1990, 1996 & 2001).

decline in the rural non-agricultural sector; the decline was found to be more in the pre-reform period.

In respect of females, the data presented in Table 4.15 reveal that the female WPRs by different age groups, both in agriculture and non-agriculture, were lower than in the case of male workers. Unlike in the case of male workers, the participation of females, particularly the youth and adults, was quite higher in the agricultural as compared to the non-agricultural sector (Figure 4.11). As in the case of male workers, the female WPRs tended to rise with rising age groups up to a certain level and then declined both in the agricultural and non-agricultural sectors. The female child labour participation rates were found to be quite higher in the agricultural sector and lower in the non-agricultural sector. The female WPRs in aggregate declined in the agricultural sector and increased in the non-agricultural sector in

Figure 4.11: Age-Profile of Female WPRs in Agriculture and Non-Agriculture in Rural India, 1993-94 and 1999-2000

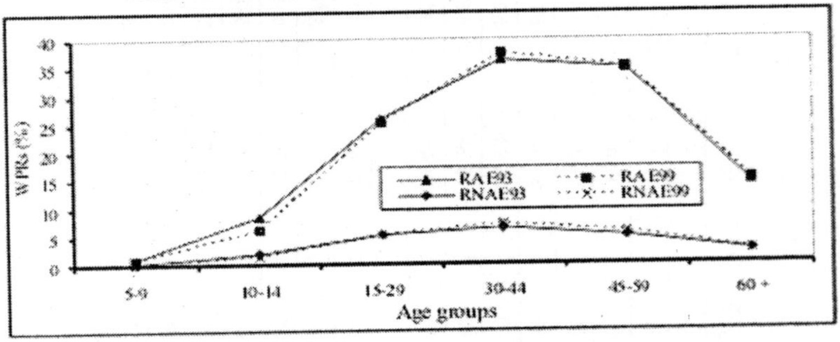

Note: RAE: rural agricultural employment; RNAE: rural non-agricultural employment.
Source: Table 4.15.

the reform regime, as compared to the pre-reform regime. However, the female child labour participation rates declined in both the sectors; while the agricultural sector witnessed a greater decline in female child labour participation rates during the pre-reform period as compared to the reform period, in the rural non-agricultural sector, it showed a greater decline during the reform period as compared to the pre-reform period. In the rural non-agricultural sector *vis-à-vis* the agricultural sector, a greater decline took place among the youth and adult WPRs during the reform period, as compared to the pre-reform period. The female WPRs in the age group of 60+ years increased in the agricultural sector and declined in the rural non-agricultural sector during both the periods.

Table 4.16 : Distribution of Female Workers by Age Groups in Agriculture and Non-Agriculture in Rural India, 1983 to 1999-2000

(Percent)

Age group (Years)	*Agriculture*				*Non-agriculture*			
	1983	*1987-88*	*1993-94*	*1999-00*	*1983*	*1987-88*	*1993-94*	*1999-00*
UPS								
5-9	1.07	0.40	0.60	0.40	0.55	0.30	0.64	0.24
10-14	8.37	5.80	4.60	3.60	7.72	6.39	5.43	4.45
15-29	37.12	37.00	34.83	33.47	36.98	40.92	39.15	36.54
30-44	31.09	33.00	34.23	36.96	30.80	31.94	33.63	36.68
45-59	18.27	19.40	20.62	19.98	18.14	16.57	16.47	17.39
60+	4.04	4.40	5.11	5.59	5.87	3.89	4.68	4.70
All	100	100	100	100	100	100	100	100
UP+SS								
5-9	1.03	1.10	0.50	0.30	0.54	0.20	0.64	0.30
10-14	8.66	6.50	4.51	3.60	7.72	6.31	5.18	4.65
15-29	37.33	37.20	35.71	34.00	36.98	40.84	39.87	37.60
30-44	30.41	32.00	34.30	37.00	30.80	32.13	33.76	36.19
45-59	18.09	18.70	19.86	19.70	18.14	16.52	15.88	16.79
60 +	4.41	4.50	5.12	5.40	5.86	4.00	4.67	4.47
All	100	100	100	100	100	100	100	100

Source: NSSO (1988, 1990, 1996 & 2001).

The data presented in Table 4.16 show that the share of female child workers (age-groups 5-9 to 10-14 years) in agriculture as well as non-agriculture was found to be quite higher than that of male workers. The share of female workers in the age-group of 60+ years in agricultural employment as well as RNAE was found to be smaller than that of their male counterparts. The data also show that a larger proportion of female children (age groups 5-9 and 10-14 years) and youth (age group 15-29 years) were more likely to take up jobs in the rural non-agricultural sector *vis-à-vis* the agricultural sector, whereas a larger proportion of female adults and aged workers (age groups

Figure 4.12 : Proportion of Female Workers by Age Groups in Agriculture and Non-Agriculture in Rural India, 1999-2000

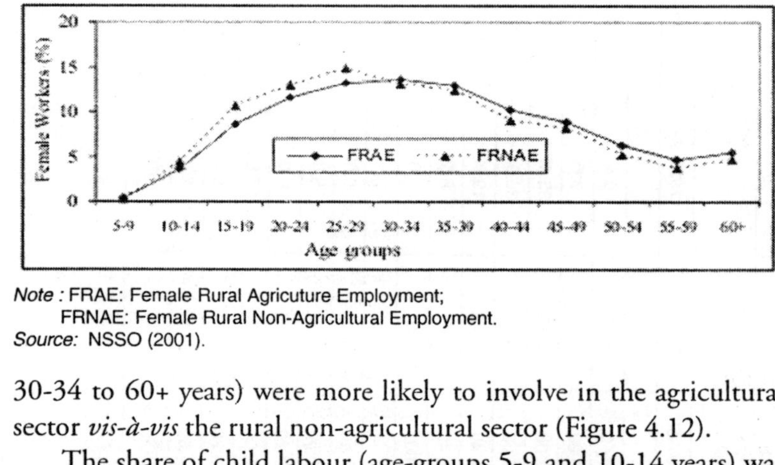

Note : FRAE: Female Rural Agricuture Employment;
FRNAE: Female Rural Non-Agricultural Employment.
Source: NSSO (2001).

30-34 to 60+ years) were more likely to involve in the agricultural sector *vis-à-vis* the rural non-agricultural sector (Figure 4.12).

The share of child labour (age-groups 5-9 and 10-14 years) was higher in the agricultural as well as non-agricultural sectors in the years of the pre-reform period, as compared to the years of the reform period. The share of child labour declined during the reform period; the decline was marginally higher in the agricultural sector *vis-à-vis* the non-agricultural sector. The data also indicate that the share of female child labour declined both in agricultural and non-agricultural sectors during the pre-reform and reform periods; a greater decline took place during the pre-reform as compared to that of the reform period, and a greater decline occurred in the agricultural sector than in the non-agricultural sector.

Similarly, the share of female youth (age-group 15-29 years) both in the agricultural and non-agricultural sectors declined and that of adults (age-group 30-44 years) increased in the years of the reform period, as compared to the years of the pre-reform period. The share of female workers in the age-group 60+ years increased in the agricultural sector and declined in the non-agricultural sector during the reform period as compared to the pre-form period. The decline in the share of youth (age-group 15-29 years) might be attributed partly to a withdrawal of some share of female youth in favour of schools/colleges and partly to a decline in employment opportunities, particularly for those who did not have the required qualifications, skills and knowledge.

RNAE by level of education

The extent of workforce participation in the agricultural and non-agricultural sectors also depends on the level of educational attainments of the workers. The workers with a higher educational background find jobs in the rural non-agricultural sector instead of in agriculture. The data provided in Table 4.17 indicate that WPRs for males, aged 15 and above, tended to decline in the agricultural sector and rise in the non-agricultural sector as the level of education rose. The proportion of male workers with higher educational attainments was found more in the non-agricultural sector than in agriculture. In other words, while agriculture had a vast majority of illiterate male workers, the non-agricultural sector had a large majority of male workers with higher levels of educational attainments. This implies that the rural non-agricultural sector began to absorb a large number of educated male workers. The male WPRs, in aggregate, declined continuously in the agricultural sector and increased in the non-agricultural sector during both the periods; the increase was much higher during the reform period *vis-à-vis* the pre-reform period.

Going by different levels of education, it was found that agriculture showed a decline in male WPRs at the level of illiterates, followed by literates up to primary and middle levels of education, during 1983 to 1999-2000; a greater decline took place during the reform period. The male WPRs relatively at higher educational

Table 4.17 : Male WPRs (15+ Years) by Level of Education in Agriculture and Non-Agriculture in Rural India, 1983 to 1999-20

(*Percent*)

Level of education	*Agriculture*			*Non-agriculture*		
	1983	*1993-94*	*1999-00*	*1983*	*1993-94*	*1999-00*
UPS						
Illiterate	77.15	75.99	72.70	13.18	15.36	16.45
Primary	64.35	64.52	61.65	23.52	25.29	25.58
Middle	47.17	48.36	48.35	22.77	25.00	26.85
Secondary	32.54	38.43	41.53	33.69	29.73	30.01
Hr. Secondary	NA	33.98	35.68	NA	29.30	31.97
Gr. & Above	21.56	28.20	30.63	55.84	51.69	49.91
All	65.51	62.15	58.89	19.49	22.41	23.99
UP+SS						
Illiterate	NA	76.33	72.93	NA	15.43	16.54
Primary	NA	65.43	62.43	NA	25.42	25.72
Middle	NA	51.63	49.81	NA	25.43	27.23
Secondary	NA	42.25	43.34	NA	30.33	30.51
Hr. Secondary	NA	37.85	38.68	NA	30.18	32.76
Gr. & Above	NA	31.31	32.96	NA	52.29	50.36
All	NA	63.69	59.85	NA	22.67	24.21

Note: NA: Not available.

Source: NSSO (1988, 1996 & 2001).

attainments (secondary, higher secondary and graduate and above) increased in agriculture during the same period; a greater increase took place during the pre-reform period. On the other hand, the male WPRs in the rural non-agricultural sector increased in almost all the levels of education categories, except for male workers with graduation and above levels of education who experienced a marginal decline during the same period. The decline in male WPRs at graduate and above level of education in the rural non-agricultural sector implies that workers with better educational qualifications were unable to secure jobs of their choice, leading to out-migration of such workers to urban areas in search of better job options. The increase in male WPRs at relatively higher levels of education in agriculture appeared to indicate that out of additional increase in the number of educated male persons, a larger share of them seemed to have reported agriculture as their main occupation due to non-availability of appropriate non-farm jobs in rural areas and due to socio-economic barriers to out-migrate.

The data presented in Table 4.18 indicate that the proportion of male workers, aged 15 and above, by level of education tended to decline rapidly with increasing levels of education both in the agricultural and non-agricultural sectors; the decline was a little faster in the latter than in the former. It is evident that agriculture had a vast majority of illiterate male workers, whereas the rural non-agricultural sector had a large majority of workers with higher educational attainments. The Achilles' heel or dominance of the peacock's feet in the Indian labour market was more discernible in the case of the agricultural sector as compared to the rural non-agricultural sector.

Table 4.18: Distribution of Male Workers (15+ Years) by Level of Education in Agriculture and Non-Agriculture in Rural India, 1983 to 1999-2000

(Percent)

Level of education	*Agriculture*			*Non-agriculture*		
	1983	*1993-94*	*1999-00*	*1983*	*1993-94*	*1999-00*
UPS						
Illiterate	58.60	50.30	45.90	33.65	28.20	25.50
Primary	27.95	28.70	27.30	34.34	31.20	27.80
Middle	9.17	12.00	14.60	14.89	17.20	19.90
Secondary	3.74	5.50	7.50	13.03	11.80	13.30
Hr. Secondary	NA	2.30	3.00	NA	5.50	6.60
Gr. & Above	0.47	1.20	1.70	4.08	6.10	6.80
All	100	100	100	100	100	100
UP+SS						
Illiterate	57.40	49.30	45.30	33.67	28.00	25.40
Primary	27.78	28.40	27.20	34.34	31.00	27.70
Middle	9.85	12.50	14.80	14.88	17.30	20.00
Secondary	4.38	5.90	7.70	13.02	11.90	13.40
Hr. Secondary	NA	2.50	3.20	NA	5.60	6.70
Gr. & Above	0.59	1.30	1.80	4.08	6.10	6.80
All	100	100	100	100	100	100

Note: NA: Not available.
Source: NSSO (1988, 1996 & 2001).

A striking feature is that the proportion of male workers who were illiterate or literate at primary level of education declined both in the agricultural and non-agricultural occupations; a greater decline took place in the latter than in the former; a greater decline started to take place during 1983 to 1993-94 as compared to 1993-94 to

1999-2000. The proportion of male workers who educated up to middle, secondary, higher secondary and graduate and above levels of education increased both in agricultural and non-agricultural occupations during 1983 to 1999-2000; a greater increase took place in the rural non-agricultural than in the agricultural sector; a greater increase occurred during 1983 to 1993-94, as compared to 1993-94 to 1999-2000.

Table 4.19 presents the data on female WPRs, aged 15 and above, at different levels of educational attainments during 1983 to 1999-2000. As in the case of male WPRs, a similar pattern was also observed in respect of female WPRs; but the female WPRs were found to be much lower than those of male WPRs, both in the agricultural and non-agricultural sectors; and the female WPRs were found to be higher in the agricultural sector than in the non-agricultural sector. The female WPRs illiterate or literate at lower levels were found to be more in the agricultural sector than in non-agricultural sector; whereas the female WPRs at higher levels of education were found to be very higher in the non-agricultural sector as compared to agriculture.

Table 4.19 : Female WPRs (15+ Years) by Level of Education in Agriculture and Non-Agriculture in Rural India, 1983 to 1999-2000

(Percent)

Level of education	*Agriculture*			*Non-agriculture*		
	1983	*1993-94*	*1999-00*	*1983*	*1993-94*	*1999-00*
UPS						
Illiterate	77.15	75.99	72.70	13.18	15.36	16.45
Primary	64.35	64.52	61.65	23.52	25.29	25.58
Middle	47.17	48.36	48.35	22.77	25.00	26.85
Secondary	32.54	38.43	41.53	33.69	29.73	30.01
Hr. Secondary	NA	33.98	35.68	NA	29.30	31.97
Gr. & Above	21.56	28.20	30.63	55.84	51.69	49.91
All	65.51	62.15	58.89	19.49	22.41	23.99
UP+SS						
Illiterate	NA	76.33	72.93	NA	15.43	16.54
Primary	NA	65.43	62.43	NA	25.42	25.72
Middle	NA	51.63	49.81	NA	25.43	27.23
Secondary	NA	42.25	43.34	NA	30.33	30.51
Hr. Secondary	NA	37.85	38.68	NA	30.18	32.76
Gr. & Above	NA	31.31	32.96	NA	52.29	50.36
All	NA	63.69	59.85	NA	22.67	24.21

Note: NA: Not available.
Source: NSSO (1988, 1996 & 2001).

The data (Table 4.20) reveal that the share of female *vis-à-vis* male workers with illiterates was found to be much higher both in the agricultural and non-agricultural sectors. But it was the agricultural sector which had a vast majority of illiterate female workers. However, such segment of female workers declined both in agricultural and non-agricultural employment during the pre-reform and reform periods; the decline was much faster in the non-agricultural sector than in agriculture. On the other hand, at higher levels of education, the share of female workers was quite lower and more so in respect of agriculture.

The beneficial change is that the proportion of female workers who were illiterate declined and those literate and literate at higher levels of education increased both in agricultural and non-agricultural employment during 1983 to 1999-2000; a greater decline took place during 1983 to 1993-94 as compared to 1993-94 to 1999-2000, and in the rural non-agricultural as compared to the agricultural sector;

Table 4.20 : Distribution of Female Workers (15+ Years) by Level of Education in Agriculture and Non-Agriculture in Rural India, 1983 to 1999-2000 **(*Percent*)**

Level of education	*Agriculture*			*Non-agriculture*		
	1983	*1993-94*	*1999-00*	*1983*	*1993-94*	*1999-00*
Males			**UPS**			
Illiterate	88.94	83.3	78.70	74.80	61.00	57.00
Primary	9.23	12.60	14.10	15.64	20.20	20.40
Middle	1.46	3.00	5.00	4.18	7.50	9.40
Secondary	0.29	0.80	1.70	4.39	5.80	6.40
Hr. Secondary	NA	0.20	0.30	NA	2.70	2.90
Gr. & Above	0.03	0.10	0.20	1.09	2.70	3.60
All	100	100	100	100	100	100
Females			**UP+SS**			
Illiterate	86.22	81.20	76.60	74.80	61.70	58.10
Primary	10.71	13.20	14.70	15.64	20.50	20.50
Middle	2.29	3.80	5.70	4.18	7.70	9.30
Secondary	0.68	1.30	2.30	4.39	5.40	5.90
Hr. Secondary	NA	0.30	0.50	NA	2.40	2.70
Gr. & Above	0.07	0.10	0.20	1.09	2.30	3.30
All	100	100	100	100	100	100

Note: NA: Not available.
Source: NSSO (1988, 1996 & 2001).

and a corresponding increase occurred in the proportion of female workers with higher levels of education.

In brief, it has been found that agriculture had a vast majority of illiterate male and female workers, aged 15 and above, as compared to those in the rural non-agricultural sector; going by gender, the share of illiterate female workers was quite significant in agriculture, followed by that of the rural non-agricultural sector, as compared to the illiterate male workers. Such segment of male and female workers declined sharply in the agricultural as well as non-agricultural sectors during both the periods; a greater decline took place in the years of the pre-reform period as compared to the reform period, in the agricultural as compared to non-agricultural sector, for male workers in agriculture, and for female workers in the non-agricultural sector. Correspondingly, the proportion of male and female workers who were literate and had higher levels of education increased both in agricultural and non-agricultural employment during the same period; a greater increase took place during the pre-reform period as compared to the reform period, and in RNAE as compared to that of in agriculture.

RNAE by size of landholdings

The extent of workers' participation in RNAE is also determined by the size of landholdings as an index of income from agriculture. It has been argued that for the smallest categories of landholdings, the rural non-agricultural sector is a major source of employment. The data (Table 4.21) show that the WPRs for males and females tended to increase in agricultural, and decline in non-agricultural, occupations as the size of landholdings rose. The workers without land or with lower size of landholdings seemed to have taken up jobs in the rural non-agricultural sector, as compared to workers with larger size of landholdings.

The data also indicate that the WPRs for males and females in agriculture declined in different categories of landholdings and more so in categories of the landless and lower sized landholdings during 1987-88 to 1993-94 as compared to 1993-94 to 1999-2000. In the case of the rural non-agricultural sector, on the other hand, the WPRs for males showed a moderate rise, particularly in the categories of lower sized landholdings, whereas the WPRs for females showed a constant decline during the same period; a greater decline took place during 1987-88 to 1993-94, as compared to 1993-94 to 1999-2000.

Table 4. 21: Male and Female WPRs (UP+SS) by Size of Landholdings in Agriculture and Non-Agriculture in Rural India, 1987-88 to 1999-2000 (*Percent*)

Land size	*Agriculture*			*Non-agriculture*		
(Ha)	*1987-88*	*1993-94*	*1999-00*	*1987-88*	*1993-94*	*1999-00*
Males						
0-0	27.80	28.60	26.60	23.90	24.50	24.20
0-1	41.20	43.33	40.22	13.10	12.62	13.05
1-2	48.10	40.70	49.50	7.20	6.40	6.80
2-4	49.22	52.00	49.70	6.05	5.50	5.90
4 +	50.41	52.70	50.90	4.49	4.90	4.90
All	40.10	41.03	37.85	13.80	14.33	15.17
Females						
0-0	21.90	22.40	20.70	8.90	8.40	7.10
0-1	29.10	30.25	26.31	4.30	3.55	3.29
1-2	29.90	31.10	31.80	2.70	1.70	1.70
2-4	29.92	33.00	31.40	2.09	1.20	1.10
4+	31.28	33.40	30.40	1.43	1.00	0.80
All	27.40	28.30	25.51	4.92	4.53	4.37

Source: NSSO (1990, 1996 & 2001).

The data on proportion of workers (Table 4.22) by different sizes of landholdings indicate that the proportion of workers participating in the agricultural as well as non-agricultural sectors tended to come down with increasing size of landholdings. The proportion of male and female workers without land was quite significant in non-agricultural sector as compared to agricultural sector. This shows that the workers who did not have productive assets such as land were more likely to take up jobs in the rural non-agricultural sector, mostly in casual/contractual types. This seemed to have lent support to the distress diversification hypothesis.

The proportion of male and female workers without land and with land but less than one hectare showed an increase both in agricultural and non-agricultural sectors; a greater rise took place in the former than in the latter. In the case of size of a landholding with one or more than one hectare, the proportion of workers, both males and females, declined continuously during 1987-88 to 1999-2000. A greater decline occurred in the agricultural than in the non-agricultural sector in rural areas, pointing to the movement of workers away from agriculture.

Table 4.22 : Distribution of Workers (UP+SS) by Size of Landholdings in Agriculture and Non-Agriculture in Rural India, 1987-88 to 1999-2000 *(Percent)*

Land size	*Agriculture*			*Non-agriculture*		
(Ha)	*1987-88*	*1993-94*	*1999-00*	*1987-88*	*1993-94*	*1999-00*
Males						
0-0	20.09	22.79	24.67	50.20	55.83	55.82
0-1	36.66	37.84	42.59	33.87	31.53	34.35
1-2	19.23	19.18	16.95	8.36	7.06	5.79
2-4	13.54	12.40	9.98	4.83	3.75	2.95
4+	10.41	7.93	5.90	2.70	2.11	1.41
All	100	100	100	100	100	100
Females						
0-0	23.51	26.69	28.91	53.42	62.94	60.21
0-1	38.53	38.24	41.52	31.82	28.23	31.57
1-2	16.96	16.79	15.70	8.52	5.75	5.10
2-4	11.71	10.79	8.95	4.58	2.45	1.90
4+	9.18	7.04	5.10	2.33	1.32	0.82
All	100	100	100	100	100	100

Source: NSSO (1990, 1996 & 2001).

Degree of Casualisation in RNAE

The Indian labour market is heterogeneous, comprising a large segment of workers in the unorganised/informal sector and a smaller proportion engaged in the organised/formal sector. With the introduction of economic reforms in India, a larger proportion of workers seemed to have been pushed out of the organised/formal sector into the unorganised/informal sector. Such change in the employment structure has resulted in what can be described as "casualisation or informalisation" of employment. Incidentally, the pace of casualisation has been gradually increasing in both agricultural and non-agricultural sectors in varying degrees across gender, age groups, level of education and size of landholdings.

The data provided in Table 4.23 show that self-employment, followed by casual employment, for both males and females constituted a major share of employment in the agricultural sector. In the rural non-agricultural sector, self-employment, followed by regular employment, constituted a major share of it. The incidence of casual employment was much higher in the agricultural sector than in non-agricultural sector; among the female workers, as compared to the male workers in agriculture; and among the male workers, as compared to the female workers in the rural non-agricultural sector.

Table 4.23 : Distribution of Workers by Status of Employment in Agriculture and Non-Agriculture in Rural India, 1983 to 1999-2000 (*Percent*)

Year	*Agriculture*			*Non-agriculture*		
	SE	*RSE*	*CWE*	*SE*	*RSE*	*CWE*
Males	**UPS**					
1983	63.07	4.32	32.61	50.22	30.77	19.01
1987-88	61.45	4.19	34.36	46.28	27.95	25.77
1993-94	60.14	1.92	37.94	47.89	28.25	23.85
1999-2000	58.05	1.8	40.15	45.71	26.93	27.36
Females						
1983	54.50	1.33	44.17	53.68	18.42	27.89
1987-88	56.93	2.28	40.79	52.17	17.64	30.19
1993-94	50.64	0.77	48.59	54.38	19.98	25.64
1999-2000	48.5	0.85	50.65	58.39	20.08	21.53
Males	**UP+SS**					
1983	64.15	4.17	31.69	50.22	30.77	19.01
1987-88	62.72	4.01	33.27	48.1	28.64	23.26
1993-94	61.08	1.86	37.06	48.22	28.04	23.74
1999-2000	58.66	1.77	39.57	45.96	26.71	27.33
Females						
1983	63.08	1.02	35.90	53.68	18.42	27.89
1987-88	62.85	1.78	35.37	55.93	15.65	28.42
1993-94	58.67	0.58	40.75	60.28	16.18	23.54
1999-2000	56.54	0.68	42.78	62.56	17.16	20.28

Note: SE: Self-employment; RSE: Regular salaried/wage employment and CWE: Casual wage employment.

Source: NSSO (1988, 1990, 1996 & 2001).

A striking trend was that, in agriculture, the share of workers in casual employment, according to UPS, increased consistently from 32.61 percent to 40.15 percent in respect of males, and from 44.17 percent to 50.65 percent in respect of females during 1983 to 1999-2000; a grater decline started to occur much before the introduction of economic reforms in the country. Such decline occurred at the cost of self-employment followed by regular employment. In the case of the rural non-agricultural sector, while the share of casual employment increased consistently from 19.01 percent to 27.36 percent for males, in respect of females, it declined from 27.89 precent to 21.53 percent during the same period; a greater increase for males, and decline for females, appeared to have taken place during the reform period. The pace of casualisation for males was relatively faster in the rural non-agricultural sector than that of in agriculture during the refom period. An important observation was that the magnitude of casualisation in RNAE had been declining for females and increasing for males during the era of globalisation.

Casualisation of RNAE by age groups of workers

The degree of casualisation for male and female workers in RNAE varied across age groups of the workers. The data presented in Table 4.24 indicate that, going by employment status, the proportion of male workers tended to increase in self-employment and decline in casual employment in the rural non-agricultural sector as the level of age groups rose. In respect of regular employment, which largely constituted a major segment of organised/formal sector employment, it was found that the proportion of male workers tended to increase as the level of age groups rose up to a certain level (age-group 45-59 years) and thereafter declined suddenly. This sudden decline in the share of male workers in the age group of 60+ years might be due to the retirement of workers, mostly in organised/formal sector employment and/or withdrawal of labour due to ageing.

The incidence of casualisation in RNAE was quite significant among the children (age-groups 5-9 and 10-14 years) and youth (age-group 15-29 years), as compared to adults and aged workers

Table 4.24 : Distribution of Male Workers by Status of Employment and Age Groups in Non-Agriculture in Rural India, 1983 to 1999-2000

Age group (Years)	*Proportion of workers Across empoloyment status*								*Proportion of workers Across age groups in each employment status*							
	UPS				*UP+SS*				*UPS*				*UP+SS*			
	1983	87-88	93-94	99-00	1983	87-88	93-94	99-00	1983	87-88	93-94	99-00	1983	87-88	93-94	99-00
SE																
5-9	38.5	36.8	51.4	27.8	38.5	39.1	57.3	43.6	0.1	0.1	0.1	0.0	0.1	0.1	0.2	0.1
10-14	47.2	45.0	46.1	45.0	47.2	50.4	50.9	48.5	2.8	2.0	1.7	1.5	2.8	2.4	2.1	1.5
15-29	48.0	43.8	46.3	41.1	48.0	46.2	46.8	41.5	35.0	35.5	35.2	32.5	35.0	36.1	35.7	36.1
30-44	46.0	43.8	45.8	46.1	46.0	45.0	45.3	46.1	32.4	34.4	36.3	39.0	32.4	33.6	35.7	38.7
45-59	54.8	49.0	47.5	45.6	54.8	50.2	47.6	45.6	20.5	19.9	18.7	18.9	20.5	19.6	18.5	19.0
60+	76.6	74.0	78.5	78.7	76.6	75.5	78.5	79.0	9.3	8.0	8.0	8.2	9.3	8.1	7.9	4.8
All	50.2	46.3	47.9	45.7	50.2	48.1	48.2	46.0	100	100	100	100	100	100	100	100
RSE																
5-9	38.5	22.2	0.0	0.0	38.5	23.2	0.0	0.0	0.2	0.1	0.0	0.0	0.2	0.1	0.0	0.1
10-14	22.6	16.3	12.1	16.4	22.6	16.2	11.4	14.6	2.2	1.2	0.7	0.9	2.2	1.3	0.8	1.7
15-29	26.8	22.3	22.2	22.0	26.8	22.9	22.0	21.7	32.0	29.9	28.6	29.5	32.0	30.1	28.9	32.9
30-44	37.8	34.5	33.0	29.6	37.8	35.4	32.9	29.5	43.4	44.8	44.3	42.5	43.4	44.5	44.0	38.5
45-59	32.6	32.8	37.3	36.9	32.6	33.6	37.2	36.9	19.9	22.1	25.0	25.9	19.9	22.1	24.8	18.6
60+	12.3	10.1	8.6	6.9	12.3	10.5	8.6	6.8	2.4	1.8	1.5	1.2	2.4	1.9	1.5	8.3
All	30.8	28.0	28.3	26.9	30.8	28.6	28.0	26.7	100	100	100	100	100	100	100	100

(Contd.)

Table 4.24 *(Contd.)*

CWE																
5-9	23.1	41.0	48.6	72.2	23.1	37.7	42.7	56.4	0.2	0.2	0.3	0.2	0.2	0.2	0.3	0.0
10-14	30.2	38.8	41.8	38.6	30.2	33.5	37.7	36.9	4.7	3.1	3.0	2.1	4.7	3.3	3.2	0.9
15-29	25.2	34.0	31.6	36.9	25.2	30.9	31.2	36.8	48.7	49.6	48.2	48.7	48.7	50.0	48.4	29.6
30-44	16.3	21.8	21.2	24.3	16.3	19.6	21.3	24.4	30.3	30.7	33.8	34.4	30.3	30.3	33.6	42.4
45-59	12.6	18.2	15.2	17.5	12.6	16.2	15.2	17.6	12.4	13.3	12.1	12.1	12.4	13.1	11.9	25.9
60+	11.1	16.0	13.0	14.4	11.1	14.0	12.9	14.3	3.5	3.1	2.7	2.5	3.5	3.1	2.7	1.2
All	19.0	25.8	23.9	27.4	19.0	23.3	23.7	27.3	100	100	100	100	100	100	100	100

Note: SE: Self-employment; RSE: Regular salaried/wage employment and CWE: Casual wage employment.
Source: NSSO (1988, 1990, 1996 & 2001).

(age-groups 30-44 to 60+ years). The youth were more likely to take up casual employment, whereas adults and aged persons took up self-employment in the rural non-agricultural sector. This follows that among the rural non-agricultural male workers, the experienced persons (adults and aged) seemed to have been engaged in self-employed jobs of challenging and competitive nature in the rural non-agricultural sector. Children from the poor households also took up jobs in the rural non-agricultural sector, which seemed to be contractual, temporary, low-end, lowly paid.

The data on proportion of workers by age-groups in each employment status shows that the share of male children and aged workers was very meagre, and that of the youth and adults was very large in the age cohort of male workers. As mentioned earlier, the share of child labour and youth was found to be more in casual employment and the share of adult male workers was found to be higher in self-employment in the rural non-agricultural sector. The data also indicate that the share of male children (age-groups 5-9 and 10-14 years) as a percentage to total male workers declined in all types of RNAE during both the periods; the decline was slightly higher during the pre-reform period, as compared to the reform period; and in casual employment, as compared to other types of RNAE. This might be due, in part, to the withdrawal of some share of male workers in the age-groups 5-9 and 10-14 in favour of educational institutions to acquire more qualifications, skills and knowledge.

The share of youth (age-group 15-29 years) increased marginally in casual employment and declined in self-employment in the rural non-agricultural sector. While the share of male workers in the age-group 60+ years also witnessed a moderate decline in casual and regular employment, it increased slightly in self-employment in the rural non-agricultural sector. The share of adult male workers, in the age group of 30-44 years, in self- and casual employment, and in the age group of 45-59 years, in regular employment, increased relatively a little faster than in any other age-group during the pre-reform and reform periods; a greater rise occurred in the former than in the latter period.

As far as females were concerned, the data presented in Table 4.25 indicate that the share of female child labourers was found to be

Table 4.25 : Distribution of Female Workers by Status of Employment and Age Groups in Non-Agriculture in Rural India, 1983 to 1999-2000

(Percent)

Age group (Years)	Proportion of workers Across empoloyment status								Proportion of workers Across age groups in each employment status							
	UPS				UP+SS				UPS				UP+SS			
	1983	87-88	93-94	99-00	1983	87-88	93-94	99-00	1983	87-88	93-94	99-00	1983	87-88	93-94	99-00
SE																
5-9	53.9	40.9	57.7	40.8	53.9	44.0	56.4	55.2	0.6	0.2	0.7	0.2	0.5	0.3	0.6	0.2
10-14	45.9	43.8	54.4	73.4	45.9	49.0	58.4	75.2	6.6	5.7	5.5	5.6	6.6	6.8	5.0	5.6
15-29	46.0	48.0	50.0	58.9	46.0	51.9	57.5	62.5	31.7	36.8	35.6	35.8	31.7	40.0	37.7	37.6
30-44	56.8	52.3	54.6	54.8	56.8	56.3	60.5	58.1	32.6	31.6	33.8	34.1	32.6	31.5	34.0	33.6
45-59	60.3	61.6	58.0	58.6	60.3	64.2	63.0	65.0	20.4	20.2	17.7	18.3	20.4	17.1	16.7	17.4
60+	73.4	66.5	75.1	67.0	73.4	70.0	75.2	77.1	8.0	5.5	6.7	6.1	8.0	4.3	6.0	5.5
All	53.7	52.2	54.4	58.4	53.7	55.9	60.3	62.6	100	100	100	100	100	100	100	100
RSE																
5-9	30.8	0.0	13.8	38.7	30.8	0.0	12.0	29.4	0.9	0.0	0.5	0.5	0.9	0.2	0.5	0.5
10-14	18.2	15.9	11.4	8.7	18.2	14.2	9.9	7.5	7.6	6.1	3.1	1.9	7.6	5.9	3.2	2.1
15-29	19.3	16.8	21.4	20.5	19.3	14.9	16.9	16.7	38.8	38.0	41.5	36.3	38.8	37.0	41.2	36.7
30-44	19.8	20.5	21.7	23.1	19.8	17.9	17.6	19.7	33.2	36.6	36.6	41.8	33.2	32.1	36.9	41.4
45-59	17.9	16.9	19.5	19.0	17.9	15.2	16.4	17.5	17.6	16.4	16.2	17.3	17.6	19.5	16.1	17.1
60+	8.2	11.9	8.6	8.6	8.2	10.2	7.2	8.7	2.6	2.9	2.1	2.3	2.6	5.4	2.2	2.3
All	18.4	17.6	20.0	20.1	18.4	15.7	16.2	17.2	100	100	100	100	100	100	100	100

(Contd.)

Table 4.25 : *(Contd.)*

CWE																
5-9	15.4	59.1	28.5	20.6	15.4	56.0	31.6	15.4	0.3	0.5	0.8	0.2	0.3	0.0	0.9	0.2
10-14	35.9	40.4	34.2	17.9	35.9	36.8	31.7	17.3	9.9	9.1	7.4	3.7	9.9	6.1	7.0	4.0
15-29	34.7	35.2	28.5	20.6	34.7	33.2	25.7	20.8	46.0	46.7	43.1	33.9	46.0	38.0	43.1	38.6
30-44	23.3	27.3	23.7	22.0	23.3	25.8	21.9	22.2	25.8	28.5	31.2	37.1	25.8	36.5	31.5	39.7
45-59	21.8	21.5	22.5	22.4	21.8	20.6	20.6	17.5	14.2	12.2	14.6	19.0	14.2	16.5	14.0	14.5
60+	18.4	21.7	16.4	24.5	18.4	19.8	17.7	14.3	3.9	3.1	3.1	6.0	3.9	2.8	3.6	3.1
All	27.9	30.2	25.6	21.5	27.9	28.4	23.5	20.3	100	100	100	100	100	100	100	100

Note: SE: Self-employment; RSE: Regular salaried/wage employment and CWE: Casual wage employment.
Source: NSSO (1988, 1990, 1996 & 2001).

very large in self-employment, as compared to regular and casual wage employment in the rural non-agricultural sector. There was a greater decline in the share of casual wage employment among the children and youth during both the periods; a greater decline occurred during the reform period. The share of female youth in casual wage employment, though it was lower next only to self-employment, declined rapidly, whereas in self-employment, it increased significantly during the same period. This indicates that a larger proportion of female youth began to shift in favour of self-employment and withdraw from casual employment in the rural non-agricultural sector; but no significant change took place in respect of regular employment. Going by the proportion of female workers in each status of RNAE, the proportion of casually employed female children and youth declined significantly, whereas the proportion of female adults and aged workers increased correspondingly during the same period.

Casualisation of RNAE by level of education

The degree of casualisation of RNAE can also be determined by the level of education of the workers. The data presented in Tables 4.26 and 4.27 show that, going by the status of employment, the share of male workers tended to decline in casual wage employment, followed by self-employment, and rise in regular employment in the rural non-agricultural sector as the level of education rose; the pace of decline was much faster in casual wage employment than in self-employment.

The share of male workers, who were illiterate or literate at lower levels of education, was quite significant in self- and casual wage employment, as compared to regular employment in the rural non-agricultural sector. The data also indicate that while the share of male workers in the categories of illiterate and literate up to primary level of education showed a decline and the share of male workers in the categories of middle, secondary, higher secondary and graduate and above levels of education showed a sharp rise in self-employment in the rural non-agricultural sector during both the periods; a greater decline took place during 1983 to 1993-94, as compared to 1993-94 to 1999-2000. On the other hand, the share of illiterate male workers declined in regular employment and increased in casual wage employment during the same period; a greater decline took place in regular employment during 1983 to 1993-94, as compared to 1993-94 to 1999-2000 and a greater increase in casual wage

Table 4.26 : Distribution of Male Workers (15+ Years) by Status of Emplayment in Non-Agriculture at Different Levels of Education in Rural India, 1983 to 1999-2000

(*Percent*)

Level of education	*UPS*			*UP+SS*		
	1983	*1993-94*	*1999-00*	*1983*	*1993-94*	*1999-00*
SE						
Illiterate	57.44	52.47	51.67	57.39	52.52	51.72
Primary	57.57	54.35	50.56	57.57	54.23	50.62
Middle	46.86	49.92	45.43	46.86	50.14	45.51
Secondary	27.34	37.29	38.77	27.34	38.35	39.33
Hr. Secondary	15.20	33.73	38.65	NA	34.41	39.73
Gr. & Above	NA	23.19	25.64	15.20	24.07	26.24
All	50.23	47.91	45.74	50.23	48.12	45.94
RSE						
Illiterate	14.11	11.76	9.35	14.10	11.70	9.36
Primary	23.13	18.82	17.27	23.13	18.80	17.21
Middle	39.96	29.93	27.87	39.96	29.60	27.61
Secondary	69.78	51.88	44.12	69.78	50.94	43.49
Hr. Secondary	NA	60.33	54.80	NA	59.45	53.79
Gr. & Above	84.04	76.05	71.56	84.04	75.17	70.58
All	31.20	28.57	27.10	31.20	28.41	26.92
CWE						
Illiterate	28.45	35.77	38.98	28.51	35.78	38.92
Primary	19.30	26.83	32.17	19.30	26.97	32.17
Middle	13.18	20.15	26.70	13.18	20.26	26.89
Secondary	2.88	10.83	17.11	2.88	10.71	17.18
Hr. Secondary	NA	5.94	6.56	NA	6.14	6.48
Gr. & Above	0.75	0.76	2.80	0.75	0.76	3.18
All	18.57	23.52	27.15	18.57	23.47	27.14

Note: NA: Not available; SE: Self-employment; RSE: Regular salaried/wage employment and CWE: casual wage employment.
Source: NSSO (1988, 1990, 1996 & 2001).

employment occurred during 1993-94 to 1999-2000, as compared to 1983 to 1993-94. This reveals that the pace of casualisation had

Table 4.27 : Distribution of Male Workers (15+ Years) by Level of Education in each Status of Employment in Non-Agriculture in Rural India, 1983 to 1999-2000

Level of education	*UPS*			*UP+SS*		
	1983	*1993-94*	*1999-00*	*1983*	*1993-94*	*1999-00*
SE						
Illiterate	38.47	30.60	28.80	38.47	30.20	28.50
Primary	39.36	35.30	30.70	39.36	34.90	30.50
Middle	13.89	18.00	19.80	13.89	18.10	19.80
Secondary	7.09	9.30	11.30	7.09	9.60	11.50
Hr. Secondary	NA	3.90	5.60	NA	4.10	5.80
Gr. & Above	1.24	3.00	3.80	1.24	3.10	3.90
All	100	100	100	100	100	100
RSE						
Illiterate	15.22	11.50	8.80	15.22	11.40	8.80
Primary	25.46	20.50	17.70	25.46	20.50	17.70
Middle	19.07	18.10	20.50	19.07	18.10	20.50
Secondary	29.14	21.70	21.70	29.14	21.60	21.70
Hr. Secondary	NA	11.70	13.40	NA	12.00	13.40
Gr. & Above	11.00	16.50	17.90	11.00	16.40	17.90
All	100	100	100	100	100	100
CWE						
Illiterate	51.54	42.50	36.60	51.68	42.20	36.30
Primary	35.69	35.50	32.90	35.69	35.60	32.80
Middle	10.56	14.80	19.60	10.56	15.00	19.80
Secondary	2.02	5.50	8.40	2.02	5.50	8.50
Hr. Secondary	NA	1.40	1.60	NA	1.50	1.60
Gr. & Above	0.17	0.20	0.70	0.17	0.20	0.80
All	100	100	100	100	100	100

Note: NA: Not available; SE: Self-employment; RSE: Regular salaried/wage employment and CWE: casual wage employment.

Source: NSSO (1988, 1990, 1996 & 2001).

been accelerated during the reform period; it was a little faster among the male workers who were illiterate and had lower levels of education. The degree of casualisation was growing at the cost of regular employment and, to some extent, at the cost of self-employment, particularly among those male workers with illiteracy and primary level of education. This implies that during the reform period, the male workers with no education or education of lower levels were pushed into casual wage employment in the rural non-agricultural sector.

The data (Table 4. 28) on proportion of male workers, aged 15 and above, by level of education in each status of employment also indicate that the share of illiterates and literate up to primary level of education was found to be highest among the casually employed male workers; the share of male workers with literates up to primary level, followed by illiterate, was highest in self-employment; and, the share of male workers with secondary, followed by middle level of education, was found to be highest in regular employment. This shows that the human capital base of the male workers in regular, followed by self-employment, was much better than that of male workers in casual wage employment in the rural non-agricultural sector.

The share of workers in the categories of illiterate and literate up to primary level of education declined significantly in each employment status of the rural non-agricultural sector during 1983 to 1999-2000; a greater decline took place in casual wage employment, followed by regular and self-employment. The share of workers in the categories of middle, secondary, higher secondary and graduate and above levels of education increased correspondingly in casual wage employment, followed by those in regular and self-employment during the same period; a greater increase took place in casual wage employment followed by regular and self-employment. This shows that male workers who were illiterate and literate up to primary level of education were not able to secure better jobs. Some proportion of the male workers, particularly in the age-groups of 5-9, 10-14, 15-19 and 20-24 years, therefore, started withdrawing from the labour market in favour of schools/colleges to acquire more qualifications, skills and knowledge.

Table 4.28 : Distribution of Male Workers (15+Years) by Status of Employment in Non-Agriculture at Different Levels of Education in Rural India, 1983 to 1999-2000 (*Percent*)

Level of education	*UPS*			*UP+SS*		
	1983	*1993-94*	*1999-00*	*1983*	*1993-94*	*1999-00*
SE						
Illiterate	58.64	58.29	64.62	58.64	63.02	67.89
Primary	50.87	63.38	63.40	50.87	68.49	66.65
Middle	41.89	50.26	55.07	41.89	59.28	60.50
Secondary	10.70	27.33	27.63	10.70	37.33	32.51
Hr. Secondary	NA	19.13	19.72	NA	26.26	25.37
Gr. & Above	7.58	9.68	12.58	7.58	15.12	16.87
All	54.20	54.36	47.97	54.20	60.38	61.95
RSE						
Illiterate	12.37	10.51	8.67	12.37	8.48	7.34
Primary	17.25	15.18	15.76	17.25	12.17	13.53
Middle	45.17	32.10	28.43	45.17	25.13	24.36
Secondary	87.86	66.78	66.67	87.86	57.22	61.95
Hr. Secondary	NA	77.34	75.84	NA	70.09	70.07
Gr. & Above	92.42	88.53	86.83	92.42	82.96	82.51
All	18.60	20.54	25.89	18.60	16.56	17.59
CWE						
Illiterate	29.00	31.20	26.71	29.00	28.50	24.77
Primary	31.88	21.44	20.85	31.88	19.34	19.81
Middle	12.94	17.65	16.50	12.94	15.59	15.14
Secondary	1.44	5.89	5.69	1.44	5.45	5.54
Hr Secondary	NA	3.53	4.44	NA	3.65	4.57
Gr. & Above	0.00	1.79	0.59	0.00	1.93	0.62
All	27.20	25.10	26.14	27.20	23.06	20.45

Note: NA: Not available; SE: Self-employment; RSE: Regular salaried/wage employment and CWE: casual wage employment.

Source: NSSO (1988, 1996 & 2001).

As far as females were concerned, it was also found that, as in the case of male workers, the share of female workers tended to decline in casual wage employment and self-employment and increase in regular employment as the level of education rose; the decline was a little steeper in respect of casual wage employment. Going by employment status, the share of illiterate female workers was found to be quite significant in self-employment, followed by casual wage employment in the rural non-agricultural sector. The share of illiterate female workers in self-employment was much higher than that of male workers. The data also reveal that the share of female workers who were illiterate or literate up primary level or above increased continuously in self-employment and declined in regular and casual wage employment during 1983 to 1999-2000. This implies that a larger share of female workers with more educational attainments was shifting to self-employment in the rural non-agricultural sector during the reform period. In other words, the growth of self-employment for females was increasing at the cost of regular and casual wage employment in the rural non-agricultural sector.

The data (Table 4.29) on educational background of the female workers, aged 15 and above, in each status of employment indicate that there was a vast majority of illiterate female workers in each status of RNAE; the casual wage employment had the highest share of illiterate female workers. The share of illiterate female workers declined significantly in each status of RNAE during 1983 to 1999-2000; a greater decline took place in self-employment in 1993-94 over 1983, and in regular employment followed by casual wage employment in 1999-2000 over 1993-94. This shows that the share of illiterate females declined more in self-employment, followed by regular, than in casual wage employment in the rural non-agricultural sector during the reform period. The share of female workers who were literate (from primary to graduate level and above) increased correspondingly in each status of RNAE during the same period. This shows that the human capital base of the female workers was also improving during the reform period.

Casualisation of RNAE by size of landholdings

The growth of RNAE is also determined by the size of landholdings.

Table 4.29 : Distribution of Female Workers (15+ Years) by Level of Education in each Status of Employment in Non-Agriculture in Rural India, 1983 to 1999-2000

(Percent)

Level of education	*UPS*			*UP+SS*		
	1983	*1993-94*	*1999-00*	*1983*	*1993-94*	*1999-00*
SE						
Illiterate	00.92	05.20	00.70	00.02	04.20	00.00
Primary	14.68	23.20	22.50	14.68	23.00	22.12
Middle	3.23	7.10	8.90	3.23	7.70	9.11
Secondary	0.87	3.00	3.10	0.87	3.40	3.10
Hr. Secondary	NA	1.00	1.00	NA	1.10	1.10
Gr. & Above	0.15	0.50	0.80	0.15	0.60	0.90
All	100	100	100	100	100	100
RSE						
Illiterate	49.74	31.10	24.02	49.74	31.50	24.22
Primary	14.50	14.70	15.72	14.50	14.90	15.82
Middle	10.15	12.00	12.91	10.15	11.90	12.91
Secondary	20.75	19.40	21.02	20.75	19.00	20.82
Hr. Secondary	NA	10.70	10.81	NA	10.70	10.71
Gr. & Above	5.42	12.10	15.52	5.42	12.00	15.52
All	100	100	100	100	100	100
CWE						
Illiterate	79.75	75.60	70.59	79.75	76.00	70.72
Primary	18.33	17.00	19.84	18.33	17.00	20.02
Middle	1.99	5.40	7.15	1.99	5.30	6.94
Secondary	0.23	1.40	1.71	0.23	1.30	1.61
Hr. Secondary	NA	0.40	0.60	NA	0.40	0.60
Gr. & Above	0.00	0.20	0.10	0.00	0.20	0.10
All	100	100	100	100	100	100

Note: SE: Self-employment; RSE: Regular salaried/wage employment and CWE: Casual wage employment.

Source: NSSO (1988, 1996 & 2001).

Among the small and marginal farmers, occupational diversification may be distress induced and residual in nature, whereas in the case of medium and large farmers, it would be growth induced and non-residual in nature. The data presented in Table 3.30 show that the proportion of male and female workers, by and large, tended to rise in regular and self-employment and decline in casual wage employment in the rural non-agricultural sector as the size of landholdings rose. In other words, the male and female workers with productive assets like land were more likely to seek employment in regular and self-employment rather than in casual wage employment in the rural non-agricultural sector. This implies that the distribution of land in favour of landless households seemed to have enabled the workers to take up self-employment or regular salaried/wage employment in the rural non-agricultural sector.

Going by employment status, however, it was found that the proportion of male and female workers without land was quite significant in self-employment, followed by those in casual and regular employment in the rural non-agricultural sector. The data also indicate that the proportion of male workers without land declined in self- and regular employment, whereas in respect of casual wage employment, it increased significantly during 1987-88 to 1999-2000; a greater decline in self-and regular, and a rise in casual wage employment occurred during the reform period. On the other hand, the share of female workers with more land increased in self-employment, followed by regular employment and declined sharply in casual wage employment in the rural non-agricultural sector during the same period; a greater decline in casual wage employment took place during the pre-reform period. This appears to indicate that the pace of "proletarianisation" was increasing in casual wage employment for male workers and in regular and self-employment for female workers in the rural non-agricultural sector. Even in the case of landed households, the share of female workers with more land declined in casual wage employment and correspondingly increased in regular and self-employment. This shows that female workers with productive assets like land seemed to have withdrawn from casual wage employment in favour of regular and self-employment in the rural non-agricultural sector.

The data on proportion of workers in each status of employment

Table 4.30 : Distribution of Workers (US+SS) by Size of Landholdings and Status of Employment in Non-Agriculture in Rural India, 1987-88 to 1999-2000

(*Percent*)

Land size (Ha)	*Proportion of workers across employment status*						*Proportion of workers in each employment status*					
	Males			*Females*			*Males*			*Females*		
	87-88	*93-94*	*99-00*	*87-88*	*93-94*	*99-00*	*87-88*	*93-94*	*99-00*	*87-88*	*93-94*	*99-00*
SE												
0-0	49.1	47.9	46.1	56.6	59.0	59.2	51.5	55.5	56.3	55.9	62.2	59.3
0-1	48.8	51.0	45.5	56.4	63.2	64.0	34.4	33.2	34.2	32.3	29.0	33.6
1-2	44.8	42.2	40.3	52.4	62.5	52.9	7.5	6.3	5.1	6.9	5.8	4.5
2-4	44.0	40.7	45.8	54.3	45.5	60.0	4.3	3.2	3.0	3.4	1.7	1.7
4 +	42.6	44.7	46.0	53.1	60.0	55.6	2.3	1.9	1.5	1.6	1.4	0.9
All	48.1	48.2	46.0	55.9	60.3	62.6	100	100	100	100	100	100
RSE												
0-0	28.5	27.7	25.3	15.7	16.9	18.3	50.3	54.9	52.2	56.3	65.7	59.1
0-1	23.6	23.3	24.8	12.8	13.2	16.9	28.1	26.0	31.4	26.7	22.4	28.5
1-2	37.3	40.6	43.3	19.1	18.8	29.4	10.5	10.3	9.2	9.2	6.4	8.0
2-4	42.7	44.4	44.1	20.7	27.3	30.0	7.1	5.9	4.8	4.7	3.9	2.8
4 +	43.0	40.4	46.0	29.9	20.0	33.3	3.9	2.9	2.5	3.2	1.7	1.6
All	28.6	28.2	26.7	15.7	16.2	17.2	100	100	100	100	100	100
CWE												
0-0	22.4	24.4	28.6	27.7	24.1	22.5	48.9	57.3	58.0	53.9	64.5	65.3
0-1	27.6	25.7	29.8	30.8	23.5	19.2	40.5	34.0	37.1	34.7	27.4	29.1
1-2	17.9	17.2	16.4	28.6	18.8	17.7	6.2	5.2	3.4	7.4	4.4	4.3
2-4	13.4	14.8	10.2	25.0	27.3	10.0	2.7	2.3	1.1	3.1	2.7	0.8
4 +	14.4	14.9	8.0	16.9	20.0	11.1	1.6	1.3	0.4	1.0	1.1	0.5
All	23.3	23.7	27.3	28.4	23.5	20.3	100	100	100	100	100	100

Note: SE: Self-employment; RSE: Regular salaried/wage employment and CWE: Casual wage employment.
Source: NSSO (1990, 1996 & 2001).

in the rural non-agricultural sector indicate that the share of male and female workers without land *vis-à-vis* those with land was found to be quite higher in all the three types of employment status, implying that the rural non-agricultural sector had a vast majority of workers belonging to landless households. The incidence of casualisation was higher among those households which did not have land, and declined rapidly as the size of landholdings increased.

Conclusions

The foregoing analysis reveals that the growth rate of RNAE had registered a continuous decline till the early 1990s, but gained momentum in the late 1990s. A significant setback occurred in respect of rural female RNAE during the reform period, as compared to the pre-reform period. It has also been found that, at the aggregate level, the growth rates of RNAE, especially for females, had largely fluctuated in accordance with the growth rates of agricultural income and foodgrains production. In other words, the growth rates of female RNAE tended to move in the direction opposite to the growth rates of agricultural income and foodgrains production, lending support to the distress diversification hypothesis in respect of female workers. The regional pattern of RNAE reveals that the growth of RNAE was not uniform across sixteen major states and none of the states witnessed a uni-linear pattern of growth rate of RNAE and almost every state had been subjected to periodic fluctuations.

The sectoral pattern of growth indicates that among the rural non-agricultural activities, which registered a positive growth since the 1970s, namely, transport, storage, communication, construction and trade, continued to show impressive progress, whereas manufacturing started to witness a lower growth rate during the reform period *vis-à-vis* the pre-reform period. A rapid decline took place in the growth of employment for males in public utilities, followed by mining and quarrying during the reform period, as compared to the pre-reform period.

In respect of females, the growth rate of employment in agriculture, as in the case of male workers, registered a positive growth rate during the pre-reform period, but a negative growth rate during the reform period. Going by activities within the rural non-agricultural sector, it was found that a negative growth rate of employment for females had been witnessed in mining and quarrying, trade, hotel and restaurant business and a positive growth in the construction, services and manufacturing sectors. However, a significant slow-down had occurred for female employment in the manufacturing and service sectors during the reform period, as compared to the pre-reform period.

It has also been found that the proportion of both male and female workers in RNAE, by and large, witnessed a sustained increase during 1972-73 to 1999-2000, except for 1987-88; the increase in the share of workers in RNAE slowed down during the reform period, as compared to the pre-reform period; and it was a little faster among the male workers as compared to that among the female workers, resulting in a widening gap between male and female RNAE. Thus, it is evident that while the share of the rural non-agricultural sector *vis-à-vis* agriculture had been on the rise, the rate at which it increased in the 1980s had not sustained itself in the 1990s.

A considerable variation in the proportion of workers in RNAE existed across states, activities and gender. While regional variation in male RNAE was largely explained by growth related factors (per capita income from agriculture), regional variation in the case of female RNAE was explained largely by distress induced factors (incidence of unemployment and poverty). The value of coefficient of variation reveals that regional disparity in male RNAE showed a decline in the years of the pre-reform period, but it again started to rise in the years of the reform period. In respect of female workers, the coefficient of variation was much higher than in respect of male workers. As in the case of male workers, the value of coefficient of variation also declined in the years of the pre-reform period and began to rise in the years of the reform period; the pace of increase was much greater than the pace of decline. This implies that during the reform period, the regional disparity in the growth of RNAE for both males and females had widened. The regional disparity in the growth of RNAE as a whole was not uniform across different activities in the rural non-agricultural sector.

The WPRs across age groups of workers show that the WPRs in the rural non-agricultural sector tended to increase with rising levels of age-groups up to a certain level (age-groups 5-9 to 30-44 years) and thereafter (age-groups 45-59 to 60+ years) declined. This constituted what could be described as the "inverted U-shaped curve" with varying contours across male and female workers. The data show that young and adult male workers were more likely to be involved in the rural non-agricultural sector, and relatively older male persons in agriculture. In respect of female workers, the young and adult were likely to be involved more in agriculture and less in non-agriculture. The child labourers and old-aged workers, both male and female, were found to be more in agricultural and less in non-agricultural occupations in rural areas. The proportion of male and female child labourers declined continuously both in the agricultural and non-agricultural sectors during the pre-reform and reform periods; a greater decline took place in agriculture during the pre-reform period. While agriculture witnessed a greater decline for youth and adult workers, the rural non-agricultural sector showed an increase during both the periods; but the pre-reform period showed a relatively greater decline in agriculture. But, in respect of female workers, the rural non-agricultural sector *vis-à-vis* the agricultural sector showed a greater decline for youth and adult workers during the reform period, as compared to the pre-reform period. The male and female workers in the age group of 60+ years declined in both the sectors; a greater decline took place in the agricultural than in the non-agricultural sector in rural areas.

The data on WPRs by level of education show that the WPRs, for both males and females, aged 15 and above, tended to decline in the agricultural and a rise in the non-agricultural sector as the level of education rose, implying that a larger share of educated workers were more likely to secure jobs in the rural non-agricultural sector. The share of educated workers in the rural non-agricultural sector was much higher than in agriculture. It was found that agriculture had a vast majority of illiterate male and female workers as compared to the rural non-agricultural sector; going by gender, the proportion of illiterate female workers was quite significant in agriculture, followed by the rural non-agricultural sector, as compared to the illiterate male workers. The proportion of illiterate workers, however, declined sharply both in the agricultural and non-agricultural sectors during both the periods; a greater decline started to take place in the years of the pre-reform period, as compared to the reform period; in agriculture, as compared to the non-agricultural sector; and for male workers in agriculture and for female workers in the non-agricultural sector in rural areas.

Similarly, the data on WPRs by size of landholding indicate that the male and female WPRs tended to increase in agriculture and decline in the non-agricultural sector as the size of landholdings increased, implying that landless workers or those with lower sized landholdings were more likely to take up jobs in the rural non-agricultural sector, as compared to the workers with larger sized landholdings (distress diversification hypothesis). The data also indicate that the WPRs for males and females declined in agriculture across different sizes of landholdings and more so in the case of workers without land or with lower sized landholdings during the pre-reform period, as compared to the reform period. In the case of the rural non-agricultural sector, on the other hand, the WPRs for males registered a moderate rise, particularly at the lower size of landholdings, whereas the WPRs for females showed a constant decline during the same period. The proportion of male and female workers without land was quite higher in the rural non-agricultural sector, as compared to agriculture. This shows that workers who did not have productive assets such as land took up jobs in the rural non-agricultural sector, mostly in casual/contractual types. This lends support to the distress diversification hypothesis. The proportion of male and female workers without land and with land less than one hectare showed a rise in both agricultural and non-agricultural employment; a greater rise took place in the former than in the latter.

As regards casualisation of RNAE, it was found that the proportion of workers, both male and female, in casual wage employment was more in the agricultural sector and less in the rural non-agricultural sector; the male workers tended to experience less incidence of casualisation, as compared to the female workers in agriculture, and *vice versa* in the rural non-agricultural sector. The pace of casualisation for males was increasing in the agricultural and rural non-agricultural sectors during both the periods; a greater increase took place in the rural non-agricultural sector and during the reform period. The pace of casualisation for females was increasing in the agricultural sector and declining in the rural non-agricultural sector during both the periods. The degree of casualisation in RNAE varied across age groups, level of education and size of landholdings.

The incidence of casualisation of employment for males in the rural non-agricultural sector tended to decline, and for females it tended to rise, as the level of age groups increased. The incidence of casualisation was increasing for males, especially among the male children and youth as compared to others. In respect of females, however, the incidence of casualisation was declining; the decline was much faster among the female children and youth as compared to others. It was found that among the casually employed persons, the incidence of casualisation tended to be more among the youth and adults, as compared to others. The youth seemed to have been involved in casual wage employment, whereas the adults and aged persons seemed to have involved in self-employment in the rural non-agricultural sector.

The proportion of male and female workers tended to decline in casual wage employment followed by those in self-employment, and increase in the case of those in regular employment in the rural non-agricultural sector as the level of education rose; the pace of decline was much faster in casual wage employment than in self-employment. This implies that education tended to reduce the proportion of male and female workers seeking employment in casual wage employment and increased sharply in regular employment. The pace of casualisation for males had accelerated during the reform period. In respect of females, it declined during the reform period; a greater decline took place among those who were illiterates and literates up to primary level of education. The growth of self-employment for females was taking place at the cost of regular and casual wage employment in the rural non-agricultural sector. The incidence of casualisation for both males and females was quite high among those who were illiterates and literates at lower levels of education. Female workers, as compared to male, in the categories of illiterates and literates at lower levels education, experienced a greater incidence of casualisation, but the share of such males and females experiencing casualisation declined more during the reform period.

The share of male and female workers with productive assets like land sought employment in self-employment and regular employment rather than in casual wage employment in the rural non-agricultural sector. The degree of casualisation tended to be higher among those workers who did not have land and declined rapidly as the size of landholdings increased. The data, over time, also show that the incidence of casualisation increased among the male workers; the increase was much faster among the workers without land or having a marginal size of land, as compared to others. In respect of females, it was declining; the decline in casualisation was much faster among the land-rich households, as compared to the land-poor households.

Appendix

Appendix 4.1 : Growth Rates of Total Workforce in Agriculture and Non-Agriculture in India, 1972-73 to 1999-2000 (*Percent/per annum*)

Period	*Agriculture*			*Non-agriculture*			*All*		
	M	*F*	*P*	*M*	*F*	*P*	*M*	*F*	*P*
Rural									
1972-73 to 1977-78	1.31	2.21	1.65	5.03	5.19	5.07	1.97	2.53	2.17
1977-78 to 1983	1.11	2.43	1.62	4.58	3.19	4.23	1.82	2.52	2.08
1983 to 1987-88	0.58	-0.15	0.29	4.28	5.09	4.48	1.44	0.55	1.12
1987-88 to1993-94	2.09	2.20	2.13	2.54	0.36	1.93	2.20	1.88	2.09
1993-94 to1999-2000	-0.13	-0.55	-0.29	2.33	1.04	2.05	0.53	-0.32	0.23
1972-73 to 1999-2000	0.95	1.15	1.03	3.52	2.61	3.30	1.52	1.33	1.45
Rural & Urban									
1072 73 to 1977-78	1.39	2.38	1.77	4.39	6.10	4.74	2.37	3.00	2.58
1977-78 to1983	1.18	2.48	1.69	4.24	3.31	4.05	2.30	2.63	2.41
1983 to 1987-88	0.56	-0.07	0.31	3.50	4.37	3.68	1.69	0.82	1.40
1987-88 to1993-94	2.14	2.11	2.13	3.61	2.18	3.32	2.76	2.13	2.55
1993-94 to1999-2000	-0.24	-0.72	-0.43	2.04	1.70	1.97	0.78	-0.15	0.48
1972-73 to 1999-2000	0.96	1.14	1.03	3.37	3.28	3.35	1.89	1.57	1.79

Note: M: Males; F: Females; P: Persons.
Source: NSSO (1981, 1988, 1990, 1996 & 2001).

NOTES

1. For more details on this pattern of development see Vyasulu (1990).
2. A phenomenal increase in the share of the service sector in national income was regarded as "excess growth" by Bhattacharya and Mitra (1990). They point out that income from the service sector is growing much in excess of the demand generated for the services by the commodity sector.
3. A study by Parthasarathy, Shameem and Reddy (1998) has estimated the total number of states with a negative growth rate of foodgrains, on the one hand, and the total number of states with a positive and rapid growth rate of workers in the rural non-agricultural sector during 1987-88, on the other. The study has found that the states, which witnessed a drastic decline in the growth rate of foodgrains, had experienced a rapid rise in the growth rate of workers in RNAE, lending support to the distress diversification hypothesis.
4. The per capita income from agriculture (cultivators and agricultural labourers) was estimated for 1983 to examine the regional variation in

RNAE (in 1983 and 1999-2000) due to variation in agricultural growth (1983). The states of abnormal variations in RNAE have not been included.

5. A greater increase in the proportion of male workers in RNAE during 1983 to 1999-2000 can be seen in terms of the level of distance between the 1983 and 1999-2000 lines, which tended to reduce along with the states with lower PCIA.
6. The states were ranked according to the rate of male literacy in 1981, as per the census data, to explore the nexus between male education and male RNAE in the pre-reform (1983) and reform (1999-2000) years.
7. With a view to capturing the distress-induced RNAE during the pre-reform and reform periods, the states were ranked on the basis of the incidence of unemployment as per CDS in 1983.
8. In order to examine the relationship between female education and female RNAE in the pre-reform (1983) and reform (1999-2000) years, the states were ranked with the rate of female literacy in 1981 as per the Census data.

5 DETERMINANTS OF RURAL NON-AGRICULTURAL EMPLOYMENT IN INDIA

Exploring Alternative Hypotheses

Introduction

The pattern of occupational diversification as analysed in the previous chapter reveals that there has been a moderate shift of workers away from the agricultural to the non-agricultural sector in rural areas. This shift has had considerable variations across gender, states and activities. It is evident from a large number of studies employing secondary data at the state/region, district and village/household levels (see, Chapter 2) that the varying shift of workers from the agricultural to the rural non-agricultural sector was attributed to a host of factors originating within and/or outside agriculture in rural and urban areas. These factors can broadly be classified into: (1) agricultural growth related; (2) distress-induced; (3) factors external to agriculture but within the rural economy; and (4) also factors originating from outside the rural economy.

It is evident from the review of literature that most of the studies attempted to identify the factors contributing to the growth of only male RNAE (Rural Non-Agricultural Employment) as a whole. But, the factors that determine the growth of male RNAE are different from those in the case of female RNAE. Consequently, the status of non-agricultural activities in which they involve themselves appears to be different, ranging from low productive, lowly paid to high productive, highly paid jobs. As the growth of male and female RNAE was mainly determined by growth related and distress-induced factors, respectively, the level and/or nature of impact of male RNAE on poverty reduction would be different from that of female RNAE. An effort identifying the different factors determining the growth of male and female RNAE separately has, therefore, been increasingly felt to be important for policy intervention.

There are few studies conducted to explore the factors contributing to the growth of sub-activities, for both males and females, in the rural non-agricultural sector; to estimate the elasticity of important determinants of male and female RNAE as a whole and also sub-activities; and to identify the important rural non-agricultural activities for both males and females which are largely growth or distress induced, or high productive or low productive. It can be hypothesised that the activities that are largely growth induced appear to have a greater impact on poverty reduction, as compared to the activities that are largely distress induced/poverty compelled. Because, the daily real income or wage earnings are relatively higher in the growth-led RNAE than in the distress induced RNAE, as it depresses the real income or wage rates due to distress. There has also been a debate on the productive dimension of RNAE at different levels of academic forums. Therefore, an attempt at identifying the activities that are relatively more productive and likely to have a greater impact on poverty and inequality reduction is the need of the hour.

Determinants of Male and Female RNAE

The growth of male and female RNAE has been attributed to several factors within and outside the rural economy (see, Chapter 2). The list of important factors contributing to the growth of male and female RNAE at the state level has been provided in Table 5.1. As is evident from the review of literature, the growth of agriculture contributes to the growth of RNAE through consumption and production linkages. Per hectare/capita net income from agriculture (PHIA/PCIA) is used to measure the agricultural prosperity and consumption and production linkages from agriculture to the rural non-agricultural sector. An increased income from agriculture raises demand for non-farm goods and services (consumption linkages). The growth of

Table 5.1 : List of Notations, Specifications and Sources of Selected Variables, 1999-2000

Sl No.	*Notation*	*Specification*	*Source*
1.	M/F RNAE	Proportion of male/female workers in RNAE	NSSO 2001
2.	M/F MF	Proportion of M/F workers in manufacturing	NSSO 2001
3.	M/F CON	Proportion of M/F workers in construction	NSSO 2001
4.	M/F THRB	Proportion of M/F workers in trade, hotel and restaurant business	NSSO 2001
5.	M TSC	Proportion of M workers in transport, storage and communication	NSSO 2001
6.	M/F CPS	Proportion of M/F workers in community and personal services	NSSO 2001
7.	M/FRAE	Proportion of M/F in rural agricultural employment	NSSO 2001
8.	PHIA	Per hectare net (NSDP) income from agriculture (current prices)	GoI 2003a, CMIE 2004
9.	PCIA	Per capita (cultivators and agricultural labourers) net income from agriculture	GoI 2003a CMIE 2004
10.	GRLH	Gini ratio for operational landholdings (estimated)	GoI 2003a CMIE 2004
11.	NFCE	Per capita non-food consumption expenditure	GoI 2002d
12.	WRLA M/F	Daily real wage rates of agricultural M/F labourers in agriculture	GoI 2003b.
13.	WRLN M/F	Daily real wage rates of agricultural M/F labourers in non-agriculture	GoI 2003b
14.	GAI	Proportion of area under Irrigation	GoI 2003a CMIE 2004
15.	PHFC	Per hectare consumption of chemical fertilizers (NPK)	GoI 2003a CMIE 2004
16.	CIMA*	Composite index of agricultural mechanisation (estimated)	GoI 1998a
17.	COM	Proportion of area under non-foodgrains to total cropped area	GoI 2003a CMIE 2004

(Contd.)

Table 5.1 : (Contd.)

18.	COMP	Proportion of area under non-foodgrains plus paddy to total cropped area	GoI 2003a CMIE 2004
19.	M/F L/hs	Rate of male /female literacy/ at higher secondary	NSSO 2001
20.	CDSUR M/F	Current daily status unemployment rate for males/females	NSSO 2001
21.	CDSURY M/F	Current daily status unemployment rate for youth (15-29) males/females	NSSO 2001
22.	UPSUR M/F	Usual principal status unemployment rate for males/females	NSSO 2001
23.	UPSURY M/F	Usual principal status unemployment rate for youth (15-29 years) males/females	NSSO 2001
24.	UPSURE M/F	Usual principal status unemployment rate for educated males/females	NSSO 2001
25.	PLLH	Proportion of landless households to total households	NSSO 2001
26.	PSMH	Proportion of small and marginal holdings to total holdings	GoI 2003a CMIE 2004
27.	ASMH	Average size of small and marginal landholdings	GoI 2003a CMIE 2004
28.	LMR	Land-man ratio (total cropped area/agricultural population)	GoI 2001 GoI 2003a
29.	PVTY	Proportion of people living below the poverty line (HCR)	GoI 2002d
30.	QM/F RNAE	Proportion of M/F workers in self and regular RNAE	NSSO 2001
31.	DoU	Degree of urbanisation (census)	GoI 2001
32.	UAGLM	Urban agglomeration (proportion of workers in manufacturing)	NSSO 2001
33.	CINFRA*	Composite index for rural infrastructure (estimated)	GoI 2003a CMIE 2004

* Estimated by using principal component analysis (PCA).

agriculture emphasising on commercial and highly market-oriented cash crops cultivation certainly promotes the growth of RNAE through its backward and forward production linkages. The commercialisation of agriculture increases the demand for agricultural inputs such as chemical fertilizers, HYVs, pesticides, insecticides, irrigation, tractors, tillers, harvesters, sprayers and such other locally manufactured agricultural equipments and non-farm services which create backward production linkages as agricultural growth picks up. It also creates forward production linkages to the processing of agricultural commodities such as paddy, maize, wheat, oilseeds, sugarcane, cotton, tobacco, areca nut and such other crops, which generate a good deal of sustained employment opportunities in the rural non-agricultural sector. Both consumption and production linkages from agriculture are expected to reinforce each other and create sustained and adequate employment as broad-based agricultural growth picks up rapidly.

The Gini ratio for operational landholdings (GRLH) is also used to measure the consumption linkages from the growth of agriculture to the growth of employment in the rural non-agricultural sector. If the Gini ratio is negatively associated with RNAE, distribution of land in favour of land-poor households increases their income levels from agriculture and generates a greater demand for non-farm goods and services mostly produced in rural or nearby cities/villages. If it is positively associated with RNAE, there will be distress diversification among the poor households.

Rural per capita non-food consumption expenditure (NFCE) is considered as the size of the market for non-farm goods and services (overall consumption linkage). If it is highly correlated with PHIA/PCIA, it reflects strong consumption linkages from agricultural growth to the growth of RNAE. The daily real wage rate of male or female agricultural labourers in agriculture (WRM\FA) can be considered as the income earning levels of the poor households, as they constitute a major segment of the rural poor, and can also be used to measure the consumption linkages of the poor households and as also an indicator reflecting distress in agriculture (when the wage rate tends to decline or stagnate in agricultural occupations or is lower than that of non-agricultural occupations). In order to measure the backward production linkages the variables such as the proportion of area under irrigation (GIA) per hectare fertilizer consumption (PHFC) and composite index for mechanisation of agriculture (CIMA)[1] are used, in respect of forward linkages, the proportion of area under non-foodgrains plus paddy (COMP) is proxied.

The production linkages from agricultural growth may create direct and indirect employment opportunities in the rural non-agricultural sector. The direct impact of backward linkages on RNAE includes creation of employment opportunities due to construction of dams/barrages (for provision of irrigation and electricity); manufacturing of materials required for dam/barrage construction (cement, steel, etc.); manufacturing of chemical fertilizers (NPK); and manufacturing of agricultural equipments such as tractors, tillers, harvesters, sprayers, etc. The indirect/derived impact of backward linkages from agricultural growth includes generation of employment opportunities due to marketing and distribution of agricultural inputs produced in urban centres and non-farm repair services. Such indirect/derived backward linkages generate employment opportunities in trade, commerce, hotel, restaurant business and services. On the other hand, the direct impact of forward linkages on RNAE includes generation of employment due to agro-processing activities located in rural areas and indirect employment creation due to distribution and marketing of processed agricultural commodities across different places in rural and urban areas.

The rate of literacy for male (RML) and female (RFL) persons is used as a surrogate to human capital. The unemployment rates according to CDS and UPS, particularly for the youth and educated persons, can be taken to mean the height of labour "reluctance" or "unwillingness" to work in agriculture but not as an indicator measuring distress in agriculture *per se*. A strong positive association, therefore, exists between the incidence of unemployment and growth in agriculture (PHIA). It can, however, be an indicator of distress in general; that is, poor performance of the entire economy instead of agricultural sector only.

The unemployment rate for illiterate persons, who are more likely to be engaged in agriculture itself, can roughly be considered as an indicator measuring distress in agriculture, as their opportunity cost tends to be much lower than those with higher levels of education. The growth of RNAE resulting from distress in agriculture is captured through the daily real wage rates of agricultural labourers in agricultural occupations, the proportion of landless households, small and marginal landholdings and the average size of small and marginal landholdings. If the proportion of landless households, small and marginal landholdings is positively, and the daily real wage rates of male and female agricultural labourers in agriculture and the average size of small and marginal landholdings are negatively associated with RNAE, the growth of RNAE can be considered as distress diversification from agriculture.

The proportion of people living below the poverty line (PVTY) and workers employed in self- and regular RNAE (QRNAE)[2] are used to measure the productive/qualitative dimension of RNAE. The PVTY is also an indicator to measure the lack of demand for non-farm goods and services. If the PVTY is negatively, and QRNAE is positively, associated with RNAE, the growth of RNAE cannot necessarily be considered as low productive, casual/contractual ones. Similarly, if the daily real wage rates of male and female labourers in the rural non-agricultural occupations are significantly and positively associated with the proportion of male and female workers in RNAE, the growth of RNAE cannot be considered as a residual sector—low productive, lowly paid jobs. The composite index for rural infrastructure (CINFRA) was constructed by considering rural roads, post-offices, banks and rural teledensity per lakh population using principal component analysis. The degree of urbanisation (DoU) and the urban agglomeration (UAGLM) were identified as the important contributing factors external to the rural economy.

As the growth of RNAE for males and females is determined by different sets of factors, a separate correlation matrix was estimated for male and female RNAE. The results presented in Table 5.2 indicate that among the selected explanatory variables, PHIA, PCIA, NFCE, WRLA, WRLN, RML, RMLhs, UPSUR, PVTY and to some extent

CINFRA emerged as the important determinants of male RNAE. A strong positive association existed between the proportion of male workers in RNAE and per hectare/capita net income from agriculture. The correlation was much stronger in the case of per hectare than in the case of per capita net income from agriculture, *albeit* both of them were significantly and positively associated with male RNAE. The net income per hectare from agriculture seemed to be a more appropriate indicator to measure the prosperity of agriculture and also consumption and production linkages from the growth of agriculture to the growth of RNAE.

A strong positive association between male RNAE and net income per hectare from agriculture implies that the growth of agriculture contributed significantly to the growth of male RNAE through consumption and production linkages. A strong positive association of rural per capita consumption expenditure on non-food items with PCIA may imply that the growth of agriculture contributed rapidly to the growth of male RNAE through consumption linkages. The Gini ratio for operational landholdings was negatively, but statistically insignificant, associated with the male RNAE. This shows that a greater access to land by the land-poor households might raise their income levels from agriculture and their demand for non-farm goods and services and thereby contributed to the growth of male RNAE.

As the growth in agricultural and non-agricultural sectors accelerates, the daily real wage rates tend to increase rapidly. When data on the income earnings of the rural poor are not available at the macro level, the daily real wage earnings of rural agricultural labourers in agricultural and non-agricultural occupations can be taken as an index of income levels of the poor households as the wage-earning households constitute a major proportion of the poor households. The daily real wage earnings of male agricultural labourers in agricultural and non-agricultural occupations had a strong positive impact on the proportion of male workers in RNAE. This implies that an increase in the income levels of the poor households might raise the demand for non-farm goods and services, mostly produced locally or in towns nearby big cities. The increase in demand by the poor for locally produced non-farm goods and services may be greater than that of the non-poor whose demand is mostly urban oriented. It is due to the fact that the degree of correlation between WRLA or WRLN and the male RNAE was much higher than the degree of correlation between PHIA, PCIA or NFCE and the male RNAE. It can generally be noticed that when the income levels of the non-poor households tend to rise due to agricultural growth, there is an increase in the demand for non-farm goods and services produced mostly in urban and semi-urban areas. As a result, an additional increase in the income levels of the non-poor households may stimulate the growth of non-agricultural employment in urban areas. That is why PCIA was found to be positively, and WRLA and WRLN were negatively, associated with the proportion of workers involved in the urban manufacturing sector (UAGLM), but statistically it was not significant.

The backward production linkages that generate employment opportunities in the rural non-agricultural sector depend on technology, scale of production and post-production activities. The use of capital-intensive technology in agriculture certainly generates demand for agricultural implements such as tractors, threshers and harvesters and consequently it displaces labourers from some agricultural operations. It certainly, however, increases the productivity of labourers in agriculture and consequently the wage rates of agricultural labourers; and the income levels of medium and large farmers tend to increase, enabling them to either reinvest in non-farm micro enterprises or to increase their household consumption expenditure on non-farm goods and services mostly produced in semi-urban and urban areas.

The direct production (backward and forward) linkages from agricultural growth to the growth of RNAE were found to be not strong. The proportion of area under irrigation was negatively, the mechanisation of agriculture and the per hectare fertilizer consumption were positively, associated with male RNAE, but it was statistically not significant. This implies that the direct backward linkages were not quite operative in the creation of employment opportunities in the rural non-agricultural sector, but they seemed to be more effective in generating employment opportunities in the manufacturing sector in urban areas (see, Table 5.2, columns 7, 8 and 11 for row 28). The indirect/derived backward production linkages, however, seemed to be more operative in generating employment opportunities in trade commerce, hotels, restaurants and services. It appears that the growth of agriculture demanding agricultural inputs/equipments may create a major share of direct employment in the urban manufacturing sector and indirect employment in rural areas. This is evident in terms of a strong positive association between PHFC/GAI and the proportion of workers in the manufacturing sector in urban areas (UAGLM), and the PHFC/CIMA/GAI and the employment for males in trade, commerce, hotel, restaurant and services in rural areas. Therefore, the urban manufacturing sector seemed to be growing at the cost of RNAE. A weak negative association existed between the proportion of male workers in RNAE and the proportion of workers in the urban manufacturing sector.

In respect of forward linkages from the growth of agriculture to the growth of employment in the rural non-agricultural sector, it has been found that the proportion of area under non-foodgrains plus paddy had a negative association with male RNAE, but it was not statistically significant. This shows that commercialisation of agriculture did not contribute to the growth of male RNAE as a whole; but it certainly contributed to the growth of employment in manufacturing mostly in agro-processing (see, Table 5.5). A weak positive association of commercialisation with the proportion of workers in the urban manufacturing sector implies that the increasing commercialisation accompanied by mechanisation of agriculture in rural areas seemed to have promoted the growth of the urban manufacturing sector (direct forward-linkage effect).

A positive but weak association between the degree of commercialisation and the proportion of male workers in trade, hotels and restaurants shows that the increasing commercialisation of agriculture seemed to have generated non-agricultural employment in distribution, supply of raw materials to urban based agro-processing units, transportation and storage of HYVs, chemical fertilizers, modern tools and machines and repair of agricultural equipments (indirect forward-linkage effect). Thus, the results show that the consumption linkages were more operative than the production linkages in the

Table 5.2 : Correlation Matrix for Male RNAE in India, 1999-2000

	1	2	3	4	5	6	7	8	9	10	11	12	13	14	15	16	17	18	19	20	21	22	23	24	25	26	27	28	29
1 RNAE	1																												
2 RAE	-0.953**	1																											
3 PHIA	0.719**	-0.690**	1																										
4 PCIA	0.629**	-0.644**	0.584*	1																									
5 GRLH	-0.102	0.127	-0.149	0.051	1																								
6 NFCE	0.668**	-0.693**	0.357	0.668**	0.252	1																							
7 GAI	-0.089	0.203	0.009	0.088	0.380	0.173	1																						
8 PHFC	0.094	-0.017	0.428	0.333	0.122	0.357	0.747**	1																					
9 COM	-0.136	0.012	-0.026	0.086	-0.255	-0.140	-0.127	-0.080	1																				
10 COMP	-0.121	0.069	0.213	0.124	-0.329	-0.479	-0.57	-0.23	0.723**	1																			
11 CIMA	0.31	-0.332	-0.740	0.083	-0.318	0.069	-0.532*	-0.579*	-0.190	-0.243	1																		
12 WRLA	0.887**	-0.933**	0.689**	0.706**	-0.056	0.804**	-0.03	252	0	-0.120	0.083	1																	
13 WRLN	0.805**	-0.851**	0.653**	0.628**	0.160	0.817**	-0.49	0.259	-0.108	-0.291	0.035	0.887**	1																
14 RML	0.789**	-0.822**	0.509*	0.628**	-0.092	0.620*	-0.428	-0.123	-0.021	-0.112	0.421	0.680**	0.771**	1															
15 RMLhs	0.697**	-0.733**	0.652**	0.551*	0.255	0.802**	0.017	0.316	-0.130	-0.330	-0.023	0.750**	0.910**	0.707**	1														
16 CDSUR	0.336	-0.396	0.592*	0.283	-0.529*	-0.099	-0.211	0.147	0.521*	0.725**	-0.054	0.325	0.197	0.341	0.105	1													
17 CDSURY	0.496	-0.543*	0.694**	0.325	-0.486	-0.007	-0.275	0.100	0.338	0.601*	0.005	0.445	0.350	0.473	0.263	0.955**	1												
18 UPSUR	0.588*	-0.622**	0.678**	0.303	-0.274	0.058	-0.403	-0.135	0.021	0.352	0.124	0.480	0.449	0.559*	0.409	0.668**	0.854**	1											
19 UPSURY	0.485	-0.507*	0.477	0.035	-0.357	-0.131	-0.394	-0.270	0.064	0.374	0.246	0.312	0.249	0.400	0.183	0.642**	0.797**	0.915**	1										
20 UPSURE	0.185	-0.249	0.415	-0.017	-0.243	-0.314	-0.505*	-0.291	-0.021	0.409	0.178	0.069	0.037	0.246	0.024	0.606*	0.717**	0.841**	0.828**	1									
21 PVTY	-0.612*	0.591*	-0.377	-0.587*	-0.310	-0.898**	-0.232	-0.394	0.145	0.434	0.030	-0.722**	-0.663**	-0.440	-0.703**	0.142	0.086	0.071	0.263	0.365	1								
22 PLLH	-0.029	0.104	0.357	0.320	0.115	-0.014	0.388	0.605*	0.390	0.550*	-0.626**	0.009	0.024	-0.013	0.121	0.524	0.420	0.142	-0.002	0.091	-0.050	1							
23 LMR	0.055	-0.101	-0.203	0.650**	0.268	0.478	0.163	0.094	0.293	-0.025	-0.014	0.238	0.219	0.211	0.080	-0.211	-0.226	-0.201	-0.356	-0.008	-0.340	0.121	1						
24 PSMH	0.126	-0.206	0.551*	-0.071	-0.315	-0.273	-0.228	0.089	0.060	0.348	0.053	0.055	0.165	0.160	0.188	0.622*	0.628**	0.629**	0.63**	0.693*	0.355	0.136	-596*	1					
25 ASMH	-0.485	0.646**	-0.599*	-0.268	0.321	-0.154	0.220	0.018	-0.366	-0.390	-0.145	-0.575*	-0.477	-0.357	-0.351	-0.652**	-0.710**	-0.680**	-0.689**	-0.522*	0.008	0.037	-0.190	0.450	1				
26 QRNAE	0.809**	-0.734**	0.820**	0.568*	-0.075	0.389	0.179	0.330	-0.039	0.174	0.014	0.740**	0.588*	0.384	0.535*	0.382	0.516*	0.575*	0.497*	0.215	-0.428	0.152	0.103	-0.566*	-0.075	1			
27 CINFRA	0.444	-0.554*	0.197	0.451	0.157	0.814**	0.173	0.099	0.072	-0.363	0.168	0.591*	.602*	0.594*	0.650**	0.020	0.042	0.029	-0.144	-0.124	-0.792**	-0.035	0.318	-0.141	0.157	0.050	1		
28 DoU	-0.022	0.059	-0.009	0.394	0.012	0.352	0.242	0.482	0.318	0.062	-0.157	0.071	0.096	0.189	0.107	0.187	0.024	-0.307	-0.400	-0.368	-0.290	0.595*	0.443	0.230	0.012	-0.197	0.375	1	
29 UAGLM	-0.210	0.230	-0.078	0.145	-0.309	0.105	0.484	0.597*	0.386	0.215	-0.307	-0.040	-0.157	-0.214	-.145	0.247	0.057	-0.328	-0.330	-0.380	-0.025	0.518*	0.258	0.097	-0.309	-0.168	0.046	0.756**	1

Note: ** and * refer to 1 and 5 percent level of significance, respectively.

promotion of male RNAE in India, as elsewhere in most of the developing economies. The consumption linkages due to an increase in the income levels of the poor households, as compared to that of the non-poor, seemed to have had a greater impact on the growth of male RNAE. As far as production linkages were concerned, the backward linkages appeared to be stronger than the forward linkages in the promotion of male RNAE as a whole.

A strong positive association existed between the proportion of male workers in RNAE and the rate of male literacy. It has also been observed that the positive association of male literacy with male RNAE tended to increase further with increasing levels of education[3]. This clearly implies that the male workers with higher educational attainments had a greater access to RNAE, as they possessed relevant information, knowledge and skills, including the ability to manage a business and adapt to changing demand patterns. As they were highly educated, they sought better employment outside agriculture. That is why there was a strong negative correlation between the proportion of male workers in agriculture (RAE) and the male unemployment rates, particularly among the educated male youth.

The residual sector hypothesis, formulated by Vaidyanathan (1986), has been widely discussed. Vaidyanathan, considering the person-day unemployment rate (CDS) as an index of distress in agriculture, found a strong positive association between the person-day male unemployment rate and the proportion of male workers in RNAE. This phenomenon was termed as "distress diversification"—a diversification into unproductive, lowly paid, non-agricultural jobs. The distress-induced RNAE was termed as the "residual sector hypothesis" by Vaidyanathan (1986).

In the present analysis, however, it was also found that a strong positive association existed between the proportion of male workers in RNAE and the male unemployment rate according to UPS only. The general and educated unemployment rates according to CDS (person-day unemployment rate) were not significantly, but yet positively, associated with male RNAE. The positive association between male RNAE and male unemployment rates, according to CDS or UPS or both, was apparently termed as distress diversification of male workers from agriculture into the rural non-agricultural sector for employment. This has been interpreted in the literature, in most cases, that due to lack of employment opportunities in agriculture, a large chunk of labourers were being pushed into the rural non-agricultural sector and the activities in which such workers involved were considered as low productive, casual/contractual, dead-end and lowly paid. If we unearth the reasons why the unemployment rates tended to be higher in those areas with higher proportion of male workers in RNAE, the growth of male RNAE could not necessarily be regarded as distress diversification resulting from lack of growth in agriculture. In agriculturally developing areas, as the manageability of remaining unemployed by the youth and educated persons from the better off households happened to be more, the incidence of unemployment was apparently found to be quite significant. As most of the unemployed workers are educated youth, they do not wish to work in agriculture; instead they prefer to wait for some time and try to secure better jobs outside agriculture. In these areas, the correlation between the growth of RNAE and unemployment rate is expected to be positive and stronger. An important question is whether it may be considered as distress diversification. If all the unemployed persons are to be illiterate or literate at lower levels of education and positively and significantly associated with male RNAE, then the hypothesis that the growth of male RNAE is distress diversification would be credible and so easily acceptable. Because, most of the persons involved in agriculture are illiterate and literate at lower levels of education, whereas most of the unemployed persons are more or less highly educated, waiting for better job options. The general unemployment rate, therefore, cannot be an appropriate indicator measuring distress in agriculture *per se*.

A strong positive association between the male unemployment rate (UPS) and the rate of male literacy (see, correlation matrix Table 5.2) shows that the educated persons form a major segment of the general unemployment pool. A rapid increase in the rate of literacy is due to, *inter alia,* rapid improvement in the agricultural income of rural households. A strong negative association of male literacy and unemployment rate with the proportion of male workers in agriculture

(RAE) shows that these labourers were educated and unwilling to work in agriculture, but it did not necessarily mean that the employment opportunities were limited and hence the workers were being pushed into the residual sector of rural non-agriculture. This phenomenon cannot necessarily be regarded as distress diversification due to lack of growth in agriculture. Youth and educated unemployment rates in rural areas, however, can be a better index to capture the poor performance of the economy, in general, and rural economy, in particular.

If the employment opportunities are very bleak, why do the daily real wage rates in agricultural occupations continue to rise? The distress in agriculture, as a matter of fact, will be better reflected through the daily real wage earnings in agricultural occupations. If the growth of RNAE is distress diversification resulting from no growth or stagnation in agriculture, we expect a strong negative association between male RNAE and the daily real wage earnings of male workers in agriculture. As observed from the foregoing analysis, there was a strong positive association between the two.

The data (Table 5.3) on the daily real wage earnings which prevailed in agricultural and non-agricultural occupations do not lend any support to the residual sector hypothesis. They reveal that the daily real wage earnings of male agricultural labourers in agricultural and non-agricultural occupations were not only increasing but also quite higher in the latter than in the former. If the growth of RNAE were to be a residual sector, the daily real wage earnings in agriculture should have declined and in non-agriculture should have been not only less than the agricultural wage rates but also declined due to influx of the labour force into the rural non-agricultural sector. An increase in the daily real wage earnings in agriculture indicates an excess demand for labour over its supply and, therefore, the question of distress movement does not arise. In view of this, Vaidyanathan (1994) has revised his hypothesis saying that the residual sector hypothesis would stand discredited if the daily real wage earnings increased in agricultural and non-agricultural occupations.

The growth of male RNAE, however, might be distress induced but not necessarily be a residual sector, meaning low productive, casual/

contractual, dead-end, lowly paid jobs. This is due to the fact that the proportion of male workers in self- and regular RNAE, as an index of quality of RNAE (QRNAE), was significantly and positively associated with male RNAE (see, correlation matrix, Table 5.2). This shows that the growth of male RNAE as a whole was not casual/contractual. Further, a strong negative association existed between the proportion of male workers in RNAE and the incidence of rural poverty, implying that the activities in which the male workers involved were not low productive, lowly paid dead-end jobs. It also implies that the reduction of poverty rather stimulated the growth of male RNAE through an increase in the demand of the poor for non-farm goods and services mostly produced in rural areas.

Table 5.3 : Daily Real Wage Earnings of Agricultural Labourers in Rural India, 1974-75 to 1999-2000 (Per day/Rs.)

	Agriculture		*Non-agriculture*	
Year	*Males*	*Females*	*Males*	*Females*
1974-75	3.60	2.52	3.63	2.36
1977-78	4.74	3.36	5.99	3.36
1983	4.03	3.04	4.38	2.54
1987-88	5.96	4.40	6.43	5.01
1993-94	7.62	5.42	9.63	5.79
1999-2000	12.90	9.20	17.51	11.01

Note: The money wage rates have been deflated by Consumer Price Index for Agricultural Labourers (CPIAL).
Source: GOI (1998b, 2003b).

An increase in the incidence of rural poverty implies a decline in the demand for non-agricultural goods and services resulting in a limited growth of employment opportunities in the rural non-agricultural sector. If the growth of male RNAE were to be low productive (residual sector), a strong positive association was expected between male RNAE and poverty. It can be seen that high productive, better-paid male RNAE was higher likely to be associated with rapid agricultural growth and higher educational attainments. The results, thus, show that the growth of male RNAE as a whole was neither distress diversification nor low productive, lowly paid, dead-end jobs. This observation requires to be verified with primary data at different levels of agricultural prosperity.

A positive but weak correlation existed between the proportion of male workers in RNAE and the composite index for rural infrastructure (such as rural roads, post-office, telecommunication facilities and banks). This implies that availability of rural infrastructure positively contributed to the growth of male RNAE. But the rural infrastructure has yet to play its fuller role in the promotion of RNAE as played in many developing countries (Oshima 1994; Lane 1996). The expansion of rural roads, transportation and communication networks encourage specialisation and division of labour by the rural households. Infrastructure facilitates exchange of agricultural inputs and products between villages and small rural towns. Expansion of trade and commerce, marketing and distributional network, establishing a linkage between farm and non-farm sectors, largely depends on the availability of good transportation and communication facilities. Infrastructure not only widens the choice of markets and inputs for rural non-farm enterprises, but also facilitates sub-contract arrangements between urban-based enterprises and those in rural areas.

A strong positive association between manufacturing activities in rural and urban areas seemed to have been established due to adequate transport and communication networks between rural and urban areas (see, Table 5.5). It may also be noted that it would affect the tastes and preferences of the rural households and orient them toward urban products. For instance, the provision of electricity in a village creates a demand for urban products such as household electrical goods, including radios, televisions and refrigerators. An increase in the income levels of the non-poor households seemed to have raised the demand for such consumer goods mostly produced in urban or semi-urban areas, which lead to development of trade and commerce between rural and urban areas. This could be one of the reasons why CINFRA was not significantly and positively associated with male RNAE as a whole, but with urban manufacturing activities. It is, thus, evident that the growth of male RNAE was determined by a set of multiple interrelated factors, the process of whose working was found to be more dynamic and complex.

An attempt has also been made to estimate the elasticity of the important determinants of male RNAE. Since many of the explanatory variables are highly correlated among themselves, a log-linear multiple regression model has been fitted, selecting the important variables which determine the growth of male RNAE.

$$\text{Log MRNAE} = \alpha + \beta_1 \log \text{PHIA} + \beta_2 \log \text{CIMA} + \beta_3 \log \text{RML} - \beta_4 \log \text{PVTY} + u \ldots (1)$$

where MRNAE is the proportion of male workers in RNAE, PHIA is the per hectare net income from agriculture, CIMA is the composite index for mechanisation of agriculture, RML is the rate of male literacy and PVTY is the proportion of people living below poverty line, α refers to the constant, β_1 to β_4 are the slope coefficients of the parameters and u is the error term. The results of the model turned out to be as follows:

$$\underset{}{\text{Log MRNAE}} = \underset{(-1.108)}{-1.163} + \underset{(2.954)}{0.452 \log \text{PHIA}^{***}} + \underset{(1.536)}{0.229 \log \text{CIMA}} + \underset{(1.796)}{0.32 \log \text{RML}^{*}} - \underset{(-2.327)}{0.325 \log \text{PVTY}^{**}}$$

$R^2 = 0.840^{***}$ Adj. $R^2 = 0.782^{***}$more $F = 14.43$

Note: Figures in brackets indicate the 't' values and ***, ** and * refer to 1, 5 and 10 percent level of significance, respectively.

The F value shows that the model was statistically significant at 1 per cent level, explaining about 78 percent variation in the proportion of male workers in RNAE. The elasticity of PHIA indicates that one percent increase in per hectare net income from agriculture led to 0.45 percent increase in the proportion of male workers in RNAE. The elasticity of CIMA shows that, although it was not statistically significant, one percent increase in agricultural mechanisation resulted in an increase of 0.23 percent male RNAE, whereas in the case of RML, it contributed to an increase of 0.32 percent. Similarly, the elasticity of PVTY indicates that one per cent decline in the proportion of people living below the poverty line contributed to an increase of 0.33 percent male RNAE, mostly raising the demand for non-farm goods and services. Among the selected correlates, the growth of agriculture, mostly through production and consumption linkages, decline in poverty and increase in education were found to be the key factors in the promotion of male RNAE.

In respect of females, it was observed that the growth process of RNAE appeared to be different and more complex. The results presented in Table 5.4 indicate that the explanatory variables such as PHIA, COMP, CDSUR, CDSURY, USUR, USURE, PSMH, GRLH were found to be important factors determining the level of female participation in RNAE. The results show that there was a strong positive correlation between the proportion of female workers in RNAE and PHIA, implying that the growth of agriculture stimulated the growth of female RNAE mostly through consumption linkages, as also the source of capital in the rural non-agricultural sector. The positive correlation was stronger in respect of PHIA as compared to that of PCIA. This may be taken to imply that increased per capita income from agriculture does not force women to work outside agriculture. A strong negative correlation existed between the proportion of females in RNAE and the Gini ratio for operational landholdings. This shows that a greater access to land by the land-poor households seemed to have stimulated the growth of RNAE mostly through consumption linkages and as also a source of capital to invest in petty shops, hotel business, tea stalls, vegetable vending, pan shops and such other micro enterprises, in which female labourers with very little knowledge were more likely to be involved. However, a negative but weak correlation between female RNAE and rural per capita non-food consumption expenditure shows that the overall consumption linkages were not strong enough to promote female RNAE.

In the case of production linkages, it has been found that the direct impact of backward linkages (PHFC and CIMA) on the growth of female RNAE was not significant, but the indirect/derived impact of backward linkages was relatively more operative in terms of creating jobs in trade, commerce, hotels and restaurants. In the case of forward linkages, there was a positive but weak association between the proportion of female workers in RNAE and the proportion of area under non-foodgrains cultivation. If we define the concept of commercialisation of agriculture by including paddy, which is also one of the important commercial and largely traded crops, the positive association existing between them tends to increase significantly. This implies that the growth of agriculture through its forward linkages contributed significantly to the growth of female RNAE, mostly in the agro-processing sector.

The forward production linkages were found to have a greater impact on the growth of female RNAE, as compared to backward production and consumption linkages. But the production (both backward and forward) linkages were found to have a greater impact on the growth of male RNAE than on female RNAE. Thus, this indicates that the backward production linkages were more effective in the promotion of male RNAE, whereas the forward production linkages were found to be more effective in the promotion of female RNAE. A strong positive association between the rate of female literacy at higher secondary level (RFLhs) and the proportion of females in RNAE shows that the provision of education to females seemed to have removed the inflexible socio-economic barriers and encouraged them to take up some jobs in the rural non-agricultural sector. In other words, illiterate females and literate ones at lower levels of education seemed to have encountered several socio-economic constraints in taking up non-farm jobs in the countryside.

A negative but statistically insignificant association existed between the proportion of female workers in RNAE and the daily real wage earnings in agricultural occupations. This shows that there was a distress situation in agriculture because of which a larger proportion of labourers, unable to get adequate employment and wages in agriculture, were forced to seek their subsistence in the rural non-agricultural sector. A negative but insignificant correlation between the proportion of female workers in RNAE and the daily real wage earnings in the rural non-agricultural occupations indicates that the participation of females tended to depress the non-agricultural wage rates. A negative but weak association of daily real wage rates of female workers in agricultural and non-agricultural occupations with the proportion of female workers in RNAE brings out the fact that the growth of female RNAE was a manifestation of distress diversification.

For validation of the distress diversification hypothesis, we expect

Table 5.4 : Correlation Matrix for Female RNAE in India, 1999-2000

		1	2	3	4	5	6	7	8	9	10	11	12	13	14	15
1	RNAE	1														
2	RAE	-0.926**	1													
3	PHIA	0.514*	-0.605*	1												
4	PCIA	0.297	-0.396	0.584*	1											
5	GRLH	-0.543*	0.674**	-0.149	0.051	1										
6	NFCE	-0.306	0.129	0.357	0.668**	252	1									
7	GAI	-0.153	0.280	0.009	0.088	0.380	0.173	1								
8	PHFC	0.052	-0.073	0.428	0.333	0.122	0.357	0.747**	1							
9	COM	0.395	-0.285	-0.026	0.086	-0.255	-0.140	-0.127	-0.080	1						
10	COMP	0.786**	-0.625**	0.213	0.124	-0.329	-0.479*	-0.530*	-0.023	0.723**	1					
11	CIMA	0.073	-0.132	-0.074	0.083	-0.318	0.069	-0.532*	-0.579*	-0.190	-0.243	1				
12	WRLA	-0.056	-0.117	0.579*	0.673**	0.233	0.747**	-0.169	0.26	-0.289	-0.264	-0.041	1			
13	WRLN	-0.068	-0.072	0.388	0.441	-0.048	0.627**	-0.044	0.077	-0.259	-0.411	0.211	0.587*	1		
14	PVTY	0.322	-0.182	-0.377	-0.587*	-0.310	-0.898**	-0.232	-0.394	0.145	0.434	0.030	-0.697**	-0.587*	1	
15	PLLH	0.384	-0.271	0.357	0.320	0.115	-0.014	0.388	-0.605*	0.390	0.550*	-0.625**	-0.071	-0.294	-0.050	1
16	PSMH	0.574*	-0.538*	0.551*	-0.071	-0.315	-0.273	-0.228	0.089	0.060	0.348	0.053	-0.122	-0.208	0.355	0.136
17	ASMH	-0.609*	0.609*	-0.599*	-0.268	0.321	-0.154	0.220	0.018	-0.366	-0.390	-0.145	-0.310	-0.185	0.008	0.037
18	LMR	-0.209	0.154	-0.114	0.657**	0.369	0.563*	0.291	0.192	0.030	-0.082	-0.149	0.476*	0.238	-0.462	0.156
19	RF L	0.284	-0.391	0.739**	0.463	-0.003	.390	-0.310	0.069	0.117	0.185	-0.127	0.533*	0.279	-0.275	0.250
20	RFLhs	0.347	-0.102	0.612*	0.174	0.130	0.306	0.000	0.207	0.160	0.130	-0.413	0.454	0.263	-0.251	0.204
21	CDSUR	0.686**	-0.767**	0.555*	0.064	-0.532*	-0.207	-0.631*	-0.202	0.251	0.474	0.181	-0.037	-0.034	0.204	0.140
22	CDSURY	0.827**	-0.843**	0.502*	0.239	-0.512*	-0.260	-0.531*	-0.179	0.498*	0.747**	0.054	-0.055	-0.119	0.269	0.377
23	UPSUR	0.558*	-0.608*	0.722**	0.570*	-0.130	0.197	-0.324	-0.054	-0.028	0.347	0.153	0.501*	0.183	-0.104	0.186
24	UPSURY	0.510*	-0.563*	0.734**	0.536*	-0.084	0.213	-0.321	-0.038	-0.036	0.319	0.081	0.514*	0.156	-0.126	0.206
25	UPSURE	0.642**	-0.592*	0.418	0.251	-0.087	-0.160	-0.264	-0.108	-0.130	0.352	0.155	0.118	-0.111	0.232	0.141
26	QRNAE	0.576*	-0.505*	0.582*	0.664**	-0.136	0.169	0.410	0.389	0.142	0.469	-0.113	0.467	0.272	-0.197	0.0.338
27	CINFRA	-0.265	0.115	0.197	0.451	0.157	0.814**	-0.173	0.099	0.072	-0.363	0.168	0.375	0.506*	-0.792**	-0.035
28	DOU	0.017	-0.054	-0.009	0.394	0.212	0.352	0.242	0.482*	0.318	0.062	-0.157	-0.100	0.018	-0.290	0.595*
29	UAGLM	0.181	-0.162	-0.078	0.145	-0.309	0.105	0.484*	0.597*	0.386	0.215	-0.307	-0.255	-0.031	-0.025	0.518*

		16	17	18	19	20	21	22	23	24	25	26	27	28	29
1	RNAE														
2	RAE														
3	PHIA														
4	PCIA														
5	GRLH														
6	NFCE														
7	GAI														
8	PHFC														
9	COM														
10	COMP														
11	CIMA														
12	WRLA														
13	WRLN														
14	PVTY														
15	PLLH														
16	PSMH	1													
17	ASMH	-0.651**	1												
18	LMR	-0.631**	0.191	1											
19	RF L	0.499*	-0.591*	-0.001	1										
20	RFLhs	0.436	-0.589*	-0.090	0.817**	1									
21	CDSUR	0.694**	-0.584*	-0.517*	0.595*	0.299	1								
22	CDSURY	0.507*	-0.537*	-0.246	0.530*	0.229	0.890**	1							
23	UPSUR	0.452	-0.579*	0.110	0.765**	0.485*	0.603*	0.647**	1						
24	UPSURY	0.492*	-0.583*	0.080	0.829**	0.572*	0.608*	0.623**	0.986**	1					
25	UPSURE	0.523*	-0.474	-0.018	0.356	0.135	0.516*	0.546*	0.739**	0.728**	1				
26	QRNAE	0.114	-0.372	0.337	0.217	0.227	0.021	0.244	0.477*	0.431	0.478	1			
27	CINFRA	-0.176	-0.141	0.340	0.375	0.227	0.050	-0.074	0.105	0.136	-0.140	0.125	1		
28	DOU	-0.307	0.260	0.393	-0.093	-0.275	-0.129	0.052	-0.189	-0.232	-0.258	-0.001	0.375	1	
29	UAGLM	-0.162	0.097	0.229	-0.293	-0.208	-0.219	-0.029	-0.344	-0.382	-0.272	0.199	0.046	0.756**	1

Note: ** and * refer to 1 and 5 percent level of significance, respectively.

the daily real wage earnings to be constant or a significant decline in agricultural as well as in non-agricultural occupations. But the data provided in Table 5.2 show that the daily real wage rates of female workers in agricultural and non-agricultural occupations have registered an increasing trend over the years. But it did not increase as much as in the case of male lobourers in both the sectors. A gradual rise in the daily real wage rates in agricultural occupations might be due to a marginal shift of female labourers in favour of the rural non-agricultural sector, resulting from distress-induced factors.

The land-man ratio can be used to measure population pressure on agriculture and may be considered as one of the indicators reflecting under-employment or lack of employment in agriculture and hence a distress situation. A negative association between the proportion of females in RNAE and land-man ratio also shows that, although there was less diversification of females in favour of RNAE, it was presumably the population pressure that pushed out some proportion of the female workers into the rural non-agricultural sector due to lack of employment opportunities in agriculture. A positive but weak association of the proportion of landless households and a strong positive association of the proportion of small and marginal farmers with the proportion of females in RNAE imply that the growth of female employment was distress diversification. This indicates that female workers belonging to the small, marginal and landless households were unable to get employment from the agricultural sector and hence they were pushed into the rural non-agricultural sector. Similarly, a strong positive association between the average size of small and marginal landholdings and female RNAE further indicates that, being unable to get adequate employment and incomes from tiny landholdings, a larger share of females seemed to have been forced to take up jobs in the rural non-agricultural sector. The nature of correlation between female RNAE and the daily real wage rates for females in agricultural and non-agricultural occupations, the proportion of landless households, the proportion of small and marginal landholdings, the average size of small and marginal landholdings and the land-man ratio, thus, show that the growth of female employment in the rural non-agricultural sector as a whole was due to distress in agriculture and thereby lending support to the distress diversification hypothesis. Such distress diversification seemed to be quite evident in the case of resource-poor households.

The youth and educated unemployment rates for females were significantly and positively associated with female RNAE, as compared to those for general unemployment rates for females. The youth and educated unemployment rates for females were also highly correlated with the rate of female literacy, which, further, was significantly and positively associated with per hectare net income from agriculture. This indicates that the need for non-agricultural employment was very essential and significant among the youth and educated females. Such growth of female RNAE, accompanied by unemployment among the youth and educated females, however, does not explain that distress in agriculture forced the females to seek employment in the rural non-agricultural sector. This explains the fact that female youth with higher educational attainments were not really willing to work in agriculture (a strong negative association existed between female employment in agriculture and female unemployment rate).

Unlike in the case of male workers, a positive but weak association between female RNAE and the rate of female literacy, however, indicates that illiterate females and those literate at lower levels, mostly from poor households, were more likely to seek jobs in the rural non-agricultural sector which were most often casual, low productive, lowly paid, dead-end jobs. Though the growth of female RNAE was found to be distress induced, the general and youth unemployment rates as per CDS, and the general and educated unemployment rates as per UPS, may not explain the distress diversification phenomenon, resulting from agricultural distress and, of course, it might be distress in general not necessarily originating from the agricultural sector *per se*.

In order to analyse the productive dimension of female RNAE, indicators such as incidence of poverty and proportion of females in self and regular RNAE were considered. It has been observed that the proportion of people living below the poverty line was positively associated with the proportion of females in RNAE, but statistically not significant. This follows that most of the activities in which they involved were not productive and highly paid. It goes without saying that when the female labourers are unable to get sufficient work in agriculture they would be forced to seek dead-end lowly paid jobs in the rural non-agricultural sector with a view to coping with the distress. In such cased, the daily real wage earnings of female workers in agricultural and non-agricultural occupations were negatively associated with the proportion of females in RNAE and the impact of the same on poverty was not on expected lines.

The above observation, however, was not supported by other indicators, which were also used to measure the productive or qualitative dimension of female RNAE. A strong positive association between the proportion of female workers in total RNAE, and the proportion of female workers only in self- and regular RNAE, however, indicates that some part of female RNAE appeared to be not casual/ contractual but less remunerative. This may be due to the fact that in the rural non-agricultural sector, a larger share of female workers began to withdraw from casual wage employment and shift to self-employment (see, Degree of Casualisation in RNAE in Chapter 4). As a result of such shift to self-employment, the income levels might have decreased and hence no greater impact might have been seen on poverty reduction. It is, thus, obvious that although the growth of female RNAE was determined by the growth as well as poverty related factors, a major part of the growth was largely due to poverty or distress-induced factors, but it was not so in the case of male RNAE. This observation requires to be tested with the primary data in regions with different levels of agricultural prosperity and availability of rural infrastructure.

The composite index for rural infrastructure was negatively associated with the proportion of female workers in RNAE, but it was statistically not significant. This shows that the provision of infrastructure facilities such as rural roads, post-offices, telecommunication and banks did not seem to have contributed positively to the growth of female RNAE. The degree of urbanisation and the growth of urban agglomeration were not significantly but positively associated with the proportion of female workers in RNAE.

Since most of the explanatory variables were highly correlated among themselves (multicollinearity), the important variables, which explained the greater variation in the case of female RNAE, were included in the model. The following log-linear multiple regression model has been specified for female RNAE.

$$\text{Log FRNAE} = \alpha + \beta_1 \log \text{COMP} - \beta_2 \log \text{ASMH} + \beta_3 \log \text{CDSURYF} + u \ldots. \quad (2)$$

where FRNAE is the proportion of female workers in RNAE, COMP is the proportion of area under non-foodgrains plus paddy (commercialisation of agriculture), ASMH is the average size of small and marginal landholdings, CDSURYF is the current daily status unemployment rate for youth females, α refers to the constant, β_1 to β_3 are the slope coefficients of the parameters and u is the error term. The results of the model are as follows:

$$\text{Log FRNAE} = \underset{(-0.599)}{-0.294} + \underset{(1.932)}{0.386 \log \text{COMP}^*} - \underset{(1.509)}{0.238 \log \text{ASMH}} + \underset{(1.883)}{0.411 \log \text{CDSURYF}^*}$$

$R^2 = 0.785^{***}$ Adj. $R^2 = 0.735^{***}$ F=14.878

Note: Figures in brackets indicate the 't' values and *** & * refer to 1 & 10 percent level of significance, respectively.

The F value of equation (2) shows that the estimated relationship was statistically significant at 1 percent level, explaining about 74 percent variation in the proportion of female workers in RNAE. The elasticity of COMP indicates that one percent increase in the proportion of area under non-foodgrains plus paddy resulted in an increase of 0.39 percent in the proportion of female RNAE mostly through forward production linkage. On the other hand, one percent decline in the average size of small and marginal landholdings led to an increase of 0.24 percent female RNAE (distress diversification from agriculture). One percent rise in the current daily status unemployment rate for female youth seemed to have resulted in an increase of 0.411 percent female RNAE, indicating the distress movement of female workers into the rural non-agricultural sector but the distress was not necessarily originating from agriculture.

It is evident from the foregoing analysis that the growth process of male and female RNAE as a whole was largely determined by several factors originating mainly from the rural economy (Chart 5.1). The

Chart: 5.1 : Determinants of Rural Non-Agricultural Employment in India

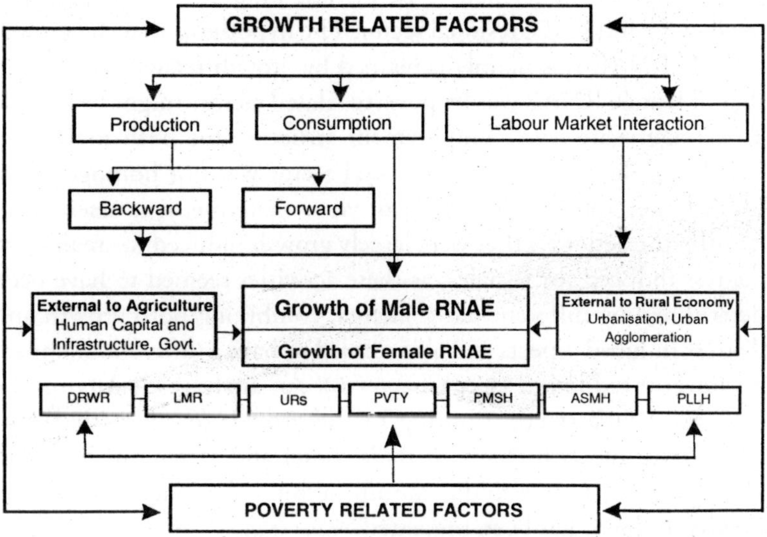

Note: DRWR: Daily real wage rate; LMR: Land-man ratio; URs: CDS/UPS unemployment rates; PVTY; Proportion of people below poverty line; PMSH: Proportion of small and marginal holdings; ASMH: Average size of small and marginal holdings; PLLH: Proportion of landless households to total households.

factors that contributed to determine the level of male participation in RNAE were not strictly the same factors that contributed to determine the level of female participation in RNAE. For the growth of male RNAE, the consumption linkages were more operative than production linkages. Further, backward production linkages were found to be stronger than forward production linkages. For the growth of female RNAE, the production linkages were little stronger than the consumption linkages and further forward production linkages were more effective than backward production linkages.

The growth of male RNAE was largely determined by growth related factors (growth induced), whereas the growth of female RNAE was mainly determined by distress/poverty-compelled factors (distress induced). The growth of RNAE for males was found to be more productive, better paid and not in the residual sector, whereas in the case of females, it seemed to be distress induced, low productive, lowly paid and in the residual sector. The hypothesis that the growth of male RNAE was mainly explained by growth related factors and that of female RNAE was by poverty related factors might be tenable at the aggregate level. Employment structure for both males and females in the rural non-agricultural sector was not homogeneous. From policy intervention point of view, therefore, it is essential to identify the activities that were largely growth induced for males and distress induced for females, as these activities seemed to have been determined by different set of factors, combining both growth and distress induced aspects and also seemed to have a differential impact on poverty reduction (see, Chapter 6). An analysis identifying the important factors contributing to the growth of sub-activities— manufacturing; construction; trade, hotel and restaurant business; and transport, storage and communication— for males and female, in what follows, has been presented.

Determinants of RNAE by Activity

Manufacturing

Manufacturing, particularly non-household, in the rural non-agricultural sector can generate a good deal of sustained employment opportunities for both males and females through consumption and production (backward and forward) linkages. The results presented in Table 5.5 show that, among the selected explanatory variables, variables such as PHFC, UAGLM, DoU, and to some extent PHIA, NFCE, WRLA, WRLN, RoL for males CDSURY and PVTY were found to be important correlates of male employment in the manufacturing sector.

This indicates that higher per hectare/capita net income from agriculture provided market support (consumption linkages) and/or became an important source of capital to invest in the manufacturing sector, mostly for fertilizer production-based manufacturing units, cottage and small-scale and agro-based industries. Similarly, the rural

Table 5.5 : Correlation Coefficients between Male and Female Employment in Manufacturing Sector and Selected Variables in Rural India, 1999-2000

Name of the explanatory variable	*Males*	*Females*
PHIA	0.473	0.365
PCIA	0.333	0.094
NFCE	0.475	-0.472
GRLH	-0.361	-0.593*
WRLA	0.440	-0.253
WRLN	0.442	-0.175
PHFC	0.630**	0.001
GAI	0.360	-0.163
CIMA	-0.128	0.051
COM	0.077	0.391
COMP	0.011	0.777**
RoL	0.393	0.145
RoLhs	0.555*	0.345
CDSUR	0.439	0.644**
CDSURY	0.417	0.772**
USPUR	'0.155	0.407
USPURY	0.087	0.360
USPURE	-0.166	0.611*
LMR	-0.033	-0.225
PLLH	0.389	0.330
PSMH	0.150	0.592*
ASMH	-0.209	-0.542*
PVTY	-0.368	0.518*
QRNAE	0.300	0.325
UAGLM	0.615*	0.245
DoU	0.510*	-0.007
CINFRA	0.295	-0.390

Note: ** and * refer to 1 and 5 percent level of significance, respectively.

per capita non-food consumption expenditure and the daily real wage rates of male labourers in agricultural and non-agricultural occupations were positively, and the Gini ratio for operational landholdings was negatively, associated with the proportion of males in manufacturing but none of them was statistically significant. This implies that there had been a limited market support for the products manufactured in rural areas (weak consumption linkages). This might be due to the fact that the local demand for manufactured goods was mostly oriented towards urban areas. A negative correlation between the incidence of poverty and the proportion of male workers in the manufacturing sector seemed to indicate that it was the lack of demand of the poor households for manufactured products that did not encourage the growth of male employment in the manufacturing sector in rural areas. The reduction of poverty, therefore, might stimulate the growth of male employment in the manufacturing sector through the expansion of demand for goods manufactured locally or in nearby villages/cities.

As far as production linkages were concerned, the degree of commercialisation of agriculture was not an important factor (weak forward production linkages), but the per hectare fertilizer consumption was found to be an important factor in the growth of male employment in the manufacturing sector (strong backward production linkages). It appears, however, that both consumption and production linkages, particularly forward linkages, were not significantly strong in the promotion of male employment in the manufacturing sector. But the production linkages seemed to be more effective than consumption linkages. The consumption linkages of rural origin appeared to be positively and strongly associated with the manufacturing sector in urban areas. An increase in the income levels of the rural households, particularly the resource-rich households, seemed to have raised the demand for urban based products, thereby creating direct employment in the urban manufacturing sector and indirect employment in trade, commerce and business in rural areas.

The rate of male literacy, particularly at higher levels, was positively and significantly associated with male employment in the manufacturing sector. In fact, the correlation between the rate of male literacy and male employment in the manufacturing sector tends to increase with increasing levels of educational attainments, implying that the male workers with higher education (mostly ITI and other technical degree holders) were more likely to be employed in the manufacturing sector, reducing the unemployment rates for educated males. That is why the unemployment rate according to UPS for educated males was negatively associated with male employment in the manufacturing sector.

The unemployment rates as per CDS and UPS, the youth unemployment rate as per UPS and the proportion of landless households, however, were positively and significantly associated with the proportion of male workers in manufacturing activities. This implies that some part of the growth in manufacturing employment was also distress induced, but not necessarily residual in nature as the incidence of poverty was negatively and the quality of RNAE positively and significantly, associated with male employment in the manufacturing sector. This distress diversification was not necessarily from distress originating in agriculture. Because, the daily real wage rate of male agricultural labourers was positively associated with male employment in the manufacturing sector.

A striking feature is that a strong positive association existed between male employment in manufacturing and urban agglomeration. This shows that the growth of the manufacturing sector in urban areas mostly through sub-contracting arrangements with those established in rural areas, contributed positively to the growth of male employment in manufacturing in rural areas. Similarly, a strong positive association existed between male employment in manufacturing sector and the degree of urbanisation. A strong positive association of urbanisation and urban agglomeration with male employment in the manufacturing sector also shows that the growth of manufacturing employment for males was largely determined by factors external to the rural economy rather than the local consumption and production linkages. The composite index for the rural infrastructure was also found to be positively associated with the proportion of males in manufacturing pursuits, but not statistically significant. The following log-linear multiple regression model has been specified for male employment in the manufacturing sector.

$$\text{Log MMF} = \alpha + \beta_1 \log \text{PHCF} + \beta_2 \log \text{RMLhs} + \beta_3 \log \text{UAGLM} + u \quad ....(3)$$

where MMF is the proportion of male workers in the manufacturing sector, PHCF is the per hectare fertilizer consumption, RMLhs is the rate of male literacy at higher secondary level, UAGLM is the proportion of workers in the urban manufacturing sector used to capture the growth of urban agglomeration, α refers to the constant, β_1 to β_3 are the slope coefficients of the parameters and *u* is the error term. The results of the model were as presented below:

$$\text{Log MMF} = -\underset{(-2.642)}{2.940} + \underset{(2.357)}{0.405} \log \text{PHCF}^{**} + \underset{(3.884)}{0.548} \log \text{RMLhs}^{***} + \underset{(2.811)}{0.491} \log \text{UAGLM}^{***}$$

$R^2 = 0.772^{***}$ Adj. $R^2 = 0.715^{***}$ $F = 13.559$

Note: Figures in parentheses indicate the 't' values and ***, ** and * refer to 1, 5 and 10 percent level of significance, respectively.

The F value of the equation (3) indicates that the estimated relationship of the model is statistically significant at 1 percent level, explaining about 72 percent variation in the proportion of male workers in manufacturing employment. The elasticity of PHFC indicates that one percent rise in consumption of chemical fertilizer per hectare resulted in an increase of 0.41 percent in male employment in the manufacturing sector. The elasticity of RMLhs also shows that one percent increase in the rate of male literacy led to 0.55 percent increase in manufacturing employment for males. Similarly, the elasticity of UAGLM shows that one percent increase in urban manufacturing employment led to 0.49 percent increase in rural manufacturing employment for males.

In respect of females, the variables such as GRLH, COMP, PSMH, ASMH, PVTY, CDSUR, CDSURY, USPURE, and to some extent PHIA were found to be important determinants. It has been found that per hectare and per capita net income from agriculture were positively associated with the proportion of females in manufacturing employment, but it was statistically not significant. It implies that there was a weak consumption linkage effect on female employment in the manufacturing sector. A strong negative association between female employment in the manufacturing sector and the Gini ratio for operational landholdings shows that there was a strong consumption linkage from the poor households. A greater access to land by the land-poor households may raise the income levels from agriculture and so also consumption levels, contributing positively to the growth of employment in the manufacturing sector.

In respect of production linkages, there had been a weak positive association between the share of female workers in manufacturing employment and per hectare fertilizer consumption and composite index for agricultural mechanisation, but it was statistically not significant. This shows that there had been a weak backward production-linkage effect of agricultural growth on female employment in the manufacturing sector. However, a strong positive association existed between female employment in manufacturing and the proportion of area under non-foodgrains plus paddy cultivation to total cropped area. This indicates that there had been a strong forward production linkage from agriculture to female employment in the manufacturing sector, mostly in agro-processing.

A negative but weak correlation between female employment in manufacturing and the daily real wage rates in agricultural and rural non-agricultural occupations, however, reveals that some growth of female employment mostly in non-agro-processing appeared to be distress diversification and residual in nature. A strong positive association between female employment in manufacturing and the proportion of small and marginal holdings, and a negative association with the average size of small and marginal landholdings, further support the fact that the participation of female workers in the manufacturing sector was distress induced, resulting from agriculture. Unable to get gainful employment from their tiny landholdings throughout the year, a large chunk of small and marginal farmers seemed to have been forced to seek employment in the rural non-agricultural sector.

The unemployment rates for females, especially for youth and educated females, were significantly and positively associated with female employment in the manufacturing sector. This illustrates that some part of female employment growth in manufacturing was induced by distress, resulting not necessarily from agriculture, among the youth and educated females. A strong positive association between the proportion of females in the manufacturing sector and the proportion of people living below the poverty line also reveals that the growth of female employment in manufacturing was little productive/remunerative (residual sector hypothesis).

Since the rate of female literacy was not significantly associated with female employment in the manufacturing sector, a larger share of illiterate females appeared to have engaged in low cadre works on daily basis. Their participation in manufacturing activities tended to increase with increasing levels of education among them. The composite index for rural infrastructure was negatively associated with the proportion of females in manufacturing employment, but it was statistically insignificant. This shows that availability of rural infrastructure did not contribute positively to the growth of female employment in the manufacturing sector. The following model was specified for female employment in the manufacturing sector.

$$\text{Log FMF} = \alpha + \beta_1 \log \text{COMP} + \beta_2 \log \text{UPSURE} + \beta_3 \log \text{PSMH} + u \text{ ....(4)}$$

where FMF is the proportion of female workers in manufacturing employment, COMP is the proportion of area under non-foodgrains plus paddy, UPSURE is the unemployment rate UPS for educated females, PSMH is the proportion of small and marginal landholdings to total landholdings, α refers to the constant, β_1 to β_3 are the slope coefficients of the parameters and *u* is the error term. The results of the model turned out to be:

$$\text{Log F MF} = \underset{(-3.337)}{-3.586} + \underset{(3.950)}{0.594 \log \text{COMP}^{***}} + \underset{(1.760)}{0.276 \log \text{UPSURE}^{*}} + \underset{(1.770)}{0.242 \log \text{PSMH}^{*}}$$

$R^2 = 0.772^{***}$ Adj. $R^2 = 0.715^{***}$ F=13.545

Note: Figures in parentheses indicate the 't' values and ***, ** and * refer to 1, 5 and 10 percent level of significance, respectively.

In respect of females, the estimated relationship of the variables included in the model was statistically significant at 1 percent level. It has been found that the proportion of area under non-foodgrains plus paddy, unemployment rate (UPS) for educated females and proportion of small and marginal landholdings explained about 72 percent variation in female employment in the manufacturing sector.

The elasticity of COMP shows that one percent rise in the proportion of area under non-foodgrains plus paddy led to an increase of 0.60 percent female employment in the manufacturing sector (strong forward-production linkage). Similarly, one percent increase in the educated unemployment rate for females as per UPS was estimated to result in an increase of 0.28 percent in manufacturing employment for females. The elasticity of PSMH indicates that one percent increase in the proportion of small and marginal landholdings caused an increase of 0.24 percent in the proportion of females in manufacturing employment (distress diversification). It, thus, appears to indicate that the growth of female employment in the manufacturing sector could be attributed partly to agricultural growth (forward production linkages) and partly to distress diversification, and it was residual in nature.

Construction

The results provided in Table 5.6 show that among the selected explanatory variables, higher rate of literacy, particularly at higher secondary level, higher per capita non-food consumption expenditure, lower incidence of poverty, higher daily real wage earnings of female agricultural labourers in agricultural and non-agricultural occupations and composite index for rural infrastructure emerged to be important determinants of male employment in construction. The formal banking institutions and government sponsored housing schemes seemed to have played an important role in financing house construction and hence a rapid increase in house construction employment. In rural areas, the role of agricultural growth seemed to have played a limited role in creating direct employment by providing investment capital; it may be important for only the rich farmers. The factors, which influenced the growth of employment in construction of houses, roads, community durable assets, etc., in most cases, appeared to be external to the agricultural sector.

A strong positive association existed between per capita non-food consumption expenditure and male employment in construction. Higher per hectare net income from agriculture was not significantly

Table 5.6 : Correlation Coeffcients between Male and Female Employment in Construction and Selected Variables in Rural India 199-2000

Name of the explantory variable	*Males*	*Females*
PHIA	0.265	-0.275
PCIA	0.232	-0.275
GRLH	0.033	-0.445
NFCE	0.764**	0.004
COM	-0.273	0.574*
COMP	-0.467	0.176
WRLA	0.797**	-0.335
WRLN	0.644**	-0.115
ROL	0.488	-0.213
ROLHS	0.574**	-0.256
PLLH	-0.340	-0.140
PSMH	-0.202	-0.090
CDSUR	-0.062	0.118
CDSURY	0.083	0.112
UPSUR	0.221	-0.344
UPSURE	-0.101	-0.289
QRNAE	0.493	-0.335
PVTY	-0.7708	0.051
DoU	0.109	0.239
UAGLM	-0.123	0.356
CINFRA	0.583*	0.286

Note: ** and* refer to 1 and 5 percent level of significance, respectively

and positively associated with male employment in construction. A strong positive association between the daily real wage earnings of male agricultural labourers in agriculture and the proportion of males in construction may indicate that an increase in the income levels of the poor households seemed to have raised demand for housing, house repair/improvement, renovation and so on. Further, a strong negative association between poverty and male employment in construction implies that an increase in the overall income levels of the poor and hence a reduction in poverty seemed to have raised their demand for housing and housing services. As a result, a good deal of employment potential was created in construction related activities in rural areas. The consumption or production linkages from agriculture were not strong enough to create employment opportunities in construction. As the daily real wage rates of male labourers in agriculture were positively and significantly associated with male employment in construction, it did not appear to be agricultural distress-induced diversification. The proportion of landless households and small and marginal landholdings were found to be negatively associated with male employment in construction, but it was statistically insignificant.

A strong positive association existed between the proportion of male workers in construction and the rate of male literacy, especially at higher secondary level. This implies that the construction seemed to have created more employment opportunities for male workers with higher secondary level of education and above in different capacities such as supervisors, store keepers, office assistant, etc. But, the rate of overall literacy was not significantly and positively associated with male employment in construction. This shows that illiterate male workers as well as literate ones at higher secondary and above levels of education were more likely to be engaged in construction employment.

It has been found that none of the unemployment rates significantly correlated with male employment in construction. It may imply that the distress in general did not seem to have contributed to the growth of male employment in construction. The factors external to the rural economy such as the degree of urbanisation and urban agglomeration seemed to have contributed negatively to the growth of male employment in construction. A strong positive association, however, existed between male employment in construction and the composite index for rural infrastructure. This implies that an increase in rural infrastructure facilities such as rural roads, banking, post-offices and telecommunication facilitated the growth of male employment in construction. The following log-linear multiple regression model has been constructed for male employment in construction.

$$\text{Log MCON} = \alpha + \beta_1 \log \text{NFCE} + \beta_2 \log \text{WRMLA} - \beta_3 \log \text{PLLH} + u \text{ ....(5)}$$

where MCON is the proportion of male workers in construction, NFCE is the rural per capita non-food consumption expenditure, WRMLA is the daily real wage rate of male agricultural labourers in agricultural occupations, PLLH is the proportion of landless households to total households, α refers to the constant, β_1 to β_3 are the slope coefficients of the parameters and *u* is the error term. The results of the model are as follows:

$$\text{Log MCON} = -1.373 + 0.366 \log \text{NFCE}^{*} + 0.490 \log \text{WRMLA}^{**} - 0.339 \log \text{PLLH}^{**}$$
$$(-1.677) \quad (1.797) \quad (2.240) \quad (-2.487)$$

$R^2 = 0.778^{***}$ Adj. $R^2 = 0.772^{***}$ $F = 13.981$

Note: Figures in parentheses indicate the 't' values and ***, ** and * refer to 1, 5 and 10 percent level of significance, respectively.

The results of equation (5) for male employment in construction show that the estimated relationship of the independent variables was statistically significant at 1 percent level, accounting for about 77 percent variation in the proportion of male workers in construction. The elasticity of NFCE indicates that one percent rise in per capita non-food consumption expenditure caused an increase of 0.37 percent in male employment in construction. Similarly, one percent increase in the daily real wage earnings of male workers in agricultural occupations led to 0.49 percent increase. PLLH caused an increase of 0.34 percent. This implies that a greater access to land by the landless or land-poor households stimulate the growth of male employment in construction, mostly by raising their demand for housing and housing services.

In respect of females, commercialisation of agriculture, defined as the proportion of area under non-foodgrains plus paddy, was found to be an important determinant of female employment in construction. The indicators capturing agricultural prosperity such as PHIA and PCIA were negatively associated with female employment in construction, but it was statistically insignificant. This implies that the growth of agriculture, resulting in higher income levels of the rural households may not force females to go for work in construction. A negative association between literacy and employment in construction shows that literate females did not seem to have sought employment in construction but only illiterate females did. The unemployment rate for females as per CDS was positively and as per UPS (general and educated) was negatively associated with female employment in construction, but it was statistically not significant. This implies that, in the short run, the shift of females to construction activities might have been driven by distress, not necessarily from agriculture; and, in the long run, females with more educational attainments seemed to have sought better jobs in the rural non-agricultural sector.

The negative but weak association of daily real wage rates of female agricultural labourers in agricultural occupations with female employment in construction seemed to have lent support to the hypothesis that the growth of female employment in construction was distress diversification from agriculture. For female workers who were unable to get adequate employment in agriculture and did not possess adequate capital and skill, construction activities appeared to have performed the function of "last resort". In other words, construction activities appeared to have acted as a "sponge" to absorb surplus female workers, mostly belonging to the poor households. The incidence of poverty was also positively associated with female employment in construction. A negative but statistically insignificant correlation between the proportion of females in self- and regular RNAE and female employment only in construction implies that construction activities for females were casual and less remunerative and thereby lending support to the residual sector hypothesis.

$$\text{Log FCON} = \alpha + \beta_1 \log \text{PHIA} - \beta_2 \log \text{GRLH} + \beta_3 \log \text{CINFRA} + u \ldots\ldots\ldots(6)$$

where FCON is the proportion of female workers in construction, PHIA is the per hectare net income from agriculture, GRLH is the Gini ratio for operational landholdings, CINFRA is the composite index for rural infrastructure, α refers to the constant, β_1 to β_3 are the slope coefficients of the parameters and *u* is the error term. The results of the model are as follows:

$$\underset{(3.104)}{\text{Log FCON} = 0.6160} - \underset{(-2.861)}{0.542} \log \text{PHIA}^{***} - \underset{(-3.207)}{0.603 \log \text{GRLH}^{***}} + \underset{(2.572)}{0.488 \log \text{CINFRA}^{***}}$$

$$R^2 = 0.601^{*} \qquad \text{Adj. } R^2 = 0.501^{*} \qquad F = 6.020$$

Note: Figures in parentheses indicate the 't' values and ***, ** and * refer to 1, 5 and 10 percent level of significance, respectively.

In respect of females, it has been found that per hectare net income from agriculture, Gini ratio for operational landholdings and composite index for rural infrastructure accounted for about 50 percent of variation in the proportion of female workers in construction. For discrediting the distress diversification hypothesis, we had expected a positive association between the growth of agriculture (PHIA) and female employment in construction. But, it turned out to be a strong negative association, providing support to the distress diversification hypothesis in the case of construction activities. As COMP was strongly and positively associated with PHIA, it was not included in the model to avoid multicollinearity problem. The elasticity of PHIA indicates that one percent decline in per hectare net income from agriculture caused an increase of 0.54 percent female employment in construction activities (distress diversification). A rapid growth of agricultural income did not seem to have encouraged females to involve in construction employment.

In times of agricultural distress, a larger share of female workers seemed to have been compelled to seek employment in construction activities available in the villages or nearby cities, towns or big villages. The elasticity of RLGH indicates that one percent decline in the Gini ratio for operational landholdings resulted in an increase of 0.60 percent in the proportion of female workers in construction employment. An increase of one percent in rural infrastructure led to an increase of 0.49 percent female employment in construction.

Trade, hotel and restaurant business

The results presented in Table 5.7 indicate that among the selected explanatory variables, PHIA, PCIA, CDSUR, CDSURY, USPUR, ASMH and WRLA were found to be important determinants of male employment in trade, hotel and restaurant business. The explanatory variables such as PHIA and PCIA were significantly and positively associated with the proportion of male workers in trade, hotel and restaurant business. It follows that the growth of the agricultural sector and thereby income levels of rural households, particularly the rich households, seemed to have increased the demand for non-farm goods produced mostly in urban or semi-urban areas and thereby promoting trade, commerce and business between urban and rural areas (derived consumption linkages). In agriculturally growing areas, we can generally see that the establishment of a large number of micro non-farm enterprises like hotels, canteen, tea stalls, sweet marts and other micro enterprises involved in trading and business of agricultural inputs and equipments and superior non-agricultural goods produced mostly in urban areas.

Table 5.7 : Correlation Coefficients between Male and Female Employment in Trade, Hotels and Restaurants and Selected Variables in Rural India, 1999-2000

Name of explanatory variables	*Males*	*Females*
PHIA	0.807**	0.540*
PCIA	0.735**	0.295
GRLH	-0.320	-0.283
WRLA	0.687**	-0.174
WRLN	0.585*	-0.210
PHFC	0.071	0.425
GAI	-0.222	0.137
CIMA	0.248	-0.159
COM	0.151	0.531*
COMP	0.374	0.640**
RoL	0.661**	0.283
RoLhs	0.751**	0.350
CDSUR	0.651**	0.438
CDSURY	0.754**	0.549*
UPSUR	0.772**	0.250
UPSURE	0.479	0.297
PLLH	0.208	0.671**
PSMH	0.383	0.603*
ASMH	-0.671**	-0.472
PVTY	-0.227	0.090
QRNAE	0.856**	0.343
CINFRA	0.150	-0.036

Note: ** and * refer to 1 and 5 percent level of significance, respectively

A striking feature was that there had been a strong positive correlation between male employment in trading, hotel and restaurant business and the daily real wage rates in agricultural and non-agricultural occupations, indicating that an increase in the income levels of the poor households contributed significantly to the growth of male employment mostly through an expansion of demand for non-farm goods such as grocery items. Thus, the growth of agriculture resulting in a rapid rise in the income levels of rural households seemed to have generated a demand for non-farm goods produced mostly in urban centres and thereby promoting trade and commerce between rural and urban areas, thereby creating a good deal of employment in trade, commerce, hotels and restaurants. No strong production linkages from agriculture were, however, found to be operating in the promotion of male employment in trade, commerce, hotels and restaurants. A weak positive association nonetheless existed between the proportion of male workers in trade, hotels and restaurants and the degree of commercialisation (forward production linkage) and per hectare fertilizer consumption (backward production linkage).

The rate of male literacy, particularly at higher secondary level, also stimulated male employment in trading, hotel and restaurant business. The male workers with higher educational attainments were more likely to gather relevant information and process it for development of their trading business in which they were involved. A strong positive correlation existed between unemployment rates according to CDS and UPS, particularly for youth, and male employment in trade, hotel and restaurant business. This implies that employment opportunities in these segments of rural non-agricultural sector were distress induced; among the youth, the distress might not have necessarily originated from agriculture. Because, there was a strong positive association between male employment in trade, hotels and restaurants and the daily real wage rates of male agricultural labourers in agricultural occupations. A strong positive association between the proportion of male workers in trade, commerce, hotel and restaurant business and the proportion of male workers in self- and regular RNAE did not seem to lend support to the residual sector hypothesis. The incidence of poverty was also negatively but insignificantly associated with male employment in trade, hotel and restaurant business, thereby discrediting the validity of residual sector hypothesis in respect of male employment in trade, commerce, hotel and restaurant business.

The distress movement of male workers, however, cannot be ruled out, *in toto*, as the proportion of landless households and small and marginal landholdings were positively and significantly, and the average size of small and marginal landholdings was negatively and significantly, associated with male employment in trade, commerce, hotels and restaurants. The rural infrastructure also positively contributed to the growth of male employment in these segments of the rural non-agricultural sector. The following log-linear multiple regression model was constructed for male employment in trade, hotels and restaurants.

$$\text{Log MTHR} = \alpha + \beta_1 \log \text{PHIA} + \beta_2 \log \text{RMLhs} - \beta_3 \log \text{ASMH} + u \text{ ....(7)}$$

where MTHR is the proportion of male workers in trade, hotel and restaurant business, PHIA is the per hectare net income from agriculture, RMLhs is the rate of male literacy at higher secondary level, ASMH is the average size of small and marginal landholdings, α refers to the constant, β_1 to β_3 are the slope coefficients of the parameters and *u* is the error term. The results of the model are as follows:

$$\text{Log MTHR} = \underset{(-3.140)}{-2.086} + \underset{(2.658)}{0.486} \log \text{PHIA}^{***} + \underset{(2.026)}{0.319} \log \text{RMLhs}^{**} - \underset{(-1.672)}{0.266} \log \text{ASMH}$$

$$R^2 = 0.781^{**} \qquad \text{Adj.}R^2 = 0.726^{***} \qquad F = 14.280$$

Note: Figures in parentheses indicate the 't' values and ***, ** and * refer to 1, 5 and 10 percent level of significance, respectively.

The results of the equation (7) show that the estimated relationship was statistically significant at 1 percent level, explaining about 73 percent variation in the proportion of male workers in trade, hotel and restaurant business. The elasticities of PHIA and RMLhs indicate that one percent rise in per hectare net income from agriculture and the rate of male literacy led to an increase of 0.49 and 0.32 percent, respectively, in male employment in trade, hotel and restaurant business. At the same time, the elasticity of ASMH also shows that one percent decline in the average size of small and marginal landholdings contributed to an increase of 0.27 percent in male employment in trade, hotel and restaurant business (distress diversification). Thus, the growth of agriculture and provision of education contributed to the growth of male employment in trade, hotel and restaurant business.

In the case of females, it has been found that PHIA, COMP,

CDSURY, PSMH and PLLH were important determinants of employment in trade, hotel and restaurant business. A strong positive correlation existed between female employment in trade, hotel and restaurant business and per hectare net income from agriculture. This follows that the growth of agriculture contributed to the growth of female employment in trade, commerce, hotels and restaurants mostly through consumption and also as a source of investment linkage. The degree of commercialisation was significantly and positively and per hectare fertilizer consumption was positively but insignificantly, associated with female employment in trade, commerce, hotels and restaurants. This implies that forward and backward production linkages were found to be more effective in the promotion of female employment in agro-processing, agro-shops, hotels, trading of agricultural inputs in which females could also be involved.

A negative but statistically insignificant association between female employment in trade, hotel and restaurant business and the daily real wage rates in agriculture may imply that that growth of female employment in these activities was partly due to distress originating from agriculture. The proportion of landless households and small and marginal landholdings was positively and significantly, and the average size of small and marginal landholdings was negatively and insignificantly, associated with female employment in trade, commerce, hotels and restaurants. This implies that the growth of female employment in these activities was due to distress in agriculture.

The rate of literacy, particularly at middle level, had a positive but statistically insignificant association with female participation in trade, hotel and restaurant business. The unemployment rates, capturing general distress, were found to be not strong enough to push female workers into these types of activities, excepting female youth. A strong positive correlation between the female unemployment rate, particularly for the youth (15-29 years), and female employment in trade, hotel and restaurant business shows that some of these activities were distress diversification among the female youth. As the incidence of poverty and quality of employment were in insignificantly and positively associated with female employment in trade hotel and restaurant business, some part of the employment generated in these segments of the rural non-agricultural sector appeared to be residual in character. The log-linear multiple regression model for female employment in trade, hotels and restaurant business was specified as follows:

$$\text{Log FTHR} = \alpha + \beta_1 \log \text{PHIA} + \beta_2 \log \text{COMP} + \beta_3 \log \text{PHFC} + u \quad \text{.............(8)}$$

where FTHR is the proportion of female workers in trade, hotel and restaurant business, PHIA is the per hectare net income from agriculture, COMP is the proportion of area under non-foodgrains plus paddy, PHFC is the per hectare fertilizer consumption, α refers to the constant, β_1 to β_3 are the slope coefficients of the parameters and u is the error term. The results of the model turned out to be as follows:

$$\text{Log FTHR} = \underset{(-3.382)}{-3.414} + \underset{(1.451)}{0.277 \log \text{PHIA}} + \underset{(3.410)}{0.589 \log \text{COMP}^{***}} + \underset{(1.713)}{0.2320 \log \text{PHFC}^{*}}$$

$$R^2 = 0.664^{***} \qquad \text{Adj. } R^2 = 0.579^{***} \qquad F = 7.890$$

Note: Figures in parentheses indicate the 't' values and ***, ** and * refer to 1, 5 and 10 percent level of significance, respectively.

It has been found that per hectare net income from agriculture, commercialisation of agriculture (including paddy) and per hectare fertilizer consumption explained about 58 percent variation in the proportion of female employment in trade, commerce, hotel and restaurant business. The elasticity of HIA, COMP and PHFC shows that one percent rise in per hectare net income from agriculture, commercialisation of agriculture and per hectare fertiliser consumption led to an increase of 0.28, 0.59 and 0.23 percent, respectively, in female employment in trading, commerce, hotel and restaurant related activities. The production, followed by consumption, linkages, resulting from agricultural prosperity seemed to have created employment opportunities for females in trading, hotel and restaurant business.

Transport, storage and communication

As the share of females in transport, storage and communication was found to be negligible, the correlates for female RNAE were not explored. The results presented in Table 5.8 show that among the selected variables, PHIA, PCIA, WRLA, RoL for males, NFCE,

PVTY, CDSURY, USPUR and CINFRA were found to be important factors determining the level of male participation in transport, storage and communication. It was found that per hectare/capita income from agriculture, daily real wage rates of male workers in agriculture and per capita non-food consumption expenditure were significantly and positively associated with male employment in transport, storage and communication activities.

Agricultural growth-led consumption linkages seemed to have raised the demand for non-farm goods mostly produced in urban areas, leading to the creation of a greater scope for the development of transport, storage and communication networks in rural areas. A

Table 5.8 : Correlation Coefficients between Male Employment in Transport, Storage and Communication and Selected Variables in Rural India, 1999-2000

Name of the explanatory variables	*Males*
PHIA	0.786**
PCIA	0.743**
WRLA	0.895**
WRLN	0.803**
NFCE	0.739**
PHFC	0.322
GAI	0.425
RoL	0.661**
RoLhs	0.741**
CDSUR	0.407
CDSURY	0.515*
UPSUR	0.511*
UPSURE	0.340
PSMH	0.081
ASMH	-0.440
PVTY	-0.695**
QRNAE	0.802**
CINFRA	0.525*

Note: ** and * refer to 1 and 5 percent level of significance, respectively.

weak positive correlation existed between male employment in transport, storage and communication and per hectare fertilizer consumption and GAI. This shows that there had been a weak production linkage effect (forward and backward) on the growth of male employment in these segments. This seems to indicate that consumption linkages were more operative than production linkages in the promotion of male employment in transport, storage and communication.

The male workers with higher educational attainments and knowledge were more likely to find jobs in transport, storage and communication activities. As the daily real wage rate of male agricultural labourers in agriculture was significantly and positively associated with male employment in transport, storage and communication, it was concluded that the growth of such employment might not be distress diversification, resulting from lack of agricultural growth. However, the unemployment rate for male youth as per CDS and the general unemployment rate for males as per UPS were found to have significantly and positively associated with male employment in transport, storage and communication. This implies that the growth of employment in these segments of the rural non-agricultural sector was due to distress in general but not necessarily distress in agriculture. The activities in which they were involved, however, were not in the residual sector. This is due to the fact that the self- and regular RNAE was significantly and positively, and poverty was also significantly and negatively, associated with male employment in transport, storage and communication. The growth of male employment in these segments of the rural non-agricultural sector was due to agricultural growth (consumption linkages) and non-residual in nature.

A strong positive association existed between the composite index for rural infrastructure and the proportion of male workers in transport, storage and communication, indicating that in rural areas, a large number of males were employed in the provision of transport and communication services. It also implies that transport and communication facilitated the growth of employment in trading, commerce, storage and the like. The following log-linear multiple regression model was specified for male employment in transport, storage and communication.

$$\text{Log MTSC} = \alpha + \beta_1 \log \text{PHIA} + \beta_2 \log \text{NFCE} + \beta_3 \log \text{RMLhs} + u \ldots\ldots\ldots(9)$$

where MTSC is the proportion of male workers in transport, storage and communication, PHIA is the per hectare net income from agriculture, NFCE is the rural per capita non-food consumption expenditure, RMLhs is the rate of male literacy at higher secondary level, α refers to the constant, β_1 to β_3 are the slope coefficients of the parameters and u is the error term. The results of the model were as presented below:

$$\underset{(-8.211)}{\text{Log MTSC} = -4.817} + \underset{(4.65)}{0.516 \log \text{PHIA}^{***}} + \underset{(3.287)}{0.399 \log \text{NFCE}^{***}} + \underset{(1.914)}{0.252 \log \text{RMLhs}^{*}}$$

$R^2 = 0.892^{***}$ Adj. $R^2 = 0.865^{***}$ $F = 32.925$

Note: Figures in parentheses indicate the 't' values and ***, ** and * refer to 1, 5 and 10 percent level of significance, respectively.

The F value of the equation (9) shows that the estimated relationship was statistically significant at 1 percent level, explaining about 87 percent variation in the proportion of male workers in transport, storage and communication by the selected explanatory variables. The elasticity of PHIA indicates that one percent rise in per hectare income from agriculture caused an increase of 0.52 percent in male employment in transport, storage and communication. Similarly, the elasticity of NFCE and RMLhs also shows that one percent rise in rural per capita consumption expenditure for non-food items and rate of male literacy at higher secondary level resulted, respectively, in an increase of 0.40 and 0.25 percent in male employment in transport, storage and communication.

Conclusions

The foregoing analysis shows that a considerable variation in the proportion of workers in RNAE across states and gender have been mainly attributed to the growth as well as poverty related factors. The factors that determined the level of male participation in RNAE were found to be not the factors that determined the level of female participation in RNAE. For the growth of male RNAE, the consumption linkages were found to be more effective than the production linkages, and further, the backward production linkages were stronger than forward production linkages. Similarly, for the growth of female RNAE, on the other hand, consumption linkages were also found to be stronger than production linkages, and further, the forward production linkages were more effective than the backward production linkages. While the growth of male RNAE was largely determined by growth related factors (agricultural growth-led diversification), the growth of female RNAE was distress/poverty-compelled factors (agricultural distress-induced diversification). The male RNAE was found to be more productive, better paid and not in the residual sector, whereas in the case of females, it was distress induced, low productive, lowly paid, and it was in the residual sector.

The important factors contributing to the growth of male RNAE were found to be higher per hectare/capita net income from agriculture, greater mechanisation of agriculture, higher daily real wage rates of agricultural labourers in agricultural and rural non-agricultural occupations, reduction in rural poverty, higher rate of literacy, particularly at higher secondary level, and availability of rural infrastructure, among others. At the same time, there had also been distress movements of male workers into the rural non-agricultural sector, but this distress movement did not lend any support to the hotly debated residual sector hypothesis, implying that the growth of male RNAE, at times, might be agricultural distress induced, but the activities need not necessarily be low productive, lowly paid, dead-end jobs. A strong negative association between male RNAE and rural poverty shows that male RNAE did not appear to be less remunerative and lowly paid. The data on daily real wage earnings of male agricultural labourers in agricultural and non-agricultural occupations in rural areas also discredited the residual sector hypothesis.

Per hectare net income from agriculture and degree of commercialisation were found to be key factors in the promotion of female RNAE mostly through production and consumption linkages. Distress-induced factors such as lower daily real wage earnings in agricultural and rural non-agricultural occupations, higher proportion of small and marginal landholdings, landless households, incidence of poverty and unemployment rates as per CDS and UPS were found to be important factors determining the level of female participation in RNAE. Further, distress-induced factors originating from agriculture and also distress in general in terms of high unemployment rates particularly among educated youth, were found to be stronger than agricultural growth-induced factors in determining the level of female participation in RNAE. The growth of female RNAE was mostly distress-driven.

A negative association existed between female RNAE and daily real wage earnings in agricultural as well as non-agricultural occupations, but it was statistically not significant. The proportion of small and marginal landholdings was significantly and positively and the average size of small and marginal landholdings was significantly and negatively associated with female RNAE. This shows that the growth of female RNAE was distress diversification resulting from agriculture. There had been a strong positive correlation between female RNAE and female unemployment rates, as per both CDS and UPS, implying that the growth of female RNAE as a whole was distress induced, but the distress which was reflected in terms of high unemployment rates, especially among the educated youth, was not necessarily distress originating from agriculture. As there was a strong positive association between female unemployment rates and agricultural growth (PHIA), the former could not be regarded as an indicator capturing distress in agriculture. A positive correlation between female RNAE and rural poverty suggests that the growth of RNAE in which the female workers were engaged was low productive and in a residual sector. In brief, the growth of female RNAE was distress induced and residual in nature — casual/contractual, low productive, lowly paid, dead-end jobs.

Thus, the data at the aggregate level reveal that the growth of male RNAE as a whole was determined largely by growth related factors and did not lend any support to the residual sector hypothesis, whereas in the case of females, it was largely owing to distress-induced factors and lent support to the residual sector hypothesis. This observation needs to be tested with primary data in regions with different levels of agricultural prosperity and availability of rural infrastructure.

At the disaggregated level, the growth process of male and female RNAE, by and large, presented similar trends. Higher per hectare fertilizer consumption, greater degree of urban agglomeration and urbanisation and, to some extent, higher per hectare net income from agriculture and daily real wage rates in agriculture contributed positively to the growth of male employment in the manufacturing sector. Factors external to the rural economy appeared to be more operative in the growth of male employment in the manufacturing sector. At the same time, it was also significantly and positively influenced by the incidence of unemployment as per CDS. In the case of female employment in the manufacturing sector, it has been found that the proportion of area under non-foodgrains plus paddy was significantly and positively associated with female employment in the manufacturing sector (strong forward production linkage). The proportion of small and marginal landholdings, average size of large holdings, incidence of unemployment among educated female youth and incidence of poverty also contributed significantly and positively to determining the level of female participation in the manufacturing sector, mostly on casual and daily basis. Female employment in the manufacturing sector, however, appeared to be distress induced and residual in nature.

An increase in per capita non-food consumption expenditure, daily real wage rates of male agricultural labourers in agriculture, rate of male literacy, reduction in the incidence of poverty and landless households and availability of rural infrastructure were found to be important factors in determining male employment in construction. Male employment in construction was not distress induced and not in a residual sector. In respect of females, the degree of commercialisation and composite index for mechanisation of agriculture were positively, and per hectare net income from agriculture was negatively, associated with female employment in construction. The level of female participation in construction was also determined by the availability of employment in agriculture. The growth of female employment in construction was found to be distress induced and lent support to the residual sector hypothesis.

Higher per hectare net income from agriculture, daily real wage earnings of agricultural labourers and rate of male literacy at higher secondary level emerged as important factors contributing to the growth of male employment in trade, hotel and restaurant business. The unemployment rates as per CDS and UPS were also significantly and positively associated with male employment in trade, hotel and restaurant business. But, male employment in these segments appeared to be agricultural growth induced, discrediting the residual sector hypothesis. In respect of females, higher per hectare net income from agriculture, degree of commercialisation and per hectare fertilizer consumption were found to be important factors. Agricultural distress related factors such as higher proportion of small and marginal farmers and lower wage rates in agriculture, on the one hand, and higher incidence of poverty and unemployment rate as per CDS, particularly among the female youth, on the other, also determined the level of female participation in trade, hotel and restaurant business. The growth of female employment in these activities was found to be distress diversification originating from within and outside agriculture and lent some credence to the residual sector hypothesis. The growth of male employment in transport, storage and communication was significantly and positively determined largely by growth related factors such as per hectare net income from agriculture, higher wage rates in agriculture, per capita non-food consumption expenditure, education at higher secondary level and availability of rural infrastructure. Although the incidence of unemployment as per CDS and UPS (distress in general) contributed to determining the level of male employment in transport, storage and communication, it was not agricultural distress induced and did not lend any support to the residual sector hypothesis. Exploring the growth process of male and female RNAE in the developing economies like India was rather complex. The sustained growth of productive RNAE largely depends, *inter alia*, on the growth of broad-based agriculture (creating backward and forward linkages), reduction of poverty, provision of need based education and demand-driven rural infrastructure.

NOTES

1. The composite index for mechanisation of agriculture (CIMA) was estimated for the year 1999-2000 by including tillers, harvesters, tractors, sprayers and other agricultural implements through principal component analysis.

2. Employment in the rural non-agricultural sector can be classified under three categories, namely, self-employment, regular employment and casual wage paid employment. Employment in the first two categories is relatively more productive, superior, non-exploitative, better paid, as compared to the last category in which the employment is casual/contractual, exploitative and lowly paid. The quality of M/F RNAE was estimated by subtracting the proportion of male/female workers in agriculture from the proportion of male/female workers involved in self- and regular employment in rural areas (GoI 2001; NSSO 2001).
3. The results of correlation coefficient between the proportion of males in RNAE and the rate of literacy at different levels of educational attainments have not been provided in Table 5.2 due to lack of space. It was, however, evident that male workers with higher educational attainments were more likely to access RNAE.

6 Impact of Rural Non-Agricultural Employment on Poverty and Income Inequality in India

Introduction

An analysis examining the impact of occupational diversification in favour of the rural non-agricultural sector on distribution of income, assets and wealth helps policy makers and development practitioners in deciding about the type of policy intervention needed to tackle the long-standing problems of poverty, inequality and unemployment. Several strategies have attempted to address these problems and strengthen the base of rural livelihood opportunities ever since the inception of the planning era in India. Implementation of Land Reform Measures and adoption of Green Revolution technologies (new-seeds, irrigations and fertilizers) were some of the important strategies till the 1970s. However, keeping in view the limited success of Land Reform Measures and the trickle-down effect of agricultural growth, it has been realised that the growth in agriculture alone was not adequate to ensure a sustained and faster decline in poverty, inequality and unemployment.

The growth-enhancing policies, thus, were supplemented with the state-managed rural development programmes such as wage based and self-employment programmes, rapid rural industrialization, entrepreneurial skill development programmes and the like which have largely focused on diversification of the employment structure in favour of the non-agricultural sector in rural areas. Consequently, the livelihood base of the resource-poor households in the rural non-agricultural sector has begun to improve. Since then the growth of employment in the rural non-agricultural sector has been on the rise in varying degrees across gender, space and over time. Such a change in the rural occupational structure is expected to have poverty reduction and income-equalising effects. It is due to the fact that the real wage earnings and income levels of the rural labour households were not only higher but also rising a little faster in the rural non-agricultural sector than in the agricultural sector over the years[1]. The expansion of RNAE has, therefore, been regarded as one of the robust policy instruments to tackle the problem of poverty, inequality and unemployment in rural areas in many developing countries of Asia, Africa and Latin America. Empirical studies in a few countries of Asia and Africa reveal that there has been an inverse association between the size of landholdings and rural non-agricultural incomes. For the smallest categories of landholdings, the rural non-agricultural sector was the major source of total household income (Samuel 1979; Liedholm and Kilby 1989). Similarly, several studies in Latin America and the Caribbean have also shown that the poor households derived a major part of their income from RNAE, contributing to a decline in the incidence of poverty (Ferreira and Lanjouw 2001; Ruben and Van den Berg 2001; Reardon, Berdegue and Escobar 2001; Lanjouw 2001; Corral and Reardon 2001).

In the Indian context too, as evident from the review of literature in Chapter 2, several studies at the aggregate level (Vaidyanathan 1986; Chadha 1986, 2001a; Chadha and Sahu 2002; Papola 1987; Dev 1990, 1994; Unni 1991; Hazell and Haggblade 1991; Bhalla 1993,1997; Visaria 1995; Parthasarathy, Shameem and Reddy 1998), drawing data from NSS and Census, have mainly focused on examining the growth structure and determinants of RNAE at different levels. A few studies, based on primary data reveal that, as in the case of other developing countries of Asia, Africa and Latin America, the income levels of the poor households involved in RNAE increased significantly and contributed to a decline in the incidence of rural poverty (Unni 1996b, 1997a; Rajasekhar and Biradar 1998; Lanjouw and Shariff 2004). Therefore, the development strategy focusing on the promotion of RNAE has been gaining importance in many developing countries of the world.

It can be noted that the rural non-agricultural activities taken up by the poor are different from those of the non-poor (Biradar and Rajasekhar 1998; Unni 1997a; Lanjouw and Shariff 2004). The poor who lack capital, education, skill, social network and other resources are more likely to take up low productive, casual/contractual, lowly paid, dead-end, non-farm jobs; whereas the non-poor, who possess adequate capital, education, skill, social network and influence are more likely to be employed in productive, regular, higher social-status oriented, better paid non-farm jobs. The working conditions and earning levels are much better in the latter than in the former. Despite having this fact in our view, in the present study, due to lack of data at the aggregate level, the RNAE and sub-activities for only sixteen major states were considered for examining their impact on poverty reduction. Against this backdrop, the present chapter attempts to address the following important questions: What is the degree of poverty across space and activities? What are the important sources of poverty reduction? How much can the growth of RNAE contribute to a decline in poverty? What can be the most effective poverty alleviation approach? What is the pattern of income distribution among different income quintiles in rural India? Can income from RNAE reduce income inequality in rural India?

Prevalence of Poverty in Rural India

The incidence of poverty continues to be widespread and persistent in India and more so in rural areas (WB 1997, 2000; Dev 2000b; Sundaram 2001a; Sundaram and Tendulkar 2003, 2004). Although the growth-led strategy based on the "trickle-down hypothesis" and multi-pronged poverty alleviation programmes were launched, India still has the world's largest number of poor. As per the Planning Commission estimates, the number of people living below the national poverty line was estimated at 260.2 million poor, accounting for 26.1 percent of the population in 1999-2000 (Table 6.1). Such a magnitude of poverty received the serious attention of researchers, policy makers and development practitioners in view of India's endeavour towards achieving one of the Millennium Development Goals of halving the

Table 6.1 : Prevalence of Poverty in India, 1973-74 to 1999-2000

Particulars	*Rural*	*Urban*	*All*
Number of poor (million)			
1973-74	261.3	60.0	321.3
1977-78	264.3	64.6	328.9
1983	252.0	70.9	322.9
1987-88	231.9	75.2	307.1
1993-94	244.0	76.3	320.3
1999-2000	193.2	67.0	260.2
Poverty ratio (%)			
1973-74	56.4	49.0	54.9
1977-78	53.1	45.2	51.3
1983	45.7	40.0	44.5
1987-88	39.1	38.2	38.9
1993-94	37.3	32.4	36.0
1999-2000	27.1	23.6	26.1
Annual growth rate (%)			
1973-74 to 77-78	0.23	1.49	0.47
1977-78 to 83	-0.95	1.88	-0.37
1983 to 1987-88	-1.65	1.18	-1.00
1987-88 to1993-94	0.85	0.24	0.70
1993-94 to1999-2000	-3.82	-2.14	-3.40
1973-74 to 1999-2000	-0.15	-0.13	-0.15

Source: Planning Commission, Government of India, New Delhi.

incidence of poverty by 2015 (WB 2001). To achieve this end, therefore, a multi-dimensional poverty alleviation approach has to be framed in the context of changing macroeconomic scenario.

Although there was a lack of clarity over the extent of poverty reduction in the year 1999-2000, the proportion of people living below the poverty line showed a sharp decline over the earlier estimates. The data presented in Table 6.1 show that a large number of people living in poverty were found to be in rural than in urban areas; three out of every four poor persons lived in rural areas in 1999-2000. The incidence of rural poverty was found to be quite higher among

SC/ST households as against Others (Thorat 2004; Biradar 2006b; Biradar and Jayasheela 2008) and among casual workers as compared to workers involved in self- or regular wage/salaried employment both in agricultural and non-agricultural occupations (Sundaram and Tendulkar 2003, 2004).

In rural areas, the annual growth rate of people in poverty registered a rapid decline in the 1980s and more so in the late1990s, the rate of decline was disputed. In rural areas, the percentage of people in poverty, by and large, declined continuously from 56.4 percent to 27.1 percent during 1973-74 to 1999-2000, whereas in the case of urban areas, it declined from 49 percent to 23.6 percent during the same period. The decline in the incidence of poverty was a little faster in rural than in urban areas. The decline in the incidence of rural poverty in the 1980s was mainly in part attributed to a faster growth of agriculture, followed by infrastructure and human resource development, particularly female literacy (WB 1997; WB 2000). But, in the late 1990s, it was hypothesised that the decline in poverty was largely due to the growth of RNAE in which the daily real wage earnings were found to be higher than those in agriculture.

Regional pattern of poverty incidence

The incidence and progress in reduction of poverty has not been uniform across the states (Table 6.2). A substantial and consistent fall in the incidence of rural poverty took place in most of the states, namely, Kerala, West Bengal, Andhra Pradesh, Gujarat, Haryana, Punjab, Karnataka, Maharashtra, Rajasthan and Tamil Nadu. No significant progress in poverty reduction, however, was discernible in states such as Bihar, Assam, Orissa, Madhya Pradesh and Uttar Pradesh. Such varied levels of poverty prevalence and rates of reduction were attributed to differential rates of growth in agricultural as well as non-agricultural sectors; the relative importance of which needed to be identified in the 1990s.

An analysis of the percentage distribution of poverty incidence across major states ranked according to the ascending order of PCIA (agricultural growth), rate of literacy (human capital) and male and female RNAE (growth of rural non-agricultural sector) in the years of the pre-reform (1983) and reform (1999-2000) periods may provide policy insights to address the problem of poverty in a multidimensional framework.

Table 6. 2 : Percentage of People Living below Poverty Line across Major States in Rural India, 1973-74 to 1999-2000

State	*1973-74*	*1977-78*	*1983*	*1987-88*	*1993-94*	*1999-00*
Andhra Pradesh	48.4	38.1	26.5	20.9	15.9	11.1
Assam	52.7	59.8	42.6	39.4	45.0	40.0
Bihar	63.0	63.3	64.4	52.6	58.2	44.3
Gujarat	46.4	41.8	29.8	28.7	22.2	13.2
Haryana	34.2	27.7	20.6	16.2	28.0	8.3
Himachal Pradesh	27.4	33.5	17.0	16.3	30.3	7.9
Karnataka	55.1	48.2	36.3	32.8	29.9	17.4
Kerala	59.2	51.5	39.0	29.1	25.8	9.4
Madhya Pradesh	62.7	62.5	48.9	41.9	40.6	37.1
Maharashtra	57.7	64.0	45.2	40.8	37.9	23.7
Orissa	67.3	72.4	67.5	57.6	49.7	48.0
Punjab	28.2	16.4	13.2	12.6	12.0	6.4
Rajasthan	44.8	35.9	33.5	33.2	26.5	13.7
Tamil Nadu	57.4	57.7	54.0	45.8	32.5	20.6
Uttar Pradesh	56.5	47.6	46.5	41.1	42.3	31.2
West Bengal	73.2	68.3	63.1	48.3	40.8	31.9
All India	56.4	53.1	45.7	39.1	37.3	27.1

Source: Planning Commission, Government of India, New Delhi.

States with higher PCIA had not only a smaller proportion of people in poverty but also witnessed a rapid progress in the reduction of poverty as compared to states with lower PCIA in 1999-2000 over 1983. In other words, the incidence of poverty tended to decline as agricultural growth increased (Figure 6.1). There were, however, some instances where poverty reduction was substantial even in those states with a poor performance in agriculture. For instance, states such as Tamil Nadu, Maharashtra and Bihar, which had made less progress in agriculture (PCIA), experienced a sharp decline in the incidence of rural poverty during the same period.

Figure 6.1 : Percentage Distribution of People in Poverty across Major States Ranked According to Ascending Order of PCIA in Rural India, 1983 and 1999-2000

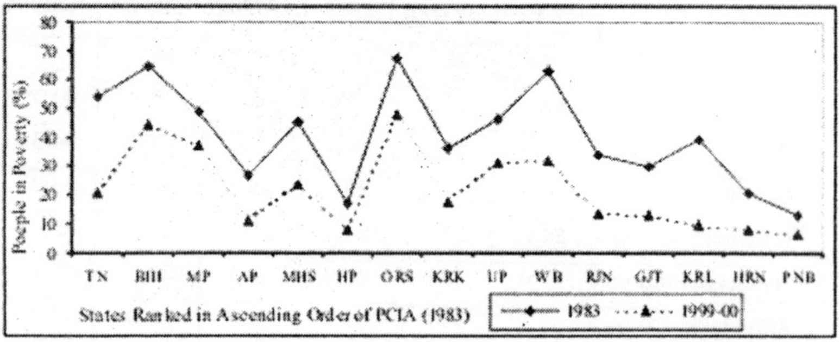

Source : Table 6.2.

Similarly, another observation is that, in rural areas, states with a lower rate of literacy *vis-à-vis* those states with higher ones, excepting for Tamil Nadu, West Bengal and Orissa, had a larger share of people in poverty. In other words, as the rate of literacy (growth of human capital) increased, in most cases, the incidence of poverty tended to decline across space and over time (Figure 6.2).

Figure 6.2 : Percentage Distribution of People in Poverty by Major States Ranked According to Ascending Order of Literacy in Rural India, 1983 and 1999-2000

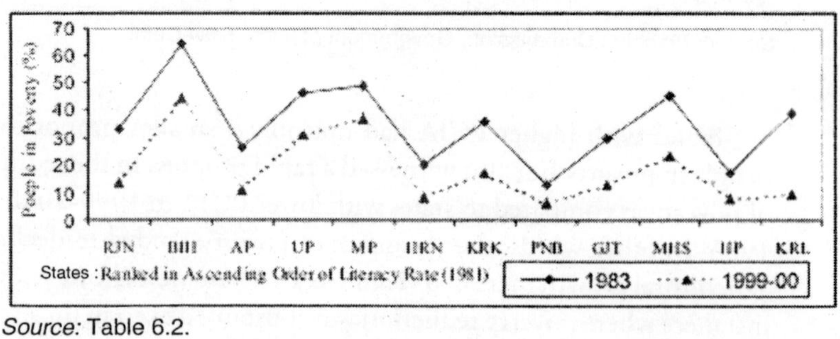

Source: Table 6.2.

A greater decline in the incidence of rural poverty took place in states with a higher rate of literacy as against states with lower ones. This indicates that the growth of human capital also played a vital role in determining the extent of workforce participation in RNAE and thereby contributing to a decline in the incidence of rural poverty during the reform period, as compared to the pre-reform period. A striking feature is that the prevalence of rural poverty was not significant in those states, which had a larger proportion of male workers in RNAE, as compared to those states with lower ones. That is to say, the incidence of rural poverty declined with an increasing share of male workers in RNAE (Figure 6.3). It has also been found that a rapid decline in the incidence of rural poverty took place in states with a higher share of male workers in RNAE *vis-à-vis* those states with lower ones, excepting for West Bengal and Orissa. This indicates that the growth of RNAE, particularly for males, contributed to a greater decline in the incidence of rural poverty in 1999-2000 over 1983.

Figure 6.3 : Percentage Distribution of People in Rural Poverty across Major States Ranked According to Ascending Order of Male RNAE, 1983 and 1999-2000

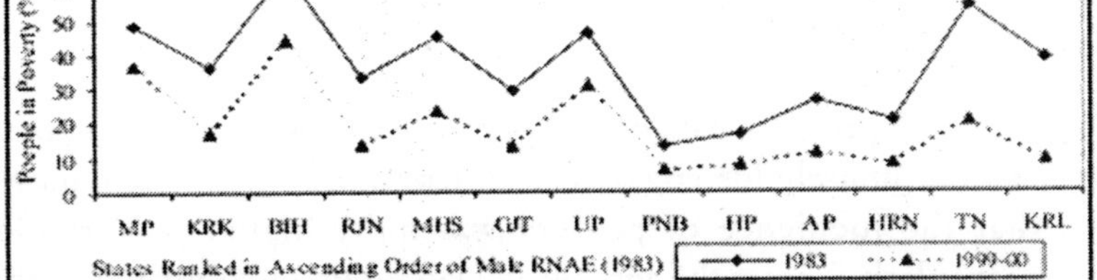

Source: Table 6.2.

Contrary to the above observation, the percentage distribution of people in poverty by states ranked according to the ascending order of female RNAE lent support to the hypothesis that the growth of female RNAE was in a residual sector. It was found that most of the states with a higher proportion of female RNAE had a higher incidence of poverty than those states which had lower proportion of female

RNAE during the pre-reform and reform periods (Figure 6.4). Although a positive association existed between them, states with a higher proportion of female RNAE experienced a relatively greater decline in the incidence of poverty than those states with a lower proportion of female RNAE in 1999-2000 over 1983. The above regional pattern of poverty incidence shows that the growth of agriculture, male RNAE and the rate of literacy, among others, were found to be important sources of poverty reduction in rural areas.

Figure 6.4 : Percentage Distribution of People in Rural Poverty across Major States Ranked According to Ascending Order of Female RNAE, 1983 and 1999-2000

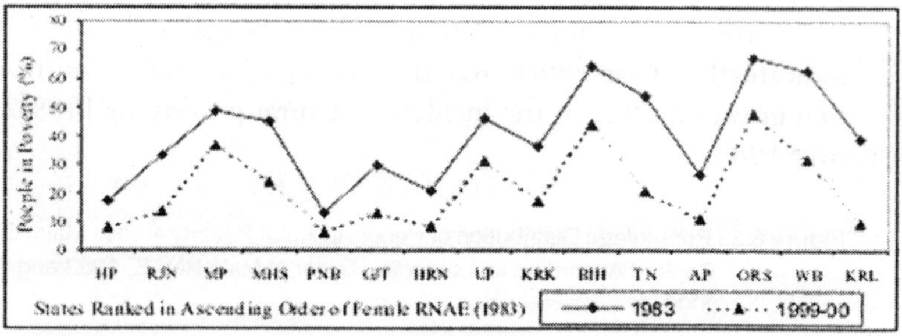

Incidence of poverty by activity in rural India

Table 6.3 provides data on the prevalence of poverty among persons in rural households engaged in different activities. The data reveal that the incidence of poverty was quite higher in agriculture, particularly among the agricultural labourers, as compared to that in the non-agricultural sector in rural areas. The incidence of poverty in the rural non-agricultural sector was lower as the daily real wage rates were found to be quite higher. Within the rural non-agricultural sector itself, the incidence of poverty was not found to be uniform. The incidence of poverty was found to be more in construction, followed by manufacturing, transport, storage and communication, trade, hotel and restaurant business and services, in that order.

Going by the major states, it was found that most of the states had a lower incidence of poverty in the rural non-agricultural sector as compared to that of the agricultural sector; only three states, namely, Assam, Haryana and Himachal Pradesh had a higher incidence of poverty in the rural non-agricultural sector than in the agricultural sector in 1999-2000. At the same time, it was observed that states with a higher share of male workers in RNAE tended to witness a lower incidence of poverty, as compared to those states with lower ones. In the case of females, however, states with a higher participation of females in RNAE showed a greater incidence of poverty, as compared to those with a lower ones. This indicates that the growth of RNAE, resulting from agricultural prosperity, seemed to be a more effective strategy for poverty alleviation; male RNAE appeared to be more effective than female RNAE in the reduction of poverty.

Nexus between RNAE and Incidence of Povorty

In the 1990s, there was a decline in the growth rate and quality of employment both in the agricultural and non-agricultural sectors, particularly in rural areas[2]. A decline in the growth rate and quality of employment in the agricultural and non-agricultural sectors in the 1990s should have caused a decline in the daily real wage earnings and income levels of the rural labour households, which depended on employment for their living, thereby resulting in further impoverishment. But this did not happen. An important and widely debated question, therefore, is: How did the incidence of rural poverty decline sharply even when the growth rate and quality of rural employment came down? In this context, we hypothesise that in spite of a decline in the growth rate and quality of rural employment, the incidence of rural poverty declined due to the shifting of rural workers away from agriculture in favour of the non-agricultural sector where they got relatively higher wage rates, thereby contributing to a decline in poverty. The data show that the growth rate of daily real wage rates and income levels were found to be relatively higher in the non-agricultural sector than in agriculture (Unni 1997b; Sharma 2001; Lanjouw and Shariff 2004; GoI 2003b).

An increase in the proportion of workers in RNAE, particularly males, was expected to raise the income levels of the rural households

Table 6.3 : Prevalence of Poverty among Persons in Rural Households Engaged in Specified Sub-Sectors by Major States 1999-2000

(*Percent*)

State	*Agriculture*	*Cultivation*	*Agrl. Labour*	*MF*	*CON*	*THB*	*TSC*	*Services*	*RNAE*
Andhra Pradesh	12.1	7.1	15.3	7.4	11.4	5.8	5.9	6.6	7.4
Assam	41.1	29.8	67.1	40.9	46.8	31.3	48.3	39.4	41.3
Bihar	44.8	31.3	59.2	47.5	58.6	30.3	34.3	41.2	42.4
Gujarat	13.3	6.4	20.2	7.0	40.4	5.3	8.7	2.8	12.8
Haryana	7.0	1.2	19.2	6.8	21.5	13.7	0.8	1.8	8.9
Himachal Pradesh	6.7	6.1	15.5	7.5	19.1	5.4	-	2.1	8.5
Karnataka	18.3	11.0	25.1	14.2	7.6	7.2	6.2	13.2	9.7
Kerala	12.1	3.7	19.4	8.9	9.2	9.7	1.7	3.9	6.7
Madhya Pradesh	38.6	27.1	53.6	39.1	48.7	25.0	12.1	15.5	28.1
Maharashtra	28.2	16.2	36.7	14.9	18.5	10.2	10.9	4.4	11.8
Orissa	54.1	40.9	63.9	44.1	63.2	25.5	13.1	12.8	31.7
Punjab	7.2	0.9	15.1	6.6	9.6	2.7	8.7	0.9	5.7
Rajasthan	13.2	11.3	27.1	4.4	22.3	7.2	11.8	8.3	10.8
Tamil Nadu	24.1	10.5	29.6	14.7	11.0	8.4	18.8	13.9	13.4
Uttar Pradesh	31.2	24.2	50.9	37.5	35.0	26.9	26.6	24.1	30.0
West Bengal	35.9	20.8	45.0	27.2	43.8	16.6	35.4	8.8	26.4
All	28.9	19.7	40.0	25.0	27.9	16.8	19.9	15.6	21.0

Note: 1. MF: Manufacturing; CON: Construction; THB: Trade, hotel and business; TSC: Transport, storage and communication; RNAE: Rural non-agricultural employment.

2. The data for workers in the organised and unorganised sectors are combined.

Source: Bhalla (2003: 21).

involved in RNAE, thereby contributing to a decline in the incidence of rural poverty. The increase in the proportion of male workers in RNAE had not affected the real wage rates as the demand for labourers in rural non-agricultural occupations had been increasing over time. Whereas in the case of females, an increase in the proportion of female workers in RNAE seemed to have raised the real wage rates in agriculture a little faster than, or caused stagnation, in the non-agricultural sector in the 1990s.

As mentioned earlier, it has been argued that the rural non-agricultural activities taken up by the poor are not the same as taken up by the non-poor; hence, the aggregate decline in the incidence of poverty cannot be strictly attributed to the aggregate growth of rural non-agricultural activities. In the absence of data on rural non-agricultural activities taken up by the poor and the non-poor, it was, however, observed that to the extent that the poor take up jobs in the rural non-agricultural sector and the real wage earnings and income levels are higher in that sector than in agriculture, the incidence of rural poverty, *a priori,* was expected to come down. In addition to this, an important observation was that due to a shift of workers away from agriculture to rural non-agriculture, the real daily wage rates in agriculture went up, thereby contributed to a further decline in poverty[3]. This is one of the important indirect benefits of the occupational diversification in favour of the non-agricultural sector in rural areas. The data show that, in the process of economic growth, the incidence of rural poverty and the proportion of workers in RNAE, particularly in respect of males, tended to move in opposite directions (Figure 6.5).

Sources of Poverty Reduction

Reduction of poverty can be achieved through several factors. Among them, the growth of agriculture (measured in terms of per capita income from agriculture, daily real wage rates of male and female agricultural labourers), human resource development (rate of male/female literacy), the growth of RNAE (proportion of male/female workers in RNAE, daily real wage rates of male and female agricultural

Figure 6.5 : Nexus between RNAE and Poverty in Rural India, 1973-74 to 1999-2000

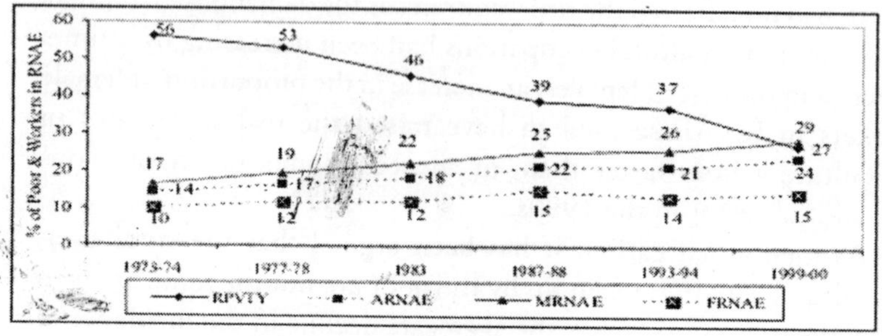

Note: RPVTY: Incidence of rural poverty; ARNAE: All rural non-agricultural employment; MRNAE: Male rural non-agricultural employment; FRNAE: Female rural non-agricultural employment.

Source: Table 6.1.

labourers in non-agriculture) and the government programmes (rural per capita public expenditure for rural development and social services) have been chosen to examine their relative importance in poverty reduction. In order to capture the indirect impact of workers shifting from agricultural to rural non-agricultural sector on poverty reduction in 1999-2000, the change in the daily real wage rates of male and female agricultural labourers in agriculture in 1999-2000 over those in 1983 was taken as a substitute in the absence of an accurate indicator. Among these sources of poverty reduction, it is imperative to measure the extent of poverty reduction due to the growth of RNAE (proportion of male and female in RNAE and the daily real wage rates in rural non-agricultural occupations). As some of these explanatory variables were highly correlated among themselves, a simple log-linear regression model was specified for each explanatory variable, considering the incidence of poverty as a dependent factor, for the years 1983 and 1999-2000. The simple log-linear regression equations for each explanatory variable have been fitted as follows:

$$\text{Log PVTY} = \alpha + \beta \log \text{PCIA} + u \qquad \text{.........(1)}$$

$$\text{Log PVTY} = \alpha + \beta \log \text{WRM/FLA} + u$$

$$\text{Log PVTY} = \alpha + \beta \log \text{M/FRNAE} + u \qquad \text{.........(2)}$$

$$\text{Log PVTY} = \alpha + \beta \log \text{WRM/FLN} + u$$

$$\text{Log PVTY} = \alpha + \beta \log \Delta \text{ WRM/FLA} + u \qquad \text{.........(3)}$$
$$\text{Log PVTY} = \alpha + \beta \log \text{RM/FL} + u \qquad \text{.........(4)}$$
$$\text{Log PVTY} = \alpha + \beta \log \text{GVTPM} + u \qquad \text{.........(5)}$$

where PVTY is the incidence of rural poverty, PCIA is per capita net income from agriculture, WRM/FLA is the daily real wage rates of agricultural male/female labourers in agricultural occupations, M/FRNAE is the proportion of male/female workers in RNAE, WRM/FLN is the daily real wage rates of agricultural male/female labourers in rural non-agricultural occupations, RM/FL is the rate of male/female literacy, GVTPM is the government programmes in terms of rural per capita government expenditure for rural development and social services and ΔWRM/FLA is the change in the daily real wage rates of male and female agricultural labourers in agricultural occupations in 1999-2000 over 1983 (used to capture the indirect impact of occupational shift in favour of RNAE). While α and β are the constant and coefficient of parameters, respectively, u is random error term in each equation.

The regression results of the fitted equations have been provided in Table 6.4. The results show that the growth of agriculture measured in terms of higher PCIA, higher daily real wage rates of male and female agricultural labourers, followed by the growth of the rural non-agricultural sector measured in terms of higher male RNAE and daily real wage rates of male and female agricultural labourers in non-agricultural occupations, human resource development measured in terms of female and male literacy were the major sources of poverty reduction during the pre-reform period.

During the reform period, on the other hand, the growth of the rural non-agricultural sector measured in terms of higher RNAE, particularly for males, and the daily real wage rates of male and female labourers, followed by the growth of agriculture, the rate of female and male literacy and government programmes measured in terms of rural per capita government expenditure for rural development and social services (GVTPM) were found to be important sources of poverty reduction. This indicates that the role of agricultural growth had been taken over by the growth of the non-agricultural sector in the reduction of poverty in rural areas during the reform period. This

Table 6. 4 : Regression Results for the Sources of Poverty Reduction in Rural India, 1983 and 1999-2000

Model	*Variable*	*1983*					*1999-2000*				
		α	β	R^2	*Adj.*R^2	*F*	α	β	R^2	*Adj.*R^2	*F*
1	PCIA	3.728 (4.934)	-0.623*** (-2.873)	0.388	0.341	8.253	4.294 (3.849)	-0.504*** (-2.635)	0.305	0.301	6.481
	WRMLA	2.032 (10.207)	-0.560** (-2.434)	0.313	0.260	5.925	3.077 (7.389)	-0.775*** (-4.423)	0.601	0.570	19.567
	WRFLA	1.816 (7.229)	-0.277 (-1.041)	0.077	0.006	1.083	3.092 (8.062)	-0.765*** (4.285)	0.585	0.553	18.553
2	MRNAE	1.830 (2.670)	-0.108 (-0.393)	0.012	-0.062	0.155	3.264 (5.489)	-0.687*** (-3.405)	0.472	0.431	11.600
	FRNAE	1.149 (6.645)	0.529** (2.247)	0.280	0.224	5.050	0.931 (3.021)	0.282 (1.058)	0.079	.0.008	1.119
	WRMLN	1.968 (8.801)	-0.460* (-1.867)	0.212	0.151	3.487	3.092 (5.661)	-0.686*** (-3.981)	0.470	0.429	11.547
	WRFLN	1.457 (8.874)	0.185 (0.677)	0.034	-0.040	0.459	2.754 (4.481)	-0.564** (-2.464)	0.318	0.266	6.002
3	WRMLA	–	–	–	–	–	2.508 (6.757)	-0.690*** (-3.435)	0.476	0.435	11.800
	WRFLA	–	–	–	–	–	2.234 (9.075)	-0.751*** (-4.095)	0.563	0.530	16.773
4	RML	1.764 (1.670)	-0.053 (-0.192)	0.030	0.001	0.637	5.948 (2.098)	-0.418* (-1.658)	0.175	—	2.750
	RFL	1.781 (5.411)	-0.185 (-0.677)	0.034	0.002	0.458	3.570 (3.405)	-0.524** (-2.220)	0.275	—	4.927
5	GVTPM	–	–	–	–	–	2.243 (2.942)	-0.342 (-1.311)	0.117	—	1.719

Note: Figures in parentheses the 't' values and ***, ** & * refer to statistical significance at 1, 5 & 10 percent level, respectively; Adj : Adjusted.

is evident from the fact that while the elasticity of PCIA declined from (-) 0.62 percent in 1983 to (-) 0.50 percent in 1999-2000, the adjusted R^2 and F values also declined during the same period. But the elasticity of WRMLA and WRFLA increased from (-) 0.56 and (-) 0.28 percent to (-) 0.78 and (-) 0.77 percent, respectively, during the same period. The impact of the daily real wage earnings of the male and female agricultural labourers on poverty reduction apparently increased; but the bulk of the increase was due to the shift of labourers to the rural non-agricultural sector, as a result of which the daily real wage earnings in agricultural occupations increased and hence there was a greater impact of the agricultural wage rates of male and female labourers on poverty reduction during the reform period. This is evident from the equation (3). However, this observation needs to be verified with primary data.

The elasticity of the variables measuring the growth of the rural non-agricultural sector also increased rapidly during the same period. The elasticity of MRNAE, WRMLN and WRFLN increased from (-) 0.11, (-) 0.46 and (+) 0.19 percent in 1983 to (-) 0.69, (-) 0.69 and (-) 0.56 percent in 1999-2000, respectively. The values of the adjusted R^2 and F also improved significantly during the same period. The increase in the values of elasticity, the adjusted R^2 and F of the variables reflecting the growth of the rural non-agricultural sector were much faster than those reflecting the agricultural growth in 1999-2000 over 1983, indicating that the growth of RNAE, particularly for males, turned out to be a key source of poverty reduction during the reform period.

Going by gender, however, it was found that a strong positive association existed between female RNAE and rural poverty, implying that the female RNAE in which they were involved appeared to be less productive, casual/contractual and lowly paid, and their participation in RNAE might have also reduced or caused stagnation in the daily real wage rates as well as income in the rural non-agricultural sector (Sharma 2001; Biradar 2006a), resulting in further impoverishment. That is why the WRFLN was also positively associated with the incidence of poverty during the same period, but statistically insignificant, (see, the beta values of FRNAE and WRFLN in 1983). During the reform period, however, the elasticity of FRNAE declined from 0.53 percent in 1983 to 0.28 percent in 1999-2000. This implies that the participation of females in low productive, casual/ contractual, lowly paid RNAE appeared to have come down significantly, though it was still positively but insignificantly associated with the incidence of poverty. Consequently, the association between RWFLN and rural poverty turned out to be strong negative during the reform period. This shows that a little reduction in the participation of female workers in low productive or lowly paid RNAE seemed to have resulted in an increase in the daily real wage rates in non-agricultural occupations, thereby partly contributing to a decline in the incidence of rural poverty during the reform period. The reduction of female participation in low productive RNAE might be attributed to a rapid increase in the rate of female literacy (human resource development).

A strong positive association between the daily real wage rates and rural poverty did not seem to have lent support to the hypothesis that the growth of female RNAE was in a residual sector— low productive, lowly paid. This may be due to the fact that the daily real wage earnings of the male or female labourers reflected only the employment situation in the casual type, not of the self-employment (or regular/salaried). The participation of females, particularly in self-employment of the rural non-agricultural sector seemed to have reduced the daily real wages/income levels, thereby resulting in a lesser or no impact on the reduction of rural poverty.

If we consider the extent of poverty reduction due to the growth of both female and male RNAE, the latter had a greater poverty reduction effect than the former (see, beta values of MRNAE and WRMLA in comparison with FRNAE and WRFLA). The hypothesis that the male RNAE contributed to a greater decline in the incidence of rural poverty, as compared to the female RNAE, might be tenable. This hypothesis needs to be verified with primary data in areas with different levels of agricultural prosperity.

In respect of literacy as an index of human capital, it was found that the elasticity of RML and RFL increased significantly during the same period. In fact, the increase was found to be much faster as compared to that of any other variable. This indicates that human resource development emerged to be one of the key factors in the reduction of rural poverty *via* facilitating the degree of accessibility to and level of male and female participation in more productive RNAE. Finally, a weak negative association was found to exist between rural poverty and rural per capita government expenditure for rural development and social services. The extent of impact on the reduction of rural poverty appeared to be insignificant, as compared to the impact of other factors. In view of a gradual decline in the pubic expenditure for rural development including poverty alleviation programmes due to the changes in the macroeconomic policies, the growth of RNAE, particularly for males, therefore, can be regarded as an important strategy for combating the long-standing problems of unemployment and poverty in rural areas. The rural development programmes, therefore, need be framed in juxtaposition with the growth of the rural non-agricultural sector.

RNAE and Income Inequality

The foregoing analysis reveals that the growth of RNAE, particularly for males, contributed significantly to a decline in the incidence of rural poverty. The impact of RNAE on income inequity is examined in what follows. As evident from a review of literature, in most of the developing countries, a large number of small and marginal farmers and landless labourers derived their major share of household income from the non-agricultural sector as compared to those of large and medium farmers (Samuel 1979; Liedholm and Kilby 1989; Rajasekhar and Biradar 1998; Elbers and Lanjouw 2001).

Traditional theories of development accorded a greater importance to the role of inter-sectoral transfer of labour from agriculture to industry as a central theme of the development process. Lewis (1954) points out that economic growth takes place when surplus labour transfers out of traditional subsistence agriculture towards a modern sector, often tacitly assumed to be industrial and urban. Fields (1980, 2000), however, demonstrates that such a process is able to generate the well known "inverted U-shaped curve" of rising and then falling income inequality, as first formulated by Kuznets (1955, 1966). It is also a matter of long-standing interest to understand how economic growth affected the distribution of income among rural households. The interaction between economic growth and distribution of income has been dynamic and divergent across different parts of the globe, depending upon the extent of accessibility and effective utilisation of resources by rural households with different income quintiles.

The contribution of income from the rural non-agricultural sector to total rural income was high, especially in those areas where unfavourable labour-to-land ratios constrained income-earning opportunities in agriculture. In land-scarce areas of third world— South and South East Asia — the rural non-agricultural sector was often viewed as a key source of income for rural households. The results of the impact of rural non-agricultural income on the distribution of income were found to be quite divergent. On the one hand, studies by Chinn (1979) and Samuel (1979) in Taiwan indicated that income from the rural non-farm sector reduced income inequality in rural areas. According to Chinn, the non-farm income benefited the poor because the share of non-farm income varied inversely with the size of the farm. On the other hand, some studies reveal different results. For instance, a few studies by Reardon, Delgado and Matlon (1992) in Burkina Faso, Collier, Radwan and Wangwe (1986) in Tanzania and Matlon (1979) in Nigeria showed that the non-farm income had a negative impact on rural income distribution. Such divergent results were perhaps due to difference in the distribution of land and educational facilities. In land scarce, labour-rich settings, like Taiwan and much of Asia, small and inadequate landholdings might tend to push poorer households out of agriculture into the rural non-agricultural sector. In these settings, the non-agricultural income might be expected to have a favourable impact on distribution of income. The obverse could hold good in land-rich settings, such as Africa, where abundant land and scarce labour might tend to keep more people in agriculture and to pull only the richer households into the rural non-agricultural sector.

In India, a study by Birthal (1997) in western Uttar Pradesh, using household survey data has found that income from the non-agricultural sector was highly inegalitarian and that accruing from livestock and agriculture was more egalitarian. While agriculture and salaried employment were found to be inequality-increasing sources, livestock, farm and non-farm wage employment were found to be inequality-decreasing sources of income. To examine the impact of RNAE on income inequality, an attempt has been made to analyse the distributional pattern of income of rural households from different sources, estimate the Gini ratios of income sources and decompose the sources of income inequality.

Distributional pattern of income from RNAE: The NCAER data

An analysis of the distribution of agricultural and non-agricultural incomes across different income quintiles and the impact of non-agricultural income sources on income inequality was mainly based on the data collected by NCAER (Lanjouw and Shariff 2004). The data were collected from a large-scale survey of 35,130 households from 1,765 villages drawn from 16 states in rural India in 1993-94.

Table 6.5 provides source-wise income shares by different income quintiles in rural India during 1993-94. The data, *prima facie*, reveal that the proportion of income from RNAE contributed, on an average, about one third (34 percent) of the total household income, whereas in respect of cultivation and agricultural wage employment, it accounted for 55 and 8 percent, respectively. It is evident that the non-agricultural sector was also an important source of income, next only to cultivation in rural India. The share of agricultural (cultivation) and non-agricultural income tended to move in opposite directions as per capita income quintiles tended to rise (Figure 6.6). This implies that a greater increase in rural non-agricultural income *vis-à-vis* agricultural income tends to reduce income inequality. The contribution of income from non-agricultural sources to total income across different per capita income quintiles shows that for the middle three income quintiles, the contribution from non-agricultural sources was close to 40 percent, while for the lowest and highest quintiles, the share was around 31 percent. This shows that, in rural India, for the middle-income quintiles, the rural non-agricultural sector was an important source of income, as compared to the lowest and highest income quintiles.

Table 6.5 : Distribution of Income from Different Sources in Rural India, 1993-94

(Percent)

Income Quintile	*Cultivation*	*AWE*	*NAWE*	*NASE*	*NARE*	*TNFE*	*OS*	*Real PCI (Rs.)*
Q1	38.2	28.2	15.8	11.4	4.4	31.6	2.0	1146
Q2	38.0	21.3	14.7	16.8	7.0	38.5	2.3	2113
Q3	45.2	13.4	10.1	16.3	11.7	38.1	3.2	3141
Q4	50.1	7.5	6.1	14.6	18.6	39.3	3.2	4712
Q5	64.5	2.1	2.0	7.9	21.1	30.9	2.5	11226
Total	54.9	8.0	5.9	11.5	17.7	34.4	2.7	4468
Gini Ratio	0.49	-0.06	0.02	0.29	0.56	0.38	0.41	0.41

Note: AWE: Agricultural wage employment; NAWE: Non-agricultural self-employment; NARE: Non-agricultural regular employment; TNAE: Total non-agricultural employment; OS: Other sources; PCI; Per capita income.

Source: Lanjouw and Shariff (2004).

In the case of the agricultural sector, while the income share from cultivation tended to rise (from 38.2 percent) with increasing levels of per capita income quintiles (to 64.5 percent), the income share from agricultural wage employment tended to decline rapidly. For the poorest income quintile, agricultural wage employment income

Figure 6.6 : Distributional Pattern of Income from Agriculture and Non-Agriculture across Different Income Quintiles in Rural India, 1993-94

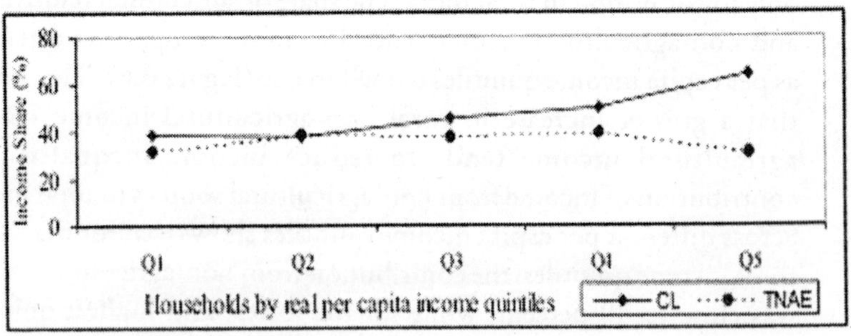

Note: CL: Cultivation; TNAE: Total non-agricultural employment.
Source: Table 6.5.

contributed as much as 28 percent of the total income, while for the highest income quintile, it accounted for only 2 percent. The income share from RNAE in total rural income was more evenly distributed, as compared to that from cultivation and agricultural wage employment across per capita income quintiles. We can see different patterns of distribution of income from different sources in the rural non-agricultural sector, namely, non-agricultural wage employment, self- and regular rural non-agricultural employment.

It was found that while income shares from non-agricultural wage employment tended to decline significantly from about 16 percent to 2 percent, the income share from non-agricultural regular employment tended to rise rapidly from about 4 percent to 21 percent, as per capita income quintiles rose. In the case of non-agricultural self-employment, for the middle three income quintiles, the contribution of income accounted for about 16 percent, while for the lowest, it accounted for about 11 percent and for the highest about 8 percent. The distribution of income from rural non-agricultural self-employment was relatively much higher than that from rural non-agricultural regular employment and higher than that from rural non-agricultural wage employment. The Gini ratios, *prima facie,* show that the inequality in the distribution of income in agriculture was highest as compared to income distribution from all rural non-agricultural sources; it was even higher than that of the total rural income of the household.

The Gini ratio for agricultural wage employment income was found to be negative, implying that income from this source amongst the rich quintiles was significantly lower, as compared to the households in the lowest income quintiles. The distribution of income within the rural non-agricultural sector indicated that the income from non-agricultural regular employment was more unequally distributed as compared to that from non-agricultural self-employment and non-agricultural wage employment; it was even highest among all types of income sources. Thus, within the rural non-agricultural sector, the income from wage employment was more equitably distributed. The Lorenz ratios for different sources of income across per capita income quintiles have been presented in Figure 6.7.

It was found that the total income from rural non-agricultural sources (TNAE) was relatively better distributed, as it was near to the line of equal distribution (LE), as compared to the distribution of income from cultivation (CL). Within the rural non-agricultural sector (TNAE), the income from rural non-agricultural self-employment (NASE), followed by non-agricultural wage employment (NAWE), was more evenly distributed than that from rural non-agricultural regular employment (NARE). Even the distribution of income from rural non-agricultural self-employment was found to be more equal than the distribution of the total per capita income from all the sources.

Figure 6.7 : Distribution of Income from Different Sources across Households by Income Quintiles in Rural India, 1993-94

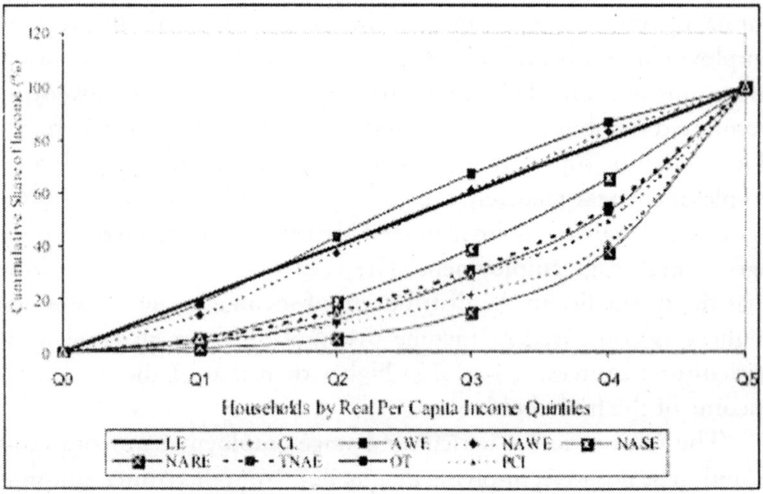

Note: LE: Line of Equal Distribution; CL: Cultivation; AWL Agricultural wage employment; NAWE: Non-agriculture wage employment; NASE: Non-agricultural self-employment; NARE: Non-agricultural regular employment; TNAE: Total non-agricultural employment; OT: Other sources; PCI: Per capita income.

Source: Table 6.3.

Regional patterns of income distribution

The data presented in Table 6.6 reveal considerable heterogeneity in distribution of income shares across per capita income quintiles in 16 major states of rural India. It was found that the income share

Table 6.6 : Percentage Distribution of Income from Different Sources across Major States in Rural India, 1993-94

Income Quintile	*CL*	*AWE*	*NAWE*	*NASE*	*NARE*	*TNAE*	*OS*	*Real PCI (Rs.)*
Andhra Pradesh								
Q1	16.7	43.0	8.9	24.3	5.1	38.2	2.1	1,397
Q2	24.9	38.6	7.4	20.7	7.3	35.4	1.1	2,449
Q3	33.3	33.8	8.2	15.1	6.4	29.8	3.2	3,503
Q4	53.2	24.2	7.0	6.2	8.6	21.8	0.8	5,141
Q5	73.1	7.8	1.9	3.8	12.8	18.5	0.6	12,709
Total	55.7	19.7	4.7	8.6	10.1	23.4	1.1	5,046
Bihar								
Q1	29.8	24.0	23.5	18.6	3.1	45.2	1.0	1,064
Q2	30.9	19.1	16.6	26.1	5.4	48.1	1.9	1,951
Q3	43.8	14.0	11.4	21.1	7.4	39.9	2.3	2,870
Q4	51.4	4.0	4.7	21.7	16.0	42.3	2.2	4,137
Q5	64.5	0.4	1.0	10.0	23.0	34.0	1.1	8,418
Total	52.8	6.7	6.4	16.5	16.0	38.9	1.6	3,690
Assam								
Q1	46.6	15.0	13.7	21.2	2.4	37.3	1.1	1,708
Q2	51.4	9.0	7.7	22.8	7.8	38.3	1.4	2,783
Q3	43.4	6.0	9.4	25.4	13.6	48.4	2.2	4,140
Q4	35.3	1.1	4.6	24.5	32.0	61.1	2.6	6,043
Q5	31.9	0.1	0.4	16.0	48.5	64.9	3.1	10,669
Total	37.7	3.3	4.6	20.7	31.3	56.5	2.5	5,071
Gujarat								
Q1	36.7	44.7	11.3	2.9	3.2	17.4	1.2	1,081
Q2	41.4	34.7	9.0	6.7	5.5	21.1	2.7	2,091
Q3	49.7	21.5	7.3	7.3	11.6	26.1	2.7	3,239
Q4	48.6	7.0	3.4	11.3	25.2	39.9	4.5	5,133
Q5	77.1	1.6	1.0	5.7	12.5	19.2	2.1	14,788
Total	63.7	9.5	3.3	6.9	13.9	24.2	2.7	5,272
Haryana								
Q1	31.3	21.9	28.5	7.0	6.2	41.8	5.0	1,755
Q2	36.2	10.3	16.8	11.6	21.9	50.2	3.3	3,252
Q3	44.4	2.6	8.0	11.6	28.5	48.1	4.9	[illegible],847
Q4	53.7	1.5	3.6	9.4	24.2	37.2	7.6	7,257
Q5	73.6	0.1	0.3	4.0	17.4	21.7	4.6	14,707
Total	58.5	3.0	5.5	7.3	20.5	33.3	5.2	6,368

(Contd..)

Table 6.6 : (Contd...)

Himachal Pradesh								
Q1	53.9	1.5	22.9	9.5	6.7	39.1	5.5	1,123
Q2	41.2	1.6	27.0	9.8	10.5	47.3	9.9	2,187
Q3	39.8	0.6	20.5	8.0	21.5	50.0	9.5	3,348
Q4	34.8	0.1	10.7	11.1	32.6	54.4	10.7	4,829
Q5	34.4	0.0	5.1	10.5	37.9	53.5	12.1	9,335
Total	37.1	0.4	12.1	10.1	29.5	51.7	10.8	4,168
Karnataka								
Q1	39.8	41.7	7.1	7.7	1.6	16.5	2.0	954
Q2	44.9	31.5	9.5	10.2	2.0	21.7	1.9	1,870
Q3	53.9	18.8	8.2	10.3	6.3	24.8	2.4	2,874
Q4	61.8	9.2	4.7	10.6	12.4	27.7	1.3	4,530
Q5	81.0	1.2	0.9	5.7	10.1	16.6	1.2	13,602
Total	69.6	8.9	3.4	7.6	9.1	20.1	1.4	4,767
Kerala								
Q1	24.1	24.9	23.8	14.5	5.4	43.6	7.3	1,670
Q2	24.9	24.0	28.6	12.1	6.5	47.2	3.9	2,974
Q3	30.3	17.6	18.4	12.9	11.2	42.5	9.6	4,041
Q4	36.4	11.7	11.8	11.0	15.3	38.1	13.8	5,859
Q5	62.3	2.5	2.1	4.9	20.4	27.4	7.8	14,288
Total	46.5	10.0	10.3	8.6	15.8	34.7	8.8	5,768
Maharashtra								
Q1	29.3	47.8	10.0	6.9	4.0	20.9	2.1	1,345
Q2	37.1	30.3	11.4	8.0	8.0	27.3	5.3	2,363
Q3	50.2	19.2	5.6	8.1	10.9	24.7	5.9	3,572
Q4	61.7	8.0	3.2	5.4	17.3	25.9	4.3	5,635
Q5	81.2	1.8	0.4	3.9	10.4	14.7	2.3	14,684
Total	66.9	10.0	3.0	5.2	11.3	19.6	3.4	5,524
Madhya Pradesh								
Q1	42.4	27.7	15.2	10.7	1.4	27.3	2.6	1,261
Q2	47.3	21.5	13.7	10.7	4.7	29.2	2.0	2,093
Q3	57.9	12.5	10.5	10.3	6.5	27.3	2.3	2,914
Q4	68.1	6.2	5.9	8.3	9.5	23.8	1.9	4,329
Q5	81.0	0.9	1.1	4.2	11.1	16.5	1.6	10,196
Total	69.4	7.3	5.6	7.0	8.9	21.4	1.9	4,159
Orissa								
Q1	44.9	25.6	18.6	7.9	0.9	27.4	2.1	743
Q2	44.6	16.9	20.2	13.0	3.6	36.8	1.6	1,454
Q3	50.6	10.2	13.2	17.8	6.2	37.1	2.0	2,223
Q4	55.6	4.3	6.3	15.0	15.7	37.1	3.1	3,413

(Contd...)

***Table 6.6** : (Contd...)*

	CL	AWE	NAWE	NASE	NARE	TNAE	OS	PCI
Q5	55.2	1.2	2.2	15.5	24.3	42.0	1.6	7,287
Total	53.1	5.9	7.3	15.1	16.6	39.0	2.0	3,027
Punjab								
Q1	26.0	24.5	27.9	11.8	5.1	44.8	4.6	1,473
Q2	28.6	9.5	16.9	17.2	21.8	55.8	6.1	2,793
Q3	37.7	4.9	9.8	17.6	26.9	54.4	3.1	4,221
Q4	51.1	2.2	4.3	8.4	27.2	39.9	6.9	6,706
Q5	70.1	0.2	1.0	4.7	20.4	26.1	3.5	16,690
Total	56.2	3.2	5.5	8.6	22.1	36.2	4.4	6,380
Rajasthan								
Q1	57.9	5.0	28.1	5.0	2.9	35.9	1.2	1,179
Q2	55.9	2.8	25.8	7.0	4.8	37.6	3.7	2,056
Q3	54.5	1.3	17.4	11.8	9.4	38.6	5.6	3,032
Q4	52.2	0.6	8.6	13.0	19.6	41.2	6.0	4,511
Q5	57.7	0.1	2.9	11.8	24.5	39.3	2.9	10,328
Total	55.9	0.9	9.8	11.2	18.2	39.2	4.0	4,227
Tamil Nadu								
Q1	26.3	51.6	11.4	7.3	1.1	19.8	2.3	1,093
Q2	27.8	27.5	18.2	14.1	10.9	43.2	1.6	2,130
Q3	32.6	21.6	15.5	18.0	10.9	44.4	1.4	3,377
Q4	35.7	14.9	14.2	7.8	23.4	45.5	4.0	5,431
Q5	42.8	5.2	7.0	9.7	33.7	50.4	1.6	12,292
Total	37.7	13.7	11.0	10.7	24.8	46.4	2.1	4,867
Uttar Pradesh								
Q1	48.4	14.2	16.3	14.2	5.1	35.7	1.7	995
Q2	9.7	15.9	19.9	9.0	44.8	1.4	1.0	2,008
Q3	4.7	9.2	17.6	19.1	45.9	2.6	3.0	3,047
Q4	1.9	5.1	15.1	24.7	44.8	2.5	8.0	4,553
Q5	61.2	0.3	2.0	8.5	26.0	36.5	2.0	10,317
Total	54.6	2.9	5.7	12.6	22.1	40.4	2.0	4,185
West Bengal								
Q1	36.3	27.4	10.2	8.0	17.7	35.9	0.4	1,124
Q2	36.5	27.2	9.9	16.1	8.8	34.9	1.4	1,858
Q3	33.6	19.3	6.6	29.7	10.3	46.6	0.5	2,471
Q4	40.2	10.0	4.3	30.9	13.4	48.5	1.2	3,534
Q5	36.1	4.0	1.3	27.6	28.9	57.8	2.0	6,788
Total	36.7	12.2	4.4	25.9	19.4	49.7	1.4	3,158

Note: CL: Cultivation; AWE: Agricultural wage employment; NAWE: Non-agricultural wage employment; NASE: Non-agricultural self-employment; NARE: Non-agricultural regular employment; TNAE: Total non-agricultural employment ;OS: Other sources; PCI: Per capita income

Source: Lanjouw and Shariff (2004).

from the non-agricultural sector *vis-à-vis* agricultural sector (cultivation) was highest in only four states, namely, Himachal Pradesh, North east- Assam, Tamil Nadu and West Bengal. In these states, the non-agricultural income sources accounted for more than 45 percent of the total income (46 percent in Tamil Nadu and 5 percent in North-east Assam). On the other hand, in states such as Gujarat, Madhya Pradesh, Andhra Pradesh, Maharashtra and Karnataka, the share of income from non-agricultural sources was less than 25 percent. In the rest, on an average, the share of non-agricultural income was around one-third.

The distribution of all non-agricultural income shares across state level income quintiles indicate that there were only four states, namely, Bihar, Haryana, Kerala and Punjab, in which households in the lowest income quintiles (Q1 and Q2) derived the largest share of income from rural non-agricultural sources. In these states, the share of income from rural non-agricultural sources tended to decline for higher income quintiles. The share of income from casual wage employment was highest for the poorest quintiles, whereas for the upper income quintiles, the share of income from non-agricultural regular employment was found to be highest in most of the states.

Gini ratios of income sources by major states

The regional patterns of Gini ratio for different income sources indicate (Table 6.7) that in states such as Uttar Pradesh, Andhra Pradesh, Karnataka, Gujarat, Maharashtra, Punjab, Kerala, Haryana, the Gini ratio of income from cultivation was quite higher than the national level ratio (ranging from 0.64 in Uttar Pradesh to 0.50 in Haryana). In the remaining states, Gini ratio was found to be lower than the national level, ranging between 0.25 in Assam and 0.49 in Tamil Nadu. This could largely be attributed to unequal distribution of landholdings.

It was also found that the Gini ratio of income from cultivation was lower than the income from rural non-agricultural sources in only four states, namely, Assam, Himachal Pradesh, Orissa and West Bengal. In the rest of the states, the reverse situation was found to

Table 6.7 : Gini Ratios of Income from Different Sources by Major States in Rural India, 1993-94

State	*CL*	*AWE*	*NAWE*	*NASE*	*NARE*	*TNAE*	*OS*	*Real PCI*
Andhra Pradesh	0.58	0.09	0.14	0.02	0.53	0.26	0.15	0.40
Bihar	0.48	-0.21	-0.16	0.22	0.59	0.31	0.29	0.37
Assam	0.25	-0.32	-0.11	0.27	0.60	0.42	0.47	0.33
Gujarat	0.56	-0.14	0.02	0.45	0.52	0.43	0.44	0.46
Haryana	0.50	-0.40	-0.28	0.21	0.36	0.23	0.39	0.38
Himachal Pradesh	0.31	-0.33	0.06	0.39	0.54	0.40	0.43	0.37
Karnataka	0.56	-0.12	0.07	0.37	0.60	0.42	0.36	0.47
Kerala	0.55	-0.02	-0.05	0.19	0.56	0.29	0.42	0.39
Maharashtra	0.56	-0.15	-0.11	0.30	0.48	0.34	0.31	0.43
Madhya Pradesh	0.48	-0.18	-0.07	0.20	0.55	0.28	0.31	0.39
Orissa	0.43	-0.14	-0.01	0.40	0.64	0.44	0.37	0.40
Punjab	0.56	-0.30	-0.15	0.19	0.45	0.29	0.37	0.43
Rajasthan	0.40	-0.26	-0.04	0.47	0.60	0.40	0.37	0.39
Tamil Nadu	0.49	0.04	0.28	0.36	0.62	0.48	0.40	0.42
Uttar Pradesh	0.64	-0.12	0.08	0.33	0.36	0.64	0.41	0.41
West Bengal	0.33	-0.05	-0.05	0.43	0.50	0.41	0.50	0.33
All India	0.49	-0.06	0.02	0.29	0.56	0.38	0.41	0.41

Note: CL: Cultivation; AWE: Agricultural wage employment; NAWE: Non-agricultural wage employment; NASE: Non-agricultural self-employment; NARE: Non-agricultural regular employment; TNAE: Total non-agricultural employment; OS : Other sources ; PCI; Per capita income.

Source: Estimated from Table 6.6.

prevail. It could also be observed that there were some states where the income from rural non-agricultural sources was more inegalitarian than that from cultivation. In the case of income from the rural non-agricultural sector, it was observed that in nine states, namely, Uttar Pradesh, Tamil Nadu, Orissa, Gujarat, Karnataka, Assam, West Bengal, Himachal Pradesh and Rajasthan, Gini ratios were quite higher than the national level, ranging from 0.64 in Uttar Pradesh to 0.40 in Rajasthan. In the remaining states, it ranged between 0.23 in Haryana and 0.34 in Maharashtra. Within the non-agricultural sector, it was found that six states had higher Gini ratios of income from rural non-agricultural regular employment and nine states from rural non-agricultural self-employment.

Decomposition of income inequality

Decomposition of income inequality by sources of income provides the contribution of each source of income to total income inequality and effects of further occupational diversification on income distribution. For decomposition of income inequality by sources of income, several techniques have been developed (Kakwani 1980; Fields 1980; Pyatt, Chen and Fei, 1980; Shorrocks 1982, 1983). Among them, the coefficient of variation was used to decompose income inequality. Using the coefficient of variation, the impact of occupational diversification on income distribution was measured as defined in the following terms.

$$\sum w_i\, c_i = 1,\ w_i = \mu_i/\mu,\ c_i = \tau_i\,(\sigma_i/\mu)/(\sigma/\mu)\ or\ c_i = \tau_i \cdot \sigma_i/\sigma \quad \text{...........(8)}$$

where μ_i and μ are the mean income from i^{th} occupation and all occupations, respectively. w_i is the contribution of ith occupation in total income. σ_i and σ are the coefficient of variation in income from i^{th} and all occupations, respectively. τ_i is the correlation coefficient between income from i^{th} occupation and total income. c_i is the relative concentration coefficient of income from i^{th} occupation in total income. If the relative concentration coefficient (c_i) associated with an occupation is more than unity, then increase in income from that occupation will raise income inequality and *vice versa;* if it is equal to unity, the existing distribution remains unaffected. The results of the decomposition of income inequality by income sources in rural India have been provided in Table 6.8.

It was found that the distribution of income from cultivation, followed by rural non-agricultural regular employment, was most inegalitarian. On the other hand, the distribution of income from rural non-agricultural wage employment, followed by non-agricultural self-employment, and agricultural wage employment, was most egalitarian. The inegalitarian distribution of income from cultivation could be attributed to unequal distribution of landholdings. Similarly, in respect of rural non-agricultural regular employment, there appears to be some entry barriers to the resource poor households. The entry barriers may be in terms of education, influence, corruption and so on, which most of the poor households do not possess. It was the rich households that had easy accessibility to such non-agricultural regular employment in government/private sector, resulting in a greater inequality in the distribution of income.

Table 6.8: Decomposition of Income Inequality by Sources of Income in Rural India, 1993-94

Income sources	*Coefficient of variation*	*Source of income weight*	*Correlation between source income and Total income*	*Relative concentration coefficient*	*Source inequality weight*
	σ	w_i	τ_i	c_i	$w_i c_i$
Cultivation	1.21	0.527	0.951	1.355	0.714
AWE	0.80	0.075	-0.208	-0.197	-0.015
NAWE	0.62	0.065	-0.127	-0.093	-0.006
NASE	0.74	0.108	0.570	0.501	0.054
NARE	1.20	0.188	0.815	1.155	0.217
TNAE	0.85	0.341	0.779	0.787	0.268
Other sources	1.27	0.037	0.611	0.918	0.034
PCI	0.84	1	–	–	1

Note: AWE: Agricultural wage employment; NAWE: Non-agricultural wage employment; NASE: Non-agricultural self-employment; NARE: Non-agricultural regular employment; TNAE: Total non-agricultural employment; PCI; Per capita income.

Source: Estimated from Table 6.6.

The relative concentration coefficient (c_i) identified cultivation and non-agricultural regular employment as inequality-increasing sources, whereas non-agricultural and agricultural wage employment, non-agricultural, self-employment and total non-agricultural employment were identified as inequality-decreasing sources of income. The proportional contribution of different sources to total income inequality has been presented in terms of source inequality weights, $w_i c_i$. The source-inequality weights show that cultivation contributed the maximum to total income inequality ($w_i c_i = 0.714$), followed by non-agricultural regular employment ($w_i c_i = 0.217$) and non-agricultural self-employment ($w_i c_i = 0.054$). The income sources from agricultural wage employment and non-agricultural wage employment contributed to a decline in overall income inequality; the contribution of the former was higher than the latter.

Conclusions

From the foregoing analysis, it may be concluded that a large number of people experiencing poverty had been concentrated in rural than in urban areas; three out of every four poor persons lived in rural areas in 1999-2000. The annual growth rate of people in rural poverty had registered a rapid decline in the 1980s and more so in the late1990s. The decline in the incidence of poverty was a little faster in rural than in urban areas. The decline in the incidence of rural poverty in the 1980s had been mainly attributed to a faster growth of agriculture, followed by infrastructure and human resource development, particularly female literacy. The incidence of poverty in the rural non-agricultural sector was found to be lower, as the daily real non-agricultural wage rates were reported to be quite higher. Within the rural non-agricultural sector, the incidence of poverty was not uniform. It was found to be higher in construction, followed by manufacturing, transport-storage-communication, trade-business-restaurant and services.

The regional pattern of poverty incidence shows that states with a higher per capita net income from agriculture, rate of literacy and male RNAE have had a smaller proportion of people in rural poverty and have also witnessed a rapid progress in poverty reduction, as compared to states with lower ones during the reform period. The growth of agriculture (measured in terms PCIA, daily real wage rates of male and female labourers in agricultural occupations) was found to have a lesser impact on the reduction of poverty as compared to the growth of the rural non-agricultural sector (measured in terms of male RNAE, daily real wage rates of males and females in rural non-agricultural occupations) during the reform period. The impact of agricultural growth on poverty reduction seemed to have slowed down in the 1990s, as compared to what it was in the 1980s. The hypothesis that the growth of the rural non-agricultural as against agricultural sector contributed to a greater decline in the incidence of rural poverty during the reform period seemed to be tenable. This hypothesis needs to be verified with primary data in regions with different levels of agricultural prosperity.

An increase in per capita net income from agriculture, the proportion of male workers in RNAE, daily real wage earnings in agriculture and non-agricultural occupations, human resource development as measured by the rate of literacy and rural per capita government expenditure for rural development and social services were found to be the important sources of poverty reduction. Among them, the growth of male RNAE and daily real wage rates of male workers in non-agricultural occupations were found to be more important sources of poverty reduction. Within the non-agricultural sector, the growth of male RNAE had a greater poverty reduction effect than that of female RNAE, as it appeared to be low productive, lowly paid. In view of a gradual decline in the proportion of government expenditure for rural development and poverty alleviation programmes, the promotion of male RNAE, encouragement of female participation in more productive non-farm jobs and an increase in agricultural growth and human resource development could be considered an effective holistic approach for poverty alleviation in the countryside.

Distribution of income from cultivation, followed by rural non-agricultural regular employment, was most inegalitarian. On the other hand, distribution of incomes from rural non-agricultural wage employment, non-agricultural self-employment and agricultural wage employment was found to be more egalitarian. The inegalitarian distribution of income from cultivation could be attributed to unequal distribution of landholdings in the country. Similarly, in respect of rural non-agricultural regular employment, there appeared to be some entry barriers to the resource poor households. The entry barriers might be in terms of lack of education, influence, social networks and so on, which most of the poor households do not possess. It was the rich households which had easy access to such non-agricultural regular employment in government/private sector, which results in a greater inequality in the distribution of income. The relative concentration coefficient identified cultivation and non-agricultural regular employment as inequality-increasing sources, whereas non-agricultural and agricultural wage employment, non-agricultural self-employment and total non-agricultural employment as inequality-decreasing sources of income. The source-inequality weights show that cultivation, followed by non-agricultural regular employment and non-agricultural self-employment, largely contributed to total income inequality. In contrast, income sources from agricultural wage employment, followed by non-agricultural wage employment, contributed to a decline in overall income inequality.

NOTES

1. The real wage rates in the rural non-agricultural occupations have been on the rise since the mid-1970s. The non-agricultural wage rates were also much higher than those of agriculture (Bhalla 1993;Vaidyanathan 1994; Unni 1997b; Sharma 2001; GoI 2003b). An increase in the proportion of workers in RNAE is, *a priori,* expected to increase the income levels of the rural nonagricultural labour households, to that extent contributing to a decline in the incidence of rural poverty.
2. The poor quality of employment can be reflected in terms of percentage of casual wage employment to total rural employment, which has been on the gradual rise. This may cause further impoverishment.
3. The rise in the daily real wage rates in agriculture due to shift of labour from the agricultural to rural non-agricultural sector can be seen in terms of a strong and positive correlation between the daily real wage rates of labourers in agricultural and non-agricultural occupations (see, for instance, correlation matrix, Tables 5.2 and 5.4 in Chapter 5). Similar types of changes have been observed by Lanjouw and Shariff (2004).
4. For instance, the incidence of unemployment and the rate of male literacy were highly correlated. In order to avoid the multicollinearity problem, the incidence of male unemployment was not included in the model, though it was one of the important factors determining the extent of male RNAE.

7 Summary of Findings and Important Suggestions

Introduction

Since the 1970s, the employment structure has started witnessing changes in different dimensions in India, especially in rural areas. One such explicit change has been diversification of employment in favour of the non-agricultural sector. An analysis of occupational diversification and its impact on distributional aspects, therefore, helps policy makers and development practitioners in deciding about the type of policy intervention needed to tackle simultaneously the triadic problems of educated youth unemployment, poverty and inequality in developing economies like India. In view of the growing population pressure on agriculture, declining land frontiers, deteriorating employment elasticity in agriculture and organised urban manufacturing sector and excess supply of labour force, on the one hand, and increasing investment opportunities due to liberalisation, privatisation and globalisation, on the other, the need for sustained growth of RNAE across different parts of the country has been well recognised since the 1990s, especially when the "educated youth unemployment" started swelling in its worst forms.

The sustained and equitable growth of RNAE has occupied a central place in the rural development programmes. In view of this, an attempt has been made to examine the dynamics of employment and unemployment by gender, areas and status; to examine the trends and patterns of RNAE by gender and activities across major states and by level of age groups, education and size of landholdings; to explore the factors contributing to the growth of RNAE and estimate the elasticity of important factors; and to study the impact of RNAE on reduction of poverty and income inequality in rural India during the reform period. The study is based on secondary data collected from NSSO for the quinquennial rounds from 1972-73 to 1999-2000. The data were analysed for sixteen major states. The simple percentages, ratios, correlation, log-linear multiple regression, coefficient of variation, principal component analysis and decomposition technique were employed for analysing the data. The important findings of the study are summarised as follows:

Dynamics of Employment and Unemployment

There has been an apprehension that the economic reforms have had an adverse impact on employment and poverty, particularly in rural India. Although it is premature to assess the impact of economic reforms on employment and poverty, the data on employment and unemployment published by NSSO (the 50th and 55th rounds) for the years 1993-94 and 1999-2000 have revealed a mixed picture. The annual growth rate of workers has been higher in urban than in rural areas throughout the period under study, implying that the employment opportunities were largely created in the urban sector. The annual growth rate of workers declined considerably during the reform period, as compared to the pre-reform period at the all-India level; the decline was greater in respect of rural workers as compared to their urban counterparts and also for females as compared to males. In rural areas, there was a considerable decline in the growth rate of male workers, whereas in respect of female workers, a negative growth rate was observed during the reform *vis-à-vis* the pre-reform period. Similarly, in urban areas, the annual growth rate of workers witnessed a sharp decline during the reform period, as compared to the pre-reform period; the decline was greater in respect of females as against males. A major decline, both in rural and urban areas, appeared to have occurred in the subsidiary status of occupation.

As regards WPRs, they began to decline marginally in the late 1980s; they sharply increased in the early 1990s; and again declined steeply in the late 1990s, both in rural and urban areas. The decline in WPRs, which started well before the reform measures, was not only higher but also accelerated in rural as against urban areas during the reform period. Although the female WPRs were found to be lower, a significant decline took place in it as compared to their male counterparts in rural as also urban areas during the reform period.

Going by age group of the workers, the WPRs tended to increase with increasing levels of age groups up to a certain level and thereafter declined for males and females both in rural and urban areas. The increase was quite sharper in the age groups 5-9, 10-14, 15-19, 20-24 and 25-29 years and became a little flatter for the higher age groups and thereafter slowly declined. This constituted the "inverted U-shaped curve" with varying contours across gender and areas. At both end sides of the curve, the WPRs tended to be lower and higher at intermediate levels. A greater decline in WPRs started to occur among the children and youth, as compared to the adults and aged workers even before the economic reforms were initiated.

The WPRs for males and females aged 15 and above, by level of education, reveal that with increasing levels of educational attainments, the WPRs tended to decline up to a certain level and increased thereafter in varying degrees across gender and areas. This implies that at both extremes of educational attainments the WPRs were higher and lower at the intermediate levels. Such a pattern of WPRs constituted the "saucer-shaped curve" with varying contours across gender and areas. In respect of urban females, the pattern of WPRs formed the "slanting U-shaped curve" as the level of education rose. This was a mirror image of the "inverted U-shaped curve" along which the ratio of females in full-time housework tended to increase as the level of education rose.

The WPRs for males aged 15 and above have been on the decline in both the periods; a greater decline took place during the reform period. Going by different levels of education, the decline in male WPRs in rural areas was higher among the workers who were illiterate or literate up to primary level of education. In the case of rural females, the decline in WPRs was greater in the pre-reform, as compared to the reform years; it increased rather marginally. Such decline was rapid among female workers educated up to graduate and above level and workers who were illiterate or literate up to primary level of education during the same period.

The male WPRs in urban areas showed a greater decline during the reform period, as compared to the pre-reform period. Although the decline in urban male WPRs was quite discernible at each level of educational attainment, a rapid decline was observed among the workers who were not literate or literate up to primary level, and even at graduate and above level of education. But, the decline was not significant during the reform period. The decline in urban female WPRs started even before the introduction of economic reforms, but a greater decline began to take place after the introduction of economic reforms. At different levels of education, the female WPRs showed a greater decline at higher levels of education, and marginal decline for illiterates or literates up to primary level of education.

As far as unemployment rates were concerned, it was found that the pattern of unemployment rates at different age groups constituted the "inverted V-shaped curve" with varying contours across males and females, in both rural and urban segments of the labour market in India. The unemployment rates were traditionally higher in urban than in rural areas. The incidence of unemployment, as per UPS and UP+SS, was found to be higher among males as compared to females in rural areas, and among females as compared to males in urban areas, and in urban as compared to rural segment of the labour market. The incidence of unemployment, for both males and females, used to be quite higher in the pre-reform years as compared to that in the reform years. Unemployment rates, as per both UPS and UP+SS, were found to be higher among male and female youth. A greater increase took place in youth unemployment rates, particularly in rural areas, in 1999-2000 over 1993-94. But, the incidence of unemployment for youth males in urban areas declined during the same period. Going by the level of education, the incidence of unemployment was quite high among highly educated workers. A marginal rise took place in the incidence of unemployment for both male and female workers who were illiterate and literate up to primary level of education, whereas in the case of workers with higher educational attainments, a greater decline occurred during 1993-94 to 1999-2000; the decline was found to be quite discernible in respect of females.

It may be concluded that the decline in WPRs mainly came from :

(i) an increase in unemployment rates among illiterates and literates up to primary level and among youth and old-aged people; (ii) withdrawal of workers from the labour market in favour of educational institutions and a corresponding increase in SPRs; and (iii) rise in the ratio of population neither in the labour market nor in educational institutions. The change in rural and urban segments of the labour market in India, especially withdrawal of the labour force in the age groups of 5-9 to 25-29 years appeared to have had a beneficial impact on the economy in terms of human capital accumulation (and also human capital deepening) so as to improve the efficiency in resource utilisation and ensure a self-sustaining growth process in the long-run.

The quality of employment, especially in the case of women in rural areas, has been significantly deteriorating in the era of globalisation, because the rate of casualisation and informalisation of formal employment has increased rapidly in recent years. The incidence of casualisation was not uniform across the different levels of education and activities both in rural and urban areas. It was found to be quite high among the illiterates, as compared to the literates, and among illiterate women, as compared to illiterate men, particularly in rural areas. As the level of education rose, the proportion of workers experiencing casualisation of employment declined in the early 1990s.

Growth, Size and Composition of RNAE

The compound annual growth rates of rural non-agricultural workers registered a constant decline till the early 1990s and gained momentum in the late 1990s. A significant setback took place in respect of rural female non-agricultural workers during the reform, as compared to the pre-reform period. It was also found, at the aggregate level, that the growth rates of non-agricultural workers fluctuated largely in accordance with the growth rates of agricultural income and foodgrains production. In other words, the growth rate of rural non-agricultural employment, particularly for females, tended to move in the opposite directions to the growth rates of agricultural income and foodgrains production, implying a distress movement of female workers into the rural non-agricultural sector. The regional dimension of the growth pattern also reveals that the growth rate of workers in RNAE was not uniform across all the sixteen major states; none of the states witnessed a uniform pattern of growth rates of workers in RNAE and almost every state was subjected to periodic fluctuations, depending upon the performance of agriculture and extent of linkages between agricultural and non-agricultural sectors in a particular area.

The sectoral pattern of growth indicates that the rural non-agricultural activities that registered positive growth rates for male workers since the 1970s, namely, transport, storage, communication; construction; and, trade, hotels and restaurants continued to show an impressive progress. On the other hand, the manufacturing sector experienced a lower growth rate during the reform *vis-à-vis* the pre-reform period. A rapid decline took place in the growth rates of male workers in public utilities, followed by mining and quarrying, during the reform period, as compared to the pre-reform period. In respect of females, a positive growth rate of employment registered in agriculture during the pre-reform period, but a negative growth rate during the reform period. Going by the activities within the rural non-agricultural sector, it was found that a negative growth rate of female workers took place in mining and quarrying; trade, hotel and restaurant business; and a positive growth continued to take place in construction; services; and manufacturing; but a significant slow-down occurred in the manufacturing and service sectors during the reform period, as compared to the pre-reform period.

The proportion of both male and female workers in RNAE, by and large, witnessed a steady increase during 1972-73 to 1999-2000, excepting 1987-88; the increase in the share of workers in RNAE slowed down during the reform period, as compared to the pre-reform period; and the slow-down was faster among the male workers than among female workers, resulting in a widening gap between males and females in RNAE. Thus, it is evident that while the share of workers in the rural non-agricultural sector *vis-à-vis* the agricultural sector had been on the rise, the rate at which it was increasing in the 1980s had not sustained itself in the 1990s.

A considerable variation in the proportion of workers in RNAE existed across states, activities and gender. The regional patterns of the proportion of workers in RNAE have been attributed to the growth as well as poverty related factors. The regional variation of male RNAE was largely explained by growth-related factors, whereas in respect of female RNAE, it was by distress-induced factors. The value of coefficient of variation reveals that the regional disparity started declining in the years of the pre-reform period, but it again began to rise in the years of the reform period. In respect of female workers, the coefficient of variation was found to be much higher than in the case of male workers. As in the case of male workers, the values of coefficient of variation also started declining in the years of the pre-reform period and began to rise in the years of reform period; the pace of increase was much greater than the pace of decline. This implies that during the reform period, the regional disparity in the growth of RNAE for both males and females had widened. The regional disparity was not uniform across different activities of the rural non-agricultural sector either.

The WPRs in the non-agricultural sector tended to increase with rising levels of age-groups up to a certain point (age-groups 5-9 to 30-44 years) and thereafter (age-groups 45-59 to 60+ years) declined. This constituted what could be described as the "inverted U-shaped curve" with varying contours across males and females. The data show that youth and adult male workers were more likely to take up jobs in the rural non-agricultural sector and relatively older male persons in agriculture. In respect of female workers, the youth and adult ones were more likely to be involved in the agricultural sector and less in the non-agricultural sector. The child labourers and aged workers, both male and female, were found to be more in the agricultural sector and less in the non-agricultural sector. The proportion of male and female child labourers declined continuously both in agricultural and non-agricultural occupations during the pre-reform and reform periods; a greater decline took place in agriculture during the pre-reform period. While agriculture witnessed a greater decline in WPRs for male youth and adult workers, the non-agricultural sector showed a rise during both the periods; but the pre-reform years showed a relatively greater decline in respect of agriculture. But, in respect of female workers, the non-agricultural sector *vis-à-vis* agriculture showed a greater decline for the youth and adult workers during the reform period, as compared to the pre-reform period. The male and female workers in the age group 60+ years declined in both the sectors; a greater decline took place in the agricultural than in the non-agricultural sector.

Going by the level of education, it was found that the WPRs, for both males and females aged 15 and above, tended to decline in the agricultural and rise in the non-agricultural sector as the level of education rose, implying that a larger share of educated workers were more likely to be involved in RNAE. The educational attainments of the workers in the non-agricultural sector were found to be much higher than those of the workers in agriculture. Agriculture had a vast majority of illiterate male and female workers, as compared to rural non-agricultural ones; going by gender, the proportion of illiterate female workers was quite significant in agriculture, followed by the non-agricultural sector, as compared to illiterate male workers. The proportion of illiterate workers, however, declined sharply both in the agricultural and non-agricultural sectors during both the periods; a greater decline started to take place in the years of the pre-reform period, as compared to the reform period; in the agricultural sector, as compared to the non-agricultural sector; and in respect of male workers in agricultural, and female workers in non-agricultural occupations.

As far as WPRs by the size of landholdings were concerned, the male and female WPRs tended to rise in agriculture and decline in the non-agricultural sector as the size of landholdings rose, implying that workers without land or with smaller size of landholdings were more likely to be involved in RNAE, as compared to workers with larger size of landholdings. The WPRs for males and females declined in all the categories of landholdings and more so in respect of landless and lower sized landholdings during the pre-reform period, as compared to the reform period. In the case of the rural non-agricultural sector, on the other hand, the WPRs for males experienced a moderate rise, particularly in the case of lower sized landholdings, whereas the WPRs for females showed a constant decline during the same period;

a greater decline took place during the pre-reform period, as compared to the reform period. The proportion of male and female workers without land was observed to be quite significant in the case of the non-agricultural sector, as compared to the agricultural sector.

This shows that workers who did not have productive assets such as land were more likely to find jobs in the rural non-agricultural sector, mostly in casual/contractual types. This lent support to the distress diversification hypothesis. The proportion of male and female workers in the categories of landless household and those with land under one hectare showed a rise both in the agricultural and non-agricultural sectors; a greater rise took place in the former than in the latter.

Degree of Casualisation in RNAE

As regards casualisation of RNAE, it was found that the proportion of workers in casual employment was higher in the agricultural and lower in the non-agricultural sector; the male workers experienced lesser incidence of casualisation as compared to the female workers in agriculture, and *vice versa* in the non-agricultural sector. The pace of casualisation for males increased in both the agricultural and non-agricultural sectors during both the periods; a greater increase took place in the non-agricultural sector during the reform period. The pace of casualisation for females was found to have increased in agriculture and declined in the non-agricultural sector during both the periods. The degree of casualisation varied across age groups, level of educational attainments and size of landholdings in rural areas.

The incidence of casualisation for male and female RNAE tended to decline as the age-group of workers went up; but it was increasing among the male children and youth much faster than in the case of others, tended to decline among the female children and youth much faster than in the case of others. However, the share of casualisation in RNAE tended to fall more among youth and adults as compared to others. The youth were more likely to take up casual employment, whereas the adults and aged persons took up self-employment in the rural non-agricultural sector.

Going by the level of education, the share of male and female workers tended to decline in casual employment, followed by self-employment and rise in regular employment in the rural non-agricultural sector, as the level of education rose; the pace of decline was much faster in casual employment than in self-employment. This implies that education reduced the share of male and female workers seeking employment in casual employment and raised their share sharply in regular salaried/wage employment in the rural non-agricultural sector. The pace of casualisation in RNAE accelerated for males during the reform period; the pace of casualisation was a little faster among workers with higher educational attainments.

In respect of females, casualisation declined during the reform period; a greater decline took place among those who were illiterate and literate up to the primary level of education, mostly due to withdrawal of labour in favour of schools and colleges. The growth of self-employment for females seemed to have taken place at the cost of regular and casual employment in the rural non-agricultural sector. The incidence of casualisation for both males and females was quite significant among those who were illiterate and literate with lower levels of educational attainments. The female workers *vis-à-vis* male workers, who were illiterate and literate at lower levels education, experienced a greater incidence of casualisation, but the share of male and female workers experiencing casualisation declined more during the reform period.

The male and female workers with larger size of landholdings were more likely to seek high quality employment such as self- and regular employment in the rural non-agricultural sector rather than casual employment in agriculture or in the rural non-agricultural sector. The degree of casualisation was found to be higher among those workers who did not have any land and it declined rapidly as the size of landholdings rose.

Determinants of Male and Female RNAE

A considerable variation in the proportion of workers in RNAE across states and gender has been attributed to the growth as well as poverty related factors. The factors that determined the level of male participation in RNAE were not the same factors that determined the level of female participation in RNAE. For the growth of male RNAE, the consumption linkages were found to be more operative than the production linkages; the backward production linkages were stronger than forward production linkages. Similarly, in the case of female RNAE, the production linkages were found to be a little stronger than consumption linkages; the forward production linkages were more effective than backward production linkages. While the growth of male RNAE was largely determined by growth related factors (growth induced), the growth of female RNAE was determined by distress/poverty-compelled factors (distress induced). The male RNAE was more productive, better paid and not in the residual sector, whereas in the case of females, it was distress induced, low productive, lowly paid and in the residual sector.

In the case of male RNAE, higher per hectare/capita net income from agriculture, mechanisation of agriculture, daily real wage rates of agricultural labourers in agricultural and rural non-agricultural occupations, reduction in the incidence of rural poverty, higher rate of literacy and availability of rural infrastructure, among others, were found to be important factors contributing to the growth of male RNAE. Growth of agriculture, decline in rural poverty and increase in male literacy contributed significantly to the growth of male RNAE through consumption and production linkages. There was also a distress movement of male workers into the rural non-farm sector, but this distress movement did not lend any support to the residual sector hypothesis, implying that though the growth of male RNAE was found to be in part distress induced, it was not entirely low productive, contractual, dead-end, lowly paid jobs. This is quite evident from the finding that there was a strong negative association between the proportion of male workers in RNAE and rural poverty. The data on daily real wage earnings of male agricultural labourers in agricultural and non-agricultural occupations also discredited the residual sector hypothesis in respect of male RNAE.

In the case of female workers, per hectare higher net income from agriculture and greater degree of commercialisation were found to be the key factors in the promotion of RNAE through consumption and production linkages. However, distress-induced factors such as lower daily real wage earnings in agricultural and non-agricultural occupations, higher proportion of small and marginal landholdings, landless households, incidence of poverty and unemployment rates, as per CDS and UPS, emerged as the important factors contributing to the growth of female RNAE. The distress-induced factors emanating from agriculture and also distress in general in terms of higher unemployment rates appeared to be stronger than agricultural growth induced factors in determining the level of female participation in RNAE.

A negative correlation existed between the proportion of females in RNAE and the daily real wage earnings in agricultural as well as non-agricultural occupations, but statistically it was not so significant. The proportion of small and marginal landholdings was significantly and positively associated with female RNAE, whereas in the case of the average size of small and marginal landholdings, a negative and significant association was found to exist. It follows from this fact that the growth of female RNAE was distress diversification resulting from agriculture. Consequently, the participation of females seemed to have brought down the daily real wage rates in the rural non-agricultural sector. A strong positive association existed between female RNAE and female unemployment rates, as per both CDS and UPS, which implies that the growth of female RNAE as a whole was distress induced, but not necessarily resulting from agriculture. In fact, there was a strong positive association between the unemployment rates and the growth of agriculture measured in terms of per hectare income from agriculture.

A positive correlation between female RNAE and rural poverty suggests that the growth of RNAE in which female workers were involved was in a residual sector. In a nutshell, the growth of female RNAE was distress-induced and residual in nature—low productive, casual, lowly paid, dead-end jobs. It may, therefore, be concluded that the growth of male RNAE as a whole was determined largely by growth related factors and did not lend any support to the residual sector hypothesis; whereas in the case of females, the growth was distress induced and lent support to the residual sector hypothesis. This observation needs to be tested with primary data in regions with different levels of agricultural prosperity and availability of rural infrastructure.

At the disaggregated level, the growth process of different activities of the rural non-agricultural sector for males and females, by and large, presented a similar picture. It was found that higher per hectare fertilizer consumption, degree of urban agglomeration, degree of urbanisation, daily real wage rates in agriculture, and to some extent, per hectare net income from agriculture, contributed positively to the growth of male employment in manufacturing sector. Factors external to the rural economy appeared to be more operative in the growth of male employment in the manufacturing sector. At the same time, it was also significantly and positively influenced by the incidence of unemployment as per CDS. In the case of female employment in manufacturing, it was found that the proportion of area under non-foodgrains plus paddy was significantly and positively associated with female employment in the manufacturing sector (strong forward production linkage). The higher proportion of small and marginal landholdings, the higher size of large holdings, greater incidence of unemployment among the educated female youth and incidence of poverty also contributed significantly and positively to determining the level of female participation in manufacturing activities, mostly on casual and daily basis. The female employment in manufacturing, however, appeared to be distress induced and residual in nature.

An increase in per capita non-food consumption expenditure, daily real wage rates of male agricultural labourers in agriculture, rate of literacy, decline in poverty and number of landless households and availability of rural infrastructure were the important factors determining the extent of male employment in construction. As a matter of fact, male employment in construction was not distress induced and not in the residual sector. In respect of females, the degree of commercialisation and composite index of mechanisation in agriculture positively, and per hectare net income from agriculture negatively, contributed to the growth of female employment in construction. Thus, the growth of male employment in construction was not distress induced, whereas in the case of females, it was distress induced and lent support to the residual sector hypothesis.

As regards male employment in trade, hotel and restaurant business, higher per hectare net income from agriculture, higher daily real wage earnings of agricultural labourers and higher rate of literacy were found to be important factors. The unemployment rates, as per CDS and UPS, were also significantly and positively associated with male employment in trade, hotel and restaurant business. The growth of male employment in these segments appeared to be growth induced and more remunerative, discrediting the residual sector hypothesis. In respect of females, the growth related factors such as per hectare net income from agriculture, degree of commercialisation and per hectare fertilizer consumption contributed to the growth of female employment in trade, hotel and restaurant business. On the other hand, distress related factors such as proportion of small and marginal farmers, incidence of poverty and unemployment rate, as per CDS, particularly for female youth and the rate of wages in agriculture also contributed to the growth of female employment in trade, hotel and restaurant business. Female employment in these sectors was essentially distress diversification, which lent some credence to the residual sector hypothesis.

Finally, the growth of male employment in transport, storage and communication was significantly determined largely by growth related factors such as higher per hectare income from agriculture, higher daily real wage rates in agriculture, reduction in the incidence of poverty, higher per capita non-food consumption expenditure, higher rate of literacy and availability of rural infrastructure. Although the incidence of unemployment, as per CDS and UPS (distress in general), influenced the level of male employment in transport, storage and communication, but it was not distress induced and it did not lend any support to the residual sector hypothesis.

RNAE and Poverty and Income Inequality

It has been found that a larger number of people experiencing poverty were concentrated in rural than in urban areas; three out of every four poor persons lived in rural areas in the 1990s. In rural areas, the annual growth rate of people in poverty registered a rapid decline in the 1980s and more so in the late 1990s. The decline in the incidence of poverty was a little faster in rural areas than in urban areas. The decline in the incidence of rural poverty in the 1980s has been mainly attributed to a faster growth of agriculture, followed by infrastructure and human resource development, particularly female literacy. The incidence of poverty in the rural non-agricultural sector was lower, as the daily real non-agricultural wage rates were found to be quite higher. The incidence of poverty was found to be not uniform across different rural non-agricultural activities.

The regional pattern of poverty incidence shows that states with a higher per capita net income from agriculture, rate of literacy and male RNAE had a smaller proportion of people in rural poverty and also witnessed a rapid progress in the reduction of it, as compared to states with lower ones during the reform period, as compared to the pre-reform period. On the other hand, the regional pattern of poverty incidence also indicates that the incidence of poverty was found to be higher in those states which had a higher share of females in RNAE during the same period, which implies that the growth of female RNAE had less impact on the reduction of poverty.

An increase in per capita net income from agriculture, proportion of male workers in RNAE, daily real wage earnings of male and female agricultural labourers in RNAE, human capital as measured by the rate of literacy and rural per capita government expenditure for rural development and social services were found to be important sources of poverty reduction. Among them, the growth of male RNAE and real daily wage rates of male agricultural labourers in agricultural and non-agricultural occupations were found to have a significant impact on poverty reduction. The growth of female RNAE did not seem to have significantly contributed to the reduction of poverty. But, the wage earnings of female agricultural labourers in agricultural and non-agricultural sectors have made a significant contribution to the reduction of poverty, but lower than that of males.

It may be concluded that occupational diversification in respect of males was mainly determined by the growth related factors, whereas in respect of females, it was by distress related factors. Occupational diversification in favour of the rural non-agricultural sector driven by growth related factors seemed to have had beneficial impact on poverty reduction as compared to the one driven by distress related factors. In view of a gradual decline in the share of public expenditure for rural development and poverty alleviation programmes, the growth of RNAE, especially in respect of males, through rapid agricultural growth and human resource development can be considered an important poverty alleviation approach in the countryside.

The distribution of income from cultivation, followed by rural non-agricultural regular employment, was found to be most inegalitarian. On the other hand, the distribution of incomes from rural non-agricultural wage employment, rural non-agricultural self-employment, and agricultural wage employment was more egalitarian. The inegalitarian distribution of income from cultivation can be attributed to unequal distribution of landholdings. Similarly, in respect of rural non-agricultural regular employment, there appeared to be some entry barriers to the resource-poor households. Lack of education, influence, social networks and so on constrained the poor households to access better quality of RNAE. It is the rich households that have an easy access to better of quality of RNAE, like regular employment in government/private sector. This may result in greater inequality in the distribution of income.

The relative concentration coefficient identified cultivation and rural non-agricultural regular employment as inequality-increasing sources, whereas rural non-agricultural and agricultural wage employment, non-agricultural self-employment and total non-agricultural employment were identified as inequality-decreasing sources of income. The source-inequality weights show that cultivation, followed by rural non-agricultural regular employment and non-agricultural self-employment contributed the maximum to total income inequality. The income from wage employment in agricultural and non-agricultural sectors contributed negatively to overall income inequality; the contribution of the former, however, was higher than the latter. This observation needs to be examined in areas with different levels of agricultural and non-agricultural prosperity and infrastructure.

Measures to Promote RNAE in Rural India

The growth rate of employment, particularly in rural areas and for educated youth, should be stepped up and sustained by the further strengthening of "inter-sectoral-growth linkages" between agricultural and non-agricultural sectors. As the quality of rural employment, particularly for females, has been deteriorating, appropriate steps should be taken to generate sustained, high quality and productive employment by providing market-oriented education, on-the-job training and imparting need-based skills. The accessibility for the poor to such employment will have a greater poverty reduction and income-equalising effect.

In view of the growing labour force with higher educational attainments, on the one hand, and declining employment elasticity in agriculture as also in the urban organised manufacturing sector, on the other, the growth of male as well as female RNAE has to be accelerated more than ever before by strengthening the multiple growth linkages between agricultural and non-agricultural sectors and rural and urban economies. As the growth of male RNAE has a greater poverty reduction effect, special emphasis should laid on the growth of male RNAE for those households, which have been living below the poverty line. At the same time, greater importance should also be accorded to the participation of female workers in more productive RNAE, as a result of which the problems of poverty and income inequality can effectively be tackled.

In the Indian labour market, one can notice that there has been a vast majority of illiterate female and male workers and the next largest category of workers are the literate ones at lower levels (primary and middle). A greater emphasis has to be accorded to the universalisation of higher secondary education and subsidisation of higher education. A larger share of the rural poor represents workers with low educational background and involved in lowly paid, dead-jobs. Hence, appropriate programmes aiming at the provision of adult education, training and orientation should be worked out and effectively implemented. To tackle the long-standing problems of unemployment and poverty, the current wage employment programmes and National Rural

Employment Guarantee Employment Scheme should be implemented comprehensively and more effectively, paying greater attention to the illiterate poor workers. For the pool of educated unemployed youth, the on-going self-employment income generation programmes should be implemented in such a manner that sustained, adequate and productive employment opportunities are generated in rural areas so that the problem of growing educated youth unemployment can be tackled.

For sustained reduction of poverty and income inequality, the growth of agricultural and non-agricultural sectors should be accorded top priority in the rural development programmes. The growth of agriculture should be promoted so as to raise the income levels of the small and marginal farmers and landless households so that the incidence of poverty and income inequality can be further reduced. If the growth rate of income of the non-poor from agriculture tends to increase much faster than that of the poor (mostly small and marginal farmers and landless labourers), the income inequality may further worsen. Therefore, the drive of land distribution in favour of landless and land-poor households may be strengthened so as to raise the income levels of the poor at much faster rate than that of the non-poor.

The rural non-agricultural sector should be promoted so as to generate sustained, productive, non-residual, remunerative employment opportunities that may be easily accessible to the poor. Provision of education, especially to the children of the poor households, may facilitate easy access to qualitative and productive RNAE. If the poor are being pushed into the low-productive, casual/contractual, dead-end, lowly paid, non-agricultural jobs of residual nature, the poverty and income inequality will further worsen. Not all non-agricultural activities tend to reduce income inequality. While the growth of regular salaried/wage employment tends to increase the income inequality, the growth of wage employment, in general, and non-agricultural wage employment in particular, followed by non-agricultural self-employment, tends to produce a good deal of income-equalising effects. Therefore, top priority should be accorded to generating wage employment followed by self-employment, particularly in the rural non-agricultural sector, by effective implementation of the current wage-based and self-employment programmes. Comprehensive and effective implementation of the National Rural Employment Guarantee programme has the potential to produce a beneficial impact on the reduction of poverty and income inequality.

The urban manufacturing sector seems to be growing at the cost of RNAE, particularly of the agro-processing type, in terms of raw materials being exported to the urban located manufacturing units. This may occur mostly due to lack of basic facilities for establishing manufacturing and commercial units in rural areas. Appropriate steps should be taken to provide all the basic facilities in rural areas so that sustained, broad-based and equitable growth of agro-processing activities in the rural non-agricultural sector itself can be promoted. Qualitative and demand-driven infrastructural facilities should be made available in the countryside so that widely dispersed and sustained non-farm employment opportunities can be generated.

Bibliography

Adams, Jr., Richard H. (1994). Non-Farm Income and Inequality in Rural Pakistan: A Composition Analysis. *The Journal of Development Studies*, 31 (1): 110-33.

Adelman, Irma (1984). Beyond Export-led Growth. *World Development*, 12 (9).

Aggarwal, Suresh Chand and J. K. Goyal (2000). Trends in Youth Unemployment in India: An Empirical Analysis. *The Indian Journal of Labour Economics*, 43 (4): 641-55.

Ahmed, R. and M. Hossain (1990). *Development Impact of Rural Infrastructure in Bangladesh. Research Report No. 83*. Washington, D.C. U.S.A.: International Food Policy and Research Institute.

Anderson, Dennis and Mark W. Leiserson (1980). Rural Non-Farm Employment in Developing Countries. *Economic Development and Cultural Change*, 28 (2): 227-248.

Basant Rakesh (1994). Economic Diversification in Rural Areas: Review of Process with Special Reference to Gujarat. *Economic and Political Weekly*, XXIX (39).

Basant, Rakesh (1998). Diversification of Economic Activities in Rural Gujarat: Some Evidence at the Micro Level. In Rakesh Basant, B. L. Kumar and R. Parthasarathy (eds.) *Non-Agricultural Employment in Rural India: The Case of Gujarat*. New Delhi: Rawat Publications.

Basant, Rakesh and B. L. Kumar (1989). Rural Non-Agricultural Activities in India: A Review of Available Evidence. *Social Scientist*, 17 (1): 11-71.

Basant, Rakesh and B. L. Kumar (1990). Non-Agricultural Employment in Rural Gujarat: A Review of Evidence. *Working Paper No. 28*. Ahmedabad: The Gujarat Institute of Area Planning.

Basant, Rakesh and B. L. Kumar (1994). Rural Non-Agricultural Employment in India: A Review of Available Evidence. In Pravin Visaria and Rakesh Basant (eds.), *Non-Agricultural Employment in India: Trends and Prospects.* New Delhi: Sage Publications.

Basant, Rakesh and B.L. Kumar (1998). Rural Non-Agricultural Employment in India and Gujarat: A Review of Secondary Data. In Rakesh Basant, B. L. Kumar and R. Parthasarathy (eds.), *Non-Agricultural Employment in Rural India: The Case of Gujarat.* New Delhi: Rawat Publications.

Basant, Rakesh and R. Parthasarathy (1998). Rural Non-Agricultural Employment in Gujarat: A Disaggregated View. In Pravin Visaria and Rakesh Basant (eds.), *Non-Agricultural Employment in India: Trends and Prospects.* New Delhi: Sage Publications.

Basu, K. and Pham Hoang Van (1998). The Economics of Child Labour. *American Economic Review*, 88 (3): 412-27.

Basu, K. (1999). Child Labour: Cause, Consequence and Cure, with Remarks on International Labour Standard. *Journal Economic Literature*, 37 (3): 1083-119.

Bell, Clive, Peter Hazell and Rodger Slade (1982). *Project Evaluation in Regional Perspective: A Study of an Irrigation Project in Northwest Malaysia.* Baltimore, MD: Johns Hopkins University Press.

Bhalla, G. S., R. K. Sharma, G. K. Chadha and S. P. Kashyap (1989). *Structural Changes in the Punjab Economy, 1969-70 to 1979-80: A Study in Input-Output Analysis.* Washington, DC. USA: International Food Policy Research Institute (Mimeo).

Bhalla, G. S. and Peter Hazell (2003). Rural Employment and Poverty: Strategies to Eliminate Rural Poverty within a Generation. *Economic and Political Weekly*, XXXVIII (33): 428-39.

Bhalla, Sheila (1993). Test of Some Propositions about the Dynamics of Changes of the Rural Workforce Structure. *The Indian Journal of Labour Economics*, 36 (3): 428-39.

Bhalla, Sheila (1997). The Rise and Fall of Workforce Diversification Processes in Rural India. In G. K. Chadha and Alakh N. Sharma (eds.), *Growth, Employment and Poverty: Change and Continuity in Rural India.* New Delhi: Vikas Publishing House Pvt. Ltd.

Bhalla, Sheila (2003). The Eclipse of Rural India in Recent

Development Processes. Paper presented at the Foundation Day Seminar on *Rural Employment with Focus on Non-Farm Sector* at National Institute for Rural Development, Hyderabad, during December 6-7 (mimeo).

Bhatt, P. R. (2003). Changes in Workforce Structure in India. *The Indian Journal of Labour Economics*, 46 (4): 877-86.

Bhattacharya, B. B. and Arup Mitra (1990). Excess Growth of the Tertiary Sector in Indian Economy. *Economic and Political Weekly*, XXV (44): 2445-50.

Bhaumik, S. K. (2003). Casualisation of the Workforce in India, 1983-2002. *The Indian Journal of Labour Economics*, 46 (4): 907-26.

Biradar, R. R. (2003). Changing Employment Structure in Rural India: Growth-led and Distress Diversification. *The Indian Journal of Labour Economics*, 46 (4): 1125.

Biradar, R. R. (2004). Human Capital Base of Labour Force among Social Groups in India: Emerging Issues and Challenges. *The Indian Journal of Labour Economics*, 47 (4): 731-48.

Biradar, R. R. (2006a). Growth of Rural Non-Farm Activities in India: Emerging Issues. In Sabanna Talawar (ed.), *Agriculture and Globalisation*. New Delhi: Serial Publications.

Biradar, R. R. (2006b). Educational Status among Social Groups in India: Emerging Issue, *at the 89th Annual Conference Volume, Indian Economic Association*: 173-85.

Biradar, R. R. (2007). Growth of Rural Non-Farm Activities in Karnataka: Emerging Issues and Prospects. In Jayasheela (ed.), *Rural Karnataka*. New Delhi: Serial Publications.

Biradar, R. R. and S. T. Bagalkoti (2001). Changing Facets of Employment in Rural India: Emerging Issues and Challenges. *Indian Journal of Agricultural Economics*, 56 (3): 538-52.

Biradar, R. R. and S. T. Bagalkoti (2004). Growth of Rural Non-Farm Activities in India: Spatial and Temporal Analysis. In P. Purushotham (ed.), *Non-Farm Employment*. Hyderabad: National Institute for Rural Development.

Biradar, R. R. and D. Rajasekhar (2007). Occupational Diversification and Poverty Alleviation in Rural India: What does Emerging Evidence Indicate? (mimeo).

Biradar, R. R. and Jayasheela (2008). Effects of Educational Inequality among Social Groups in Rural India.*Journal ot Rural Development,* 26 (3): 379-401.

Birthal, Pratap Singh (1997). Occupational Diversity and Income Distribution: Some Implications for Rural Development. *Journal of Rural Development*, 16 (3): 465-75.

Centre for Monitoring Indian Economy (CMIE) (2004): *Agriculture*, Economic and Intelligence Service, Centre for Monitoring Indian Economy, February.

Chadha, G. K. (1984). *Dynamics of Rural Transformation: A Study of Punjab (India) Since 1950.* ICSSR Project Report. New Delhi: Indian Council for Social Science Research (mimeo).

Chadha, G. K. (1986). Agricultural Growth and Rural Non-farm Activities: An Analysis of Indian Experience. In Yang-Boo Choe and Fu-Chen Lo (eds.), *Rural Industrialisation and Non-farm Activities of Asian Farmers.* Korea: Rural Economics Institute, Asian and Pacific Development Centre.

Chadha, G. K. (1994). *Employment, Earnings and Poverty: A Study of Rural India and Indonesia.* New Delhi: Sage Publications.

Chadha G. K. (1997). Access of Rural Households to Non-Farm Employment: Trends, Constraints and Possibilities. In G. K. Chadha and Alakh N. Sharma (eds.), *Growth, Employment and Poverty: Change and Continuity in Rural India.* New Delhi: Vikas Publishing House Pvt. Ltd.

Chadha, G. K. (2000). Youth Unemployment in India: Past Policy Gaps and Future Strategic Options. *The Indian Journal Labour Economics,* 43 (4): 591-624.

Chadha, G. K. (2001a). Impact of Economic Reforms on Rural Employment: No Smooth Sailing is Anticipated. *Indian Journal Agricultural Economics*, 56 (3): 491-525.

Chadha, G. K. (2001b). Education, Skills, and Working Environments of Workers in India's Rural Industry. *International Journal of Occupational and Environmental Health*, 7 (2).

Chadha, G. K. (2002). Rural Non-Farm Employment in India: What Does Recent Data Experience Teach Us. *The Indian Journal of Labour Economics*, 45 (4): 663-94.

Chadha, G. K. (2004). Human Capital Base of the Indian Labour Market: Identifying Worry Spots. *The Indian Journal Labour Economics*, 47 (1): 3-38.

Chadha, G. K. and P. P. Sahu (2002). Post-Reform Setbacks in Rural Employment: Issues That Need Further Scrutiny. *Economic and Political Weekly*, XXXVII (21): 1998-2026.

Chand, Ramesh (1996). Agricultural Diversification and Farm and Non-Farm Employment in Himachal Pradesh. *The Indian Journal of Labour Economics*, 39 (4): 841-51.

Chandrasekhar, C. P. (1993). Agrarian Change and Occupational Diversification: Non-Agricultural Employment and Rural Development in West Bengal. *The Journal of Peasant Studies*, 20 (2): 205-70.

Chaubey, P. K. (2000). Youth and Unemployment in Perspective: The Indian Context. *The Indian Journal of Labour Economics*, 43 (4): 625-40.

Chenery, B. H. and M. Syrquin (1975). *Patterns of Development 1950-1970*. New York: Oxford University Press.

Chinn, Dennis (1979). Rural Poverty and the Structure of Farm Household Income in Developing Countries: Evidence from Taiwan. *Economic Development and Cultural Change*, 27 (2): 283-301.

Clark, Colin (1940). *The Conditions of Economic Progress*, London: Macmillan.

Collier, P., Radwan, S. and S. Wangwe (1986). *Labour and Poverty in Rural Tanzania*. Oxford: Clarendon Press.

Corral, L. and T. Reardon (2001). Rural Nonfarm Incomes in Nicaragua. *World Development, Special Issue*, 29 (3): 427-42.

Deshpande, Sudha (2003). Changing Employment Structure in Large States of India: What do the NSSO Data Show? *The Indian Journal of Labour Economics*, 46 (4): 811-44.

Dev, S. Mahendra (1990). Non-Agricultural Employment in Rural India: Evidence at a Disaggregate Level. *Economic and Political Weekly*, XXV (28): 1526-36.

Dev, S. Mahendra (1994). Some Aspects of Non-Agricultural Employment in Rural India: Evidence at a Disaggregate Level. In Pravin Visaria and Rakesh Basant (eds.), *Non-Agricultural Employment in India: Trends and Prospects*. New Delhi: Sage Publications.

Dev, S. Mahendra (2000a). Economic Liberalisation and Employment in South Asia – I. *Economic and Political Weekly*, XXXV (1 & 2): 40-58.

Dev, S. Mahendra (2000b). Economic Reforms Poverty, Income Distribution and Employment. *Economic and Political Weekly*, XXXV (10): 823-35.

Eapen, Mridul (1994). Rural Non-Agricultural Employment in Kerala: Some Emerging Tendencies. *Economic and Political Weekly*, XXIX (21): 1285-96.

Eapen, Mridul (1995). Rural Non-Agricultural Employment in Kerala: Inter-District Variations. *Economic and Political Weekly*, XXX (12): 634-38.

Eapen, Mridul (2001). Rural Non-Farm Employment: Agricultural Versus Urban Linkages—Some Evidence from Kerala State, India. *The Journal of Peasant Studies,* 28 (3): 67-88.

Elbers, C. and P. Lanjouw (2001). Intersectoral Transfer, Growth, and Inequality in Rural Ecuador. *World Development, Special Issue*, 29 (3): 481-96.

Escobal, J. (2001). The Determinants of Nonfarm Income Diversification in Rural Peru. *World Development, Special Issue,* 29 (3): 497-508.

Fei, John, C. H. and Gustav Ranis (1961). A Theory of Economic Development. *American Economic Review,* 51 (4): 533-65.

Ferreira, F. H.G. and P. Lanjouw (2001). Rural Non-farm Activities and Poverty in the Brazilian Northeast. *World Development, Special Issue*, 29 (3): 509-628.

Fields, G. (1980). *Poverty, Inequality and Development.* New York: Cambridge University Press.

Fields, G. (2000). *Distribution and Development: A New Look at the Developing World.* Cambridge, MA: Russel Sage Foundation and MIT Press.

Fisher, A. G. P (1939). Production— Primary, Secondary and Tertiary. *Economic Record,* 15 (6): 24-38.

Ghose, A. K. (2004). The Employment Challenges in India. *Economic and Political Weekly*, XXIX (48): 5105-16.

Ghosh, Arun (1989). Mystery of a Declining Capital-Output Ratio. *Economic and Political Weekly,* XXIV (46): 2527-30.

Gill, Amerita Sher and Upinder Sawhney (2003). Inter-state Variations in Non-farm Rural Employment: Some Conclusions from 2001 Census Data. *The Indian Journal of Labour Economics*, 46 (4): 961-69.

Government of India (GoI) (1981). *Census of India, General Population, Series 1, Part II-B, Primary Census Abstract.* New Delhi: Office of the Registrar General, Ministry of Home Affairs.

GoI (1991). *Census of India, General Population, Series 1, Part II-B, Primary Census Abstract.* New Delhi: Office of the Registrar General, Ministry of Home Affairs.

GoI (1998a). *Indian Livestock Census – 1992, Volume II, Part 2, Poultry and Agricultural Implements and Machinery.* New Delhi: Directorate of Economics and Statistics, Government of India.

GoI (1998b). *Rural Labour Enquiry: A Report on Wages and Earnings of Rural Labour Households.* New Delhi, Shimla: Labour Bureau, Ministry of Labour.

GoI (1999). *The Ninth Five Year Plan, 1997-2002. Vol. I,* Planning Commission. New Delhi: Government of India.

GoI (2000) *Report of the Study Group on Labour Statistics.* New Delhi: Ministry of Labour.

GoI (2001). *Census of India 2001, Series-1, Provisional Population Totals, Paper-1.* New Delhi: Office of the Registrar General, Ministry of Home Affairs.

GoI (2002a). *Report of the Study Group on Skill Development, Training and Workers' Education.* New Delhi: National Commission on Labour, Ministry of Labour.

GoI (2002b). *Planning Commission Report on Labour and Employment.* New Delhi: ECONOMIC India, Info-services.

GoI (2002c). *Report of the Special Group on Targeting Ten Million Employment Opportunities Per Year over the 10th Plan Period.* New Delhi: Planning Commission, Government of India.

GoI (2002d). *National Human Development Report 2001.* New Delhi: Oxford University Press.

GoI (2003a). *Statistical Abstract, India 2003.* New Delhi: Central Statistical Organisation, Ministry of Statistics and Programme Implementation.

GoI (2003b). *Rural Labour Enquiry Report on Wages and Earnings of Rural Labour Households 1999-2000*, 55th Round of NSS. New Delhi, Shimla: Ministry of Labour, Government of India.

Gumber, Anil (2000). Correlates of Unemployment Among Rural Youth in India: An Inter-State Analysis. *The Indian Journal Labour Economics*, 43 (4): 657-71.

Haggblade, Steven, Peter Hazell and James Brown (1989). Farm-Nonfarm Linkages in Rural Sub-Saharan Africa. *World Development*, 17 (8): 1173-201.

Harriss, Barbara (1987). Regional Growth Linkages from Agriculture and Resources Flows in Non-Farm Economy. *Economic and Political Weekly*, XXII (1&2): 31-46.

Hazell, Peter B. R. and Steven Haggblade (1991). Rural-Urban Growth Linkages in India. *Indian Journal Agricultural Economics*, 46 (4): 515-29.

Hazell, Peter B. R. and Steven Haggblade (1993). 'Farm-Nonfarm Growth Linkages and the Welfare of the Poor. In M. Lipton and J. van dir Gaag (eds), *Including the Poor.* Washington, D.C.: World Bank.

Hazell, Peter B. R., and Alisa Roell (1983). Rural Growth Linkages: Household Expenditure Patterns in Malaysia and Nigeria. *Research Report No. 41.* Washington, D.C.: International Food Policy Research Institute.

Hirschman, A. O. (1958). *The Strategy of Economic Development.* New Haven CT: Yale University Press.

Hirway, Indira (2002). Employment and Unemployment Situation in 1990s: How Good Are NSS Data? *Economic and Political Weekly*, XXXVII (21): 2027-36.

Honakeri, P. M. (2002). Pattern of Rural Non-farm Employment in Rural Areas of Gadag District in Karnataka: A Case Study. *The Indian Journal of Labour Economics*, 45 (4): 981-2.

Hossain, M. (1988). Nature and Impact of the Green Revolution in Bangladesh. *Research Report No. 67.* Washington, D.C.: International Food Policy and Research Institute.

ILO (1985). *A Contribution to a Discussion Paper on Entry of Youths into Working Life in Asia and Pacific.* Melbourne: Tenth Conference of Asian and Pacific Labour Ministers.

Islam, Nurul (1997). The Nonfarm and Rural Development: Review of Issues and Evidence. *Food, Agriculture and the Environment Discussion Paper 22.* Washington. D.C.: International Food Policy Research Institute.

Janvry, A. de and E. Sadoulet (1993). Rural Development in Latin America: Relinking Poverty to Growth. In M. Lipton and J van dir Gaag (eds.), *Including the Poor.* Washington, D.C.: World Bank.

Janvry, A. de and. E. Sadoulet (2001). Income Strategies Among Rural Households in Mexico: The Role of Off-farm Activities. *World Development, Special Issue,* 29 (3): 467-80.

Jayadevan, C. M. (1996). Casualisation of Workforce in India: An Analysis of Spatial Variations. *The Indian Journal of Labour Economics,* 39 (4): 763-84.

Jayaraj, D. (1989). Determinants of Rural Non-agricultural Employment. *Working Paper No. 90.* Madras: Madras Institute of Development Studies.

Johnston, Brace F. and Peter Kilby (1975). *Agriculture and Structural Transformation: Economic Strategies in Late-Developing Countries.* London: Oxford University Press.

Kakwani, N. C. (1980). *Poverty, Inequality and Development.* New York: Cambridge University Press.

Kaur, Kuldip (2002). Determinants of Rural Nonfarm Employment in India: An Inter-State Analysis. *The Indian Journal of Labour Economics,* 45 (4): 974.

Krishnamurty, J. (1984). Changes in the Indian Workforce, *Economic and Political Weekly.* XIX (50): 2121-28.

Kumar, B. L. (1996). Changes in the Composition of Rural Labour Force: Some Evidence from a Village Survey. *The Indian Journal of Labour Economics,* 39 (4): 809-16.

Kung, James K. S. and Yiu-fai Lee (2001). So What if There is Income Inequality? The Distributive Consequence of Non-Farm Employment in Rural China. *Development and Cultural Change,* 50 (1): 19-46.

Kuznets, Simon (1955). Economic Growth and Income Inequality. *American Economic Review,* 45 (2): 1-28.

Kuznets, Simon (1966). *Modern Economic Growth: Rate, Structure and Spread*. Cambridge: Yale University Press.

Lane, D. W. (1996). Political Base of rural Industrilisation: Korea and Taiwan. In *Towards the Rural-Based Development of Commerce and Industry: Selected Experience from East Asia*. Washington, D.C.: World Bank.

Lanjouw, J. O. and P. Lanjouw (1995). Rural Non-farm Employment: A Survey. *Policy Research Working Paper No. 1463*. Washington, D.C.: World Bank.

Lanjouw, Peter (1999). Rural Nonagricultural Employment and Poverty in Ecuador. *Economic Development and Cultural Change,* 48 (1): 91-122.

Lanjouw, Peter (2001). Non-farm Employment and Poverty in Rural El Salvador. *World Development,* 29 (3): 529-47.

Lanjouw, Peter and Abusaleh Shariff (2004). Rural Non-Farm Employment in India: Access, Income and Poverty Impact. *Economic and Political Weekly*, XXIX (40): 4429-46.

Leones, J. P. and S. Feldman (1998). Nonfarm Activity and Rural Household Income: Evidence from Philippine Micro Data. *Economic Development and Cultural Change,* 46 (4): 789-806.

Lewis, W. A. (1954). *Economic Development with Unlimited Supplies of Labour,* The Manchester School, 22: 139-91.

Liedholm, Carl and Peter Kilby (1989). The Role of Nonfarm Activities in the Rural Economy. In G. Jeoffrey William and V. R. Panchamukhi (eds.), *The Balance Between the Industry and Agriculture in Economic Development: Sector Proportions*, Vol. 2. London: ILO/Macmilan Press.

Madheswaran, S. and Amita Dharmadikhary (2000). Income and Employment in Service Sector in India. *The Indian Journal of Labour Economics*, 43 (4): 835-64.

Mathur, Ashok and Rajendra P., Mamgain (2002). Technical Skills, Education and Economic Development in India. *The Indian Journal of Labour Economics*, 45 (4): 1015-45.

Matlon, P. (1979). Income Distribution among Farmers in Northern Nigeria: Empirical Results and Policy Implications. *Africa Rural Economy Paper No. 18.* Michigan State University.

Mecharla, Prasada Rao (2002). Determinants of Inter-district Variations in Rural Non-farm Employment in Andhra Pradesh. *The Indian Journal of Labour Economics*, 45 (4): 807-20

Mellor, Johan and Uma J. Lele (1973). Growth Linkages of the New Foodgrain Technologies. *Indian Journal of Agricultural Economics*, 28, (1): 35-55.

Mellor, John (1976). *The New Economics of Growth: A Strategy for India and the Developing World.* Ithaca, New York: Cornell University Press.

Minhas, B. S. and Grace Majumdar (1987). Unemployment and Casual Labour in India: An Analysis of Recent NSS Data. *Indian Journal of Industrial Relations,* 22 (3).

Nagaraj, R. (1990). Growth Rate of India's GDP, 1950-51 to 1987-88: Examination of Alternative Hypothesis. *Economic and Political Weekly*, XXV (26): 1396-403.

Narayanamoorthy, A., Queenie Rodrigues and Ashwini Phadnis (2002). Determinants of Rural Non-farm Employment: An Analysis of 256 Districts. *The Indian Journal of Labour Economics,* 45 (4): 759-69.

Nath, G. B. (1996). Farm/Off-farm Labour Market Participation and Workforce Restricting in Rural India: A Micro Study. *The Indian Journal of Labour Economics,* 39 (4): 787-93.

National Sample Survey Organisation (NSSO) (1981). Report on the Second Quinquennial Survey on Employment and Unemployment, (All India), NSSO 32nd Round, *Sarvekshana,* 5 (1 & 2).

NSSO (1988). Report on the Third Quinquennial Survey on Employment and Unemployment, All India. *Sarvekshana,* 11 (4).

NSSO (1990). Results of the Fourth Quinquennial Survey on Employment and Unemployment, All India, NSSO 43rd Round (July 1987-June 1988). *Sarvekshana, Special Number.* New Delhi: Government of India.

NSSO (1996). Key Results on Employment and Unemployment in India 1993-94, Fifth Quinquennial Survey, NSSO 50th Round (July 1993-June 1994). *NSSO Report No.406.* New Delhi: Government of India.

NSSO (2001). Employment and Unemployment Situation in India 1999-2000, Part-I, NSSO 55th Round (July 1999-June 2000).

NSSO Report No. 458. New Delhi: Ministry of Statistics and Programme Implementation, Government of India.

Oshima, H. T. (1994). *The Significance of Off-Farm Employment and Incomes in Post-War East Asian Growth*. Manila: Asian Development Bank.

Papola, T. S. (1987). Rural Industrialisation and Agricultural Growth: A Case Study of India. In R. Islam (ed.) *Rural Industrialisation and Employment in Asia.* New Delhi: ILO/ARTEP.

Papola, T. S. (1999). Labour Studies in India: Some Reflection on Developments in the Post-Independence Period. *The Indian Journal of Labour Economics,* 42 (2): 175-85.

Parthasarathy, G. and K. A. Nirmala (2000). Employment and Unemployment of Youth. *The Indian Journal Labour Economics,* 43 (4): 683-701.

Parthasarathy, G., Shameem and B. Sambi Reddy (1998). Determinants of Rural Non-Agricultural Employment: The Indian Case. *Indian Journal of Agricultural Economics*, 53 (2): 139-54.

Pyatt, G., Chen, C. N. and J. Fei (1980). The Distribution of Income by Factor Components. *Quarterly Journal of Economics*, 95 (3): 451-73.

Rajasekhar, D (1995). Rural Non-farm Employment in Karnataka: A Disaggregated Analysis at the District Level. *Agricultural Situation in India,* August.

Rajasekhar, D. and R. R. Biradar (1998). *Rural Non-Farm Activities in Karnataka.* Bangalore: Institute for Social and Economic Change (Mimeo).

Rangarajan, C. (1982). Agricultural Growth and Industrial Performance in India. *Research Report No. 33*. Washington, D.C.: International Food Policy Research Institute.

Ranjan, Sharad (1996). Regional Rural Non-Farm Employment in Uttar Pradesh: Some Emerging Trends. *The Indian Journal of Labour Economics*, 39 (2): 365-74.

Reardon, T., Delgado, C. and P. Matlon (1992). Determinants and Effects of Income Distribution Among Farm Households in Burkina Faso. *Journal of Development Studies,* 28 (2): 264-96.

Reardon, T., J. Berdegue and G. Escobar (2001). Rural Non-farm Employment and Incomes in Latin America: Overview and Policy Implications. *World Development, Special Issue*, 29 (3): 395-409.

Reardon, T., Stamoulis, K., Cruze, M.E., Balisacan, A., Berdegue, J. A. and Banks, B. (1998). Rural Nonfarm Income in Developing Countries. In FAO (ed.), *The State of Food and Agriculture 1998.* Rome: Food and Agricultural Organisation.

Rosenzweig, M. (1988). Labor Markets in Low Income Countries. In H. Chenery and T.N. Srinivasan (eds.), *Handbook of Development Economics, Vol. 1.* North Holland, Amsterdam.

Ruben, R. and M. Van den Berg (2001). Nonfarm Employment and Poverty Alleviation of Rural Farm Households in Honduras. *World Development, Special Issue,* 29 (3): 549-60.

Sahu, P. P. (2003). Casualisation of Rural Workforce in India: An Analysis of Recent Trends. *The Indian Journal of Labour Economics*, 46 (4): 927-44.

Saleth, R. Maria (1996). Rural Non-Farm Employment and Income in Tamil Nadu: A Quantitative Analysis at the Household Level. *The Indian Journal of Labour Economics*, 39 (2): 335-51.

Samuel, P. S. Ho (1979). Decentralised Industrialisation and Rural Development: Evidence from Taiwan. *Economic Development and Cultural Change*, 28 (1): 77-96.

Sanghvi, P. (1969). Surplus Manpower in Agriculture and Economic Development. New York: Asia Publishing House.

Sankaranarayana, V. (1980). *Inter-State Variation in Rural Non-Agricultural Employment: Some Tentative Results. Working Paper No. 104.* Trivandrum: Centre for Development Studies.

Saraswat, S. P., Ranveer Singh and Prem Singh Daliya (1995). Structural Changes in Farm and Non-Farm Employment in Himachal Pradesh: A Case Study of Kot Village. *Indian Journal of Agricultural Economics*, 50 (3): 430-36.

Shariff, Abusaleh and Anil Gumber (1999). Employment and Wages in India: Pre- and Post-Reform Scenario. *The Indian Journal of Labour Economics,* 42 (2): 195-214.

Sharma, H. R. (2001). Employment and Wage Earnings of Agricultural Labourers: A State-wise Analysis. *The Indian Journal of Labour Economics,* 44 (1): 27-38.

Sharma, H. R., Virendar Kumar and R.K. Sharma (1999). Rural Non-farm Employment in Himachal Pradesh, 1971-1991: A District Level Study. *The Indian Journal of Labour Economics*, 42 (2): 251-63.

Shorrocks, A. F. (1982). Inequality Decomposition by Factor Components. *Econometrica*, 50 (1): 193-211.

Shorrocks, A. F. (1983). The Impact of Income Components on the Distribution of Family Income. *Quarterly Journal of Economics*, Vol. 98:311-26.

Shukla, Vibhooti (1991). Rural Non-Farm Activity: A Regional Model and Its Empirical Application to Maharashtra. *Economic and Political Weekly*, XXVI (45): 2587-96.

Shukla, Vibhooti (1992). Rural Non-Farm Employment in India: Issues and Policy. *Economic and Political Weekly*, XXVII (28): 1477-88.

Shylendra, H. S. and P. Thomas (1995). Non-farm Employment: Nature, Magnitude and Determinants in a Semi-Arid Village of Western India. *Indian Journal of Agricultural Economics*, 50 (3): 410-16.

Sidhu, H. S. and Jasdeep Singh Toor (2002). Nature and Determinants of Rural Non-farm Activities in Punjab. *The Indian Journal of Labour Economics*, 45 (4): 871-81.

Singh, Ajit Kumar (1994). Changes in the Structure of Rural Workforce in Uttar Pradesh: A Temporal and Regional Study. In Pravin Visaria and Rakesh Basant (eds.), *Non-Agricultural Employment in India: Trends and Prospects*, New Delhi: Sage Publications.

Singh, C. S. (2003). Skill, Education and Employment: A Dissenting Essay. *Economic and Political Weekly*, XXXVIII (31): 3271-76.

Singh, Harpal (1993). Non-Agricultural Working Force in Rural Haryana: 1991 (Spatial Analysis). *Man Power Journal*, 29 (2): 37-44.

Srivastava, Nirankar and Amaresh Dubey (2002). Rural Non-farm Employment in India: Spatial Variations and Temporal Change. *The Indian Journal of Labour Economics*, 45 (4): 745-58.

Sundaram, K. (2001a). Employment and Poverty in 1990s: Further Results from NSSO 55th Round Employment-Unemployment

Survey 1999-2000. *Economic and Political Weekly*, XXXVI (32): 3039-49.

Sundaram, K. (2001b). Employment– Unemployment Situation in the Nineties: Some Results from NSSO 55th Round Survey. *Economic and Political Weekly*, XXXVI (11): 931-40.

Sundaram, K. and S. D. Tendulkar (2003). Poverty among Social and Economic Groups in India in 1990s. *Economic and Political Weekly*, XXXVII (50): 5263-76.

Sundaram, K. and S. D. Tendulkar (2004). The Poor in the Indian Labour Force Scenario in the 1990s. *Economic and Political Weekly*, XXXIX (48): 5125-32.

Thorat, Sukhadeo (2004). Persistence of Poverty Why do Scheduled Castes and Scheduled Tribes Stay Chronically Poor? London: Working Paper for Centre for Chronic Poverty.

Tulpule, Bagaram (1997). Rural Labour, Poverty and Development. *The Indian Journal of Labour Economics*, 40 (1): 15-23.

Unni, Jeemol (1991). Regional Variation in Rural Non-Agricultural Employment: An Exploratory Analysis. *Economic and Political Weekly*, XXVI (3): 109-22.

Unni, Jeemol (1996a). Diversification of Economic Activities and Non-Agricultural Employment in Rural Gujarat. *Economic and Political Weekly*, 31 (33): 2243-51.

Unni, Jeemol (1996b). Non-Agricultural Employment and Rural Livelihoods: Macro vis-à-vis Micro View. *The Indian Journal of Labour Economics*, 39 (4): 795-807.

Unni, Jeemol (1997a). Non-agricultural Employment, Livelihoods and Poverty in Rural India. *Working Paper No. 88*. Ahmedabad: Gujarat Institute of Development Research.

Unni, Jeemol (1997b). Employment and Wages Among Rural Labourers: Some Recent Trends. *Indian Journal of Agricultural Economics*, 52 (1): 59-72.

Unni, Jeemol (1998). Non-Agricultural Employment and Poverty in Rural India: A Review of Evidence. *Economic and Political Weekly*, XXXIII (13): A-36-44.

Unni, Jeemol (2001). Gender and Informality in Labour Market in South Asia. *Economic and Political Weekly*, XXXVI (26): 2360-70.

Unni, Jeemol and Uma Rani (2000). Globalisation, Information

Technology Revolution and Service Sector in India. *The Indian Journal of Labour Economics*, 43 (4): 803-27.

Vaidyanathan, A. (1986). Labour Use in Rural India: A Study of Spatial and Temporal Variations. *Economic and Political Weekly*, XXI (52): A-130-46.

Vaidyanathan, A. (1994). Employment Situation: Some Emerging Perspective. *Economic and Political Weekly,* XXIX (50): 3147-56.

Vaidyanathan, A. (2001). Employment in India, 1977-78 to 1999-2000: Characteristics and Trends. *Journal Indian School of Political Economy*, 13 (2): 217-54.

Vanamala, M. (2001). Informalisation and Feminisation of a Formal Sector Industry: A Case Study. *Economic and Political Weekly*, XXXVI (26): 2378-89.

Verma, Binoy N. and Neelam Verma (1995). Distress Diversification from Farm to Non-Farm Rural Employment Sector in the Eastern Region. *Indian Journal of Agricultural Economics*, 50 (3): 422-29.

Visaria, Pravin (1995). Rural Non-Farm Employment in India: Trends and Issues for Research. *Indian Journal of Agricultural Economics*, 50 (3): 398-409.

Visaria, Pravin (1996). The Structure of the Indian Workforce, 1961-1994. *The Indian Journal of Labour Economics,* 39 (4).

Visaria, Pravin (1998). Unemployment among Youth of India; Level, Nature and Policy Implications. *Employment and Training Paper 36*. Geneva: ILO.

Vyas, V. S. and G. Mathai (1978). Farm and Non-Farm Employment in Rural Areas: A Perspective for Planning. *Economic and Political Weekly*, XIII (6 &7): 333-47.

Vyasulu, Vinod (1990). The Non-Farm Sector in the Indian Economy: A Discussion Note on Research Priorities. *Indian Journal of Social Science, July-September.*

World Bank (WB) (1978). *Rural Enterprise and Non-Farm Employment.* Washington, D.C.: Oxford University Press.

WB (1997). *India: Achievements and Challenges in Reducing Poverty.* Washington, D.C.: Oxford University Press.

WB (2000). *India: Reducing Poverty, Accelerating Development.* Washington, D.C.: Oxford University Press.

WB (2001). *Global Economic Prospects and the Developing Countries.* Washington, D.C.: Oxford University Press.

Index